Ecotoxicology of Marine Organisms

T0199705

Editors

Bernardo Duarte

and

Isabel Caçador

MARE – Marine and Environmental Sciences Centre
Faculty of Sciences of the University of Lisbon
Campo Grande, Lisbon
Portugal

CRC Press
Taylor & Francis Group
Boca Raton London New York

CRC Press is an imprint of the
Taylor & Francis Group, an **informa** business

A SCIENCE PUBLISHERS BOOK

CRC Press
Taylor & Francis Group
6000 Broken Sound Parkway NW, Suite 300
Boca Raton, FL 33487-2742

First issued in paperback 2020

ISBN-13: 978-1-138-03549-2 (hbk)
ISBN-13: 978-0-367-77923-8 (pbk)

Library of Congress Cataloging-in-Publication Data

Names: Duarte, Bernardo, 1983- editor.
Title: Ecotoxicology of marine organisms / editors, Bernardo Duarte and Isabel Caçador, MARE (Marine and Environmental Sciences Centre), Faculty of Sciences of the University of Lisbon, Campo Grande, Lisbon, Portugal.
Description: Boca Raton : CRC Press, Taylor & Francis Group, [2019] | "A science publishers book.» | Includes bibliographical references and index.
Identifiers: LCCN 2019027243 | ISBN 9781138035492 (hardcover)
Subjects: LCSH: Marine organisms--Toxicology. | Environmental toxicology.
Classification: LCC QH91.5 .E26 2019 | DDC 578.77--dc23
LC record available at https://lccn.loc.gov/2019027243

Visit the Taylor & Francis Web site at
http://www.taylorandfrancis.com

and the CRC Press Web site at
http://www.crcpress.com

Acknowledgements

The Editors would like to thank all the authors for their important contributions.

The Editors would also like to acknowledge the effort and help provided by the Reviewers.

The Editors would also like to thank Fundação para a Ciência e Tecnologia (FCT) for supporting the Editors activity throughout the projects PTDC/CTA-AMB/30056/2017 (OPTOX), UID/MAR/04292/2013 and SFRH/BPD/115162/2016.

The Editors would also like to thank to Miguel P. Pais, Sofia Henriques and António Teixeira for their kind donation of the photos present in the cover.

Preface

Open oceans, coastal zones and transitional waters are key areas for the planet ecosystem functioning, mostly due to its high productivity and reclining ability. For decades, these ecosystems have been threatened by multiple stressors, such as climate change, habitat displacement, bioinvasions, eutrophication and contaminant disposal. These stressors consequently have severe impacts on all marine organisms, from bacteria and microalgae to fishes and top predators.

Over recent decades the world has experienced the adverse consequences of uncontrolled development of multiple human activities (Gavrilescu et al. 2015). Emerging pollutants include a wide range of man-made chemicals (such as pesticides, cosmetics, personal and household care products, pharmaceuticals), which are of worldwide use (Thomaidis et al. 2012). Recent EUROSTAT statistics published in 2013 showed that between 2002 and 2011, over 50 percent of the total production of chemicals is composed by environmentally harmful compounds, of which over 70 percent have significant environmental impact. It is also striking that the pace of chemical discovery is growing rapidly, with the Chemicals Abstracts Service (CAS REGISTRY) reporting in May 2011 the registration of the 60 million chemical substance. Following CAS REGISTRY 50 million substance registration in only 2009, this second major milestone highlights the continued acceleration of synthetic chemical innovation globally (Gavrilescu et al. 2015). The relationship between increased human population density and environmental change in coastal regions is well known. Coastal water is the ultimate sink for sewage and other by-products of human activities. The EU's Task Group recommended that the assessment of achievement of "Good Environmental Status" under Descriptor 8 should be based upon monitoring programs covering the concentrations of chemical contaminants and also biological measurements relating to the effects of pollutants on marine organisms. They also concluded that the combination of conventional and newer, effect-based, methodologies, with the assessment of environmental concentrations of contaminants provides a powerful and comprehensive approach.

Ecotoxicology emerged from this need to understand how contaminants affect the different organisms. The term ecotoxicology was coined by René Truhaut in 1969 who defined it as "the branch of toxicology concerned with the study of toxic effects, caused by natural or synthetic pollutants, to the constituents of ecosystems, animal (including human), vegetable and microbial, in an integral context" (Truhaut 1977). This becomes particularly important if focusing marine ecosystems and organisms, highly prone to contaminant inputs.

Most of the previous publications (with some rare exceptions) focused on terrestrial or freshwater, rather than marine ecosystems. Moreover, the already existing publications are mostly concerned with traditional contaminants and model organisms. Thus, the editors of this book saw the need for a book dedicated to Marine Ecotoxicology focusing not only non-model but key marine organisms (from microalgae, to seagrasses, fishes, invertebrates and marine reptiles) and the effects of several contaminant classes (trace elements, metalloids, pharmaceutical residues, biotoxins, etc.) on these organisms. With this, the editors intend to provide reference and recent information that can be used by undergraduate and postgraduate students, but also by senior researchers, particularly related to future research needs.

This book, which is divided into nine chapters, was developed with the cooperation and input of several senior researchers' experts in ecotoxicology. The chapters in this book proceed address different levels of organism complexity and different contaminant addressing the wider scope possible.

The Editors hope that you, the readers, learn as much as we did from these nine chapters and use the knowledge provided herein to further advance the science of marine ecotoxicology.

References

Gavrilescu, M., Demnerová, K., Aamand, J., Agathos, S. and Fava, F. 2015. Emerging pollutants in the environment: Present and future challenges in biomonitoring, ecological risks and bioremediation. N. Biotechnol. 32: 147–156. https://doi.org/10.1016/j.nbt.2014.01.001.
Thomaidis, N.S., Asimakopoulos, A.G. and Bletsou, A.A. 2012. Emerging contaminants: A tutorial mini-review. Glob. Nest J. 14: 72–79.
Truhaut, R. 1977. Ecotoxicology: Objectives, principles and perspectives. Ecotoxicol. Environ. Saf. 1: 151–173.

Bernardo Duarte and Isabel Caçador
MARE - Marine and Environmental Sciences Centre
Faculty of Sciences of the University of Lisbon
Campo Grande 1749-016 Lisboa, Portugal

Contents

1 Environmental Disturbance Caused by Stressors
Changing the Focus from Individuals to Habitats

Cristiano V.M. Araújo,[1,2,*] *Matilde Moreira-Santos*[2]
and Rui Ribeiro[2]

INTRODUCTION

The recognition that contamination is a worldwide problem and that a great number of ecosystems suffer some level of (anthropogenic) disturbance gave rise to environmental risk assessment (ERA) schemes and resulting environmental decisions to be strongly based on ecotoxicological assays, designed to accurately evaluate the effects of the contaminants on organisms (Luoma 1996, Sherman 2000). Therefore, although the concept of environmental vulnerability needs to be expanded to long-term effects on ecosystems (De Lange et al. 2010), the traditional ecotoxicological approach, including mesocosm studies, is focused on how toxic a compound or an environmental sample is; and toxicity is assayed, that is, measured, from biological responses at the individual and, in mesocosm studies, at higher levels (Chapman 1995, Schmitt-Jansen et al. 2008). The fundamental assumption sustaining this approach is that organisms in ecosystems are directly exposed to contamination and that, after a specific exposure period, a biological effect is produced. The methodological procedures to perform such ecotoxicological assays consists in exposing organisms to compound(s) or environmental sample(s) in a forced exposure test system, in which organisms are continuously in contact with the stressful agent. However,

[1] Institute of Marine Sciences of Andalusia (CSIC), Department of Ecology and Coastal Management, Puerto Real, Spain.
[2] CFE-Centre for Functional Ecology, Department of Life Sciences, University of Coimbra, Coimbra, Portugal.
   Emails: matilde.santos@zoo.uc.pt; rui.ribeiro@zoo.c.pt
* Corresponding author: cristiano.araujo@icman.csic.es

considering that laboratory ecotoxicological assays intend to simulate the real field situation, by using a forced exposure system in which organisms are mandatorily exposed to contaminants, their natural biological possibility to escape from a toxic effect is not taken into account (Lefcort et al. 2004).

It is a fact that a scenario of forced exposure to contamination under a real situation is unavoidable for organisms with no ability to move, neither actively nor passively (e.g., drift). However, theoretically, the same is not expected for mobile organisms. In fact, many studies have shown the ability of a great number of organisms to detect environmental disturbers and escape towards less impacted environments (Araújo et al. 2016a, Tierney 2016). To cite some few examples, species of amphipods (Kravitz et al. 1999, Hellou et al. 2009), copepods (Ward et al. 2013, Araújo et al. 2014a), cladocerans (Lopes et al. 2004, Rosa et al. 2008, 2012), snails (Hellou 2010, Araújo et al. 2016b), decapods (Richardson et al. 2001, Araújo et al. 2016c), amphibians (Araújo et al. 2014b,c), and fish (Pedder and Maly 1986, Moreira-Santos et al. 2008, Araújo et al. 2014d) were able to avoid disturbances caused by various types of contamination. However, contrarily to these observations, some field studies have reported that under particular circumstances, organisms seem to prefer inhabiting disturbed areas, rather than avoiding them. For instance, along a mercury gradient in a shallow coastal lagoon in Portugal, higher density, biomass and growth productivity of the estuarine mudsnail *Peringia ulvae* were observed in the area with intermediate contamination, probably due to resource availability and presence of refuges (Cardoso et al. 2013). Also, the euryhaline fish *Gasterosteus aculeatus*, when exposed simultaneously to a predator chemical signal and toxic cyanobacteria chose feeding in a potentially toxic environment to reduce the risk of predation (Engström-Öst et al. 2006). In the same way, tilapia fry performed intermittent displacement to previously avoid fish farming effluent concentrations when food availability was increased (Araújo et al. 2016d). A study performed in the estuarine zones of New South Wales (Australia) showed that fish larvae were most abundant in disturbed and contaminated areas, probably due to nutrient enrichment (McKinley et al. 2011). Most probably, in all the latter cases, the final choice for inhabiting the disturbed habitat was headed by the balance between avoiding the exposure to contamination or staying in the disturbed habitat to avoid the most aversive/costly environment (e.g., higher predation risk, limited food resources).

The non-forced exposure approach allows a new vision about the role of contaminants as habitat disturbers within scenarios where toxic effects at the individual level are not necessarily expected to occur. In this context, the present chapter aims to show how risky environmental disturbances can be, if the focus encompasses also habitats, particularly regarding stressor-driven spatial displacement. Firstly, since the traditional ecotoxicological methods used to assess the toxic effects of contaminants on organisms do not consider this approach, we present an exposure system that allows changing the exposure paradigm: the free-choice, non-forced exposure system. Secondly, we discuss the ecological relevance of the spatial avoidance response for assessing the role of the contaminants as habitat disturbers. Then, an additional very important view-point addressed and discussed here is how ecotoxicological assays can contribute to assess the habitat selection process (avoidance/preference) under scenarios in which the co-occurrence of contamination and other aversive or attractive factors play a crucial role for organisms to decide whether to move or not from the contaminated habitat. A brief discussion with reference to moribundity (stupefaction), here defined as the loss of the ability of organisms to detect and avoid contamination, is also provided as contaminants can also impair the organism mobility, which in turn can slow down or even prevent their spatial avoidance. Moreover, the results of a short experiment with marine shrimps to assess the effect of post-larval stage on avoidance response are presented. Given that species dispersion patterns in contaminated areas may result, not only from avoidance, but also from preference responses (Van den Brink 2008), recolonization is, therefore, an essential response for the recovery of disrupted habitats, which has been neglected in ecotoxicological studies, but that can potentially be incorporated into ERA schemes by using the non-forced exposure system. In this context, the importance of understanding to what extent contaminants regulate organisms' displacement to avoid contaminant-disturbed habitats and recolonize recovering habitats and the inclusion of both responses in ERA, to increase ecological realism and decrease uncertainties in environmental decision making, are also discussed. In this chapter we did not consider avoidance measured as impairment in the ability to move, nor studies in which free-choice was restricted to two compartments (one treated and one control). Only studies using a free-choice, gradient, non-forced exposure multi-compartmented system were considered.

A Free-choice, Gradient, Non-forced Exposure System

As previously commented, ecotoxicological assays assume that organisms are forcedly exposed to contamination and consequently, that the possibility to avoid unfavourable environments is improbable. Indeed, traditional exposure systems consist in confined environments used to verify the biological effect on organisms and, in mesocosm studies, on communities and ecosystems resulting from a continuous and forced exposure to the stressful agent. However, an alternative tool to this approach, that allows organisms to select different environments, has been recently developed: the free-choice, gradient, non-forced exposure system (Fig. 1). This exposure system simulates a gradient or patches of contamination within a chamber divided into different compartments, through which organisms can freely move. In this non-forced exposure system, a physiological response cannot be directly related to a given concentration, as organisms can move across all the different concentrations or patches being assayed. However, two responses linked to habitat selection, and thus species dispersion, can be measured: avoidance and recolonization. An avoidance assay simulates the introduction of a stressor in the habitat inhabited by a population, assuming that unobstructed connectivity with an undisturbed/clean/ reference habitat exists. Organisms are exposed to all levels of the stressor along a gradient (from 0 percent—clean compartment—to 100 percent—the opposite extremity of the system) and may respond by moving away from the most disturbed compartments. Organisms are, thus, initially distributed along the interconnected compartments in an uniform manner (equal number of organisms per compartment) and their final spatial distribution, at the assay end, is registered, allowing one to calculate the stressor intensity eliciting a fixed amount of avoidance, such as the

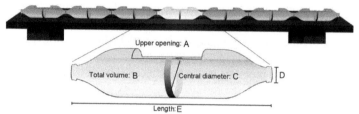

Fig. 1. Schematic representation of a free-choice, non-forced static exposure system.

median avoidance contaminant concentration (i.e., the contaminant concentration avoided by 50 percent of tested organisms; AC_{50}). A recolonization assay simulates a recovering habitat being made available to a population inhabiting a clean habitat. A similar gradient of the stressor is present along the interconnected compartments, but all the tested organisms are initially introduced in the clean extremity. Those who are insensitive to the stressor will occupy adjacent compartments. Their final spatial distribution, along the system, at the end of the assay, grant the possibility of computing the stressor intensity which was undetected and, thus, colonized by a given proportion of organisms, such as the median recolonization contaminant concentration (i.e., the contaminant concentration undetected and colonized by 50 percent of tested organisms: RC_{50}). These responses do not measure the effects of contaminants at the physiological level, but rather at the habitat level, because the free-choice, gradient, non-forced exposure system permits individuals to circumvent toxic effects resulting from continuous exposure to the stressful agent (for more details see next section).

To ensure that the system design does not influence the avoidance/recolonization response of the organisms, two factors are of concern: (i) the maintenance through the time of the linear or patchy gradient across the various compartments during the entire exposure period and (ii) the absence of a preferential distribution of the organisms when no contaminant (control) is present. The goal of the system calibration stage is to ensure that, in the absence of organisms, the system is able to maintain the different samples inside the interconnected compartments with a minimal mixture. An easy method to calibrate the free-choice, non-forced system is to establish a gradient of different concentrations of a reference compound and assess if and how fast the mixture occurs. Using sodium chloride as the reference compound, its concentration can be easily and accurately indirectly determined by measuring conductivity directly inside each compartment with no necessity to take samples; successive conductivity measurements along the selected exposure period can be made to verify the pattern of mixture inside the system. This procedure will allow verifying the maximum exposure period during which the system maintains the established gradient differences across the compartments. Even if the calibration indicates the capacity of the system to maintain the gradient across the compartments, samples can always be taken at the end of the exposure period to confirm the final concentration in each compartment. The temporal pattern of contaminant homogenization will depend on the system architecture regarding the degree of connectivity among compartments. Small distances and large interface areas between adjacent compartments will dictate a faster mixing. Obviously, the mixture of the contaminant concentrations among compartments will be accentuated when organisms are introduced into the system and it will be more intense the more turbulence the organisms provoke, which is a function of number, speed, size, shape, surface malleability and other hydrodynamical features. Therefore, the exposure period may be different for each species, to be able to maintain the desired contamination gradient at the end of the assay.

Regarding the test to prove the non-preferential distribution of the organisms along the system, only control medium is used to fill the system, to verify the possible occurrence of false positive avoidance/recolonization triggered by factors (confounding factors) other than the contaminant(s) being assayed, such as non-homogeneous temperature and electromagnetic radiation, including light, inclination, noise and other mechanical vibrations. These confounding factors resulting from the laboratory physical conditions can be evaluated and easily eliminated. Furthermore, behavioural factors (e.g., social attraction, such as schooling or repulsion) can also generate spatial distribution distortions, which can be compensated through load decrease and/or replication increase. The former approach reduces the probability of social interactions among individuals within each replicate, while the latter strategy eliminates or, at least, strongly smoothes discrepancies among replicates when data is pooled for the computation of the contaminant concentration eliciting a fixed amount of avoidance.

For whole-sediment experiments, in which only effects caused by sediment are targeted, different systems have been used, simulating contamination gradients (Araújo et al. 2012) or patches (Araújo et al. 2016b), although the goals were similar. Experiments tackling patchy distributions of sediment contamination have been performed in dishes divided into fields, so that, in each field, a natural or spiked sediment sample is deployed. To avoid desiccation, clean water should be carefully introduced above the sediment. There is no separation among sediment samples, and, thus, organisms can move freely among the fields.

Contaminants as Habitat Disturbers in Marine/Estuarine Environments

From way back, contaminants have been considered of concern due to the direct effects they can cause on the health status of organisms. Various potential changes in the health status of organisms have been proposed as test endpoints, to be considered in the assessment of risks due to contamination (Newman and Unger 2001, Walker et al. 2001). This rationale arises from the assumption that once contaminants are discharged into ecosystems, organisms are exposed to them and, then, a passive absorption process takes place. As a consequence of this absorption, a cascade of effects is expected to start taking place, beginning with sub-organismal effects, such as molecular alterations, and then effects succeed to the cell level, with metabolic and physiological disruptions, and follow to reach the whole-population level (behaviour, fecundity, etc.), and eventually community and ecosystem levels (Newman and Unger 2001, Gerhardt 2007). Although the velocity and intensity (sensitivity) of the response at high levels of biological organization can change in function of various factors (e.g., contaminant's mode of action, concentration, exposure time, species detoxification ability), such traditional approaches generally assume that the contact between organisms and the contaminant is mandatory (Lefcort et al. 2004). This assumption is correct for organisms that are not able to move, but if organisms with mobile skills are able to detect aversive agents and interpret such information as dangerous, it is expected that they try to avoid the stressful agent, escaping towards more favourable habitats (Hellou 2010, Tierney 2016). This response, here called avoidance, may prevent organisms to be continuously exposed to contamination and, therefore, direct deleterious effects at the individual level lose relevance, i.e., ecological meaning.

The displacement of individuals escaping from contamination brings serious consequences for the population, because avoidance consequences are similar to the death of the organisms: populations may totally, or partially, disappear (Lopes et al. 2004). Moreover, by triggering avoidance, contaminants can pose serious environmental risks even when concentrations are not supposed to adversely affect organisms' physiology (Rosa et al. 2008). In this context, contaminants, together with many other stressful agents, may act like habitat disturbers by triggering avoidance responses that affect species distribution patterns. The most appropriate way to verify the contaminant-driven spatial displacement of organisms is by using the free-choice, non-forced exposure system. This novel exposure approach allows for assessing how the spatial distribution of the organisms would be affected by the presence of any stressful agent, either along a contamination gradient or in a patchy contamination distribution (Araújo et al. 2016a). In a coastal habitat, for instance, if contaminants are discharged from a specific point-source, it is expected that a linear gradient contamination is formed. Therefore, the closer the organisms are to the contaminant source, the higher the environmental risk; though in time, as contamination is diluted with the dispersion, the repulsiveness caused by contaminants will be reduced and ultimately disappears. Under such scenarios, it is expected that organisms able to avoid contamination move to less disturbed areas, in which effects caused by the exposure to contamination are minimal or inexistent. Theoretically, the preferential spatial distribution of the organisms may be determined by the contamination gradient, so that lesser biodiversity would be expected where contamination is higher. Different studies with salmon (*Salmo salar*) and trout (*S. trutta*) have shown that contamination can trigger an early downstream migration (Saunders and Sprague 1967) and induce a preferential distribution, as the fish avoid disturbed and contaminated habitats (Åtland and Barlaup 1995, Woodward et al. 1995, Thorstad et al. 2005).

A contamination-driven spatial distribution can also occur in scenarios of patchy gradients of contamination. If contaminants are not uniformly and linearly dispersed, but rather form patches with different contamination levels, avoidance can occur resulting in the changing of the spatial rearrangement of the population (Spromberg et al. 1998). This is the typical contamination scenario expected for sediments, given that a homogeneous distribution of the contaminant is not expected to occur, but instead, patches with different contamination levels may be formed. These patches, not only create inhabitable areas, but may also prevent the spatial displacement of the organisms for large distances for less or even uncontaminated patches (Ares 2003). Recently, a few studies with coastal benthic species to assess the avoidance response in whole-sediment avoidance assays were reported. In a first study, the estuarine mudsnail *P. ulvae* was exposed to a linearly-disposed contaminated estuarine sediment by using a non-forced exposure approach (Araújo et al. 2012). Four different contamination levels were tested from mixing different proportions of test and reference sediment samples. Results showed that mudsnails avoided the most contaminated sediment samples and were able to move towards the reference sediment. The authors also observed that under extreme toxicity conditions, mudsnails retracted into their shells and became inactive instead of actively moving to more favourable locations along the contamination gradient. As discussed above, the contaminant spatial distribution may be spatially heterogeneous (Swartz et al. 1982, Lefcort et al. 2004), with the presence of uncontaminated sediment patches favouring the rearrangement of a population that will use those patches as refuge zones (Swartz et al. 1982). Motivated by this assumption, an experiment with the same mudsnail species was designed, aiming to assess its ability to detect different contamination levels in a heterogeneous contamination scenario with patchy sediment distribution (Araújo et al. 2016b). It was confirmed that the organisms were able to recognize the uncontaminated patches in that complex and heterogeneous scenario, changing their spatial distribution in function of the uncontaminated sediment distribution.

Recently, the approach using the avoidance and recolonization responses was employed to assess if the spatial distribution could be predicted by non-forced assays (Araújo et al. 2017). Previous visual observations of the estuarine gastropod *Olivella semistriata* along the Ecuadorian coast in Manta (Ecuador) indicated that organisms were not found close to two zones of untreated effluent discharges. Then organisms were exposed in the laboratory to different sediment samples and assessed by habitat preference by using avoidance and recolonization responses. In general, the results obtained indicated that the laboratory responses were in accordance with the spatial distribution of the snails densities observed in the field.

To date, the vast majority of the studies on avoidance in non-forced exposure systems have been carried out with freshwater species (Araújo et al. 2016a). Studies using estuarine or marine species are scarce. Alongside *P. ulvae* and *O. semistriata*, two other estuarine/marine species were investigated: whiteleg shrimp *Litopenaeus vannamei* larval stages and cobia fish *Rachycentron canadum* fries were exposed to a copper contamination gradient using a free-choice, gradient, non-forced exposure system (Araújo et al. 2016c). Concentrations between 0.1 and 1.0 µg Cu L^{-1} caused avoidance by around 80 percent of whiteleg shrimp larvae and by 50 percent of cobia fries.

Avoidance Alongside Preference as Measurements of Habitat Selection

Although contamination can play a crucial role in the habitat selection process by organisms, many other factors may influence an organism decision of escaping or not to another environment. Understanding this process is of environmental concern; however, once again, the forced exposure approach prevents this type of analysis since the possibility of the organism moving away does not exist. Besides, the free-choice, non-forced exposure system opens a complementary alternative to apply a new approach that includes not only avoidance and recolonization responses, but also habitat selection. An assay on habitat selection simulates the spatial arrangement of interconnected adjacent habitats with different characteristics including different degrees of disturbance (e.g., water pH, food, suspended particles, sediment type, predator chemical cues, salinity, anthropogenic compounds...), mimicking real or potential field scenarios of a metapopulation. Field-collected samples, or their laboratory imitations, are deployed along the system as spatially placed in the field. A highly disturbed habitat may occur between two reference sites and constitute a barrier against migration, configuring a habitat fragmentation outline. Such a situation can be mimicked and its level of obstruction can be preliminarily evaluated with this novel approach. Organisms are initially distributed along the interconnected compartments in an uniform manner (equal number of organisms per compartment) and their final spatial distribution, at the assay end, will no longer reflect, the mere intensity of a specific fully-controlled stressor, but, instead, all the antagonistic and synergistic cues that may be present.

If mobile organisms are able to detect contaminants and escape towards less disturbed areas, then in an area with the different contamination levels it would be expected a preferential spatial distribution of organisms in habitats with less disturbance (such as found by Araújo et al. 2017). However, curiously, this expectation does not always agree with the observed pattern of species distribution. In a shallow coastal lagoon in Portugal with a mercury gradient, the mudsnail *P. ulvae* seemed to prefer inhabiting an area with an intermediate contamination level (Cardoso et al. 2013). Similarly, in New South Wales, Australia, the highest abundance of fish larvae was observed in the most contaminated estuaries (McKinley et al. 2011). Although extraordinary, this behavior could be explained by the presence of other factors than the contamination that attracts organisms to that particular habitat. For instance, the fish *G. aculeatus* when exposed simultaneously to two different environments, one with a chemical predator signal and the other one with toxic cyanobacteria, preferred feeding in the potentially toxic environment to reduce the risk of predation (Engström-Öst et al. 2006). Similarly, the fish *Danio rerio* when exposed to environmentally relevant herbicide concentrations preferred to move into a plume of contaminants (mixture of four herbicides) due to an attractive effect caused by herbicides (Tierney et al. 2011). And recently, a study with tilapia fry (*Oreochromis* sp.) showed that the intensity of the avoidance response to an effluent sample was reduced when food availability was higher in the environments with higher contamination levels (Araújo et al. 2016d).

The possibility to assess the habitat preference selection process by organisms in an ecotoxicological context changes the paradigm of the exposure focus from individuals to habitats. The ecological risk posed by contaminants is not only linked to the effects they can cause on organisms—contaminants as toxicants—but also to the disturbance they can cause in the dynamic patterns of organisms migration—contaminants as habitat disturbers—and thus on populations' spatial distribution. Moreover, this approach will contribute to understand the interactions between contaminants and other environmental factors and, then, to assess how determining such interactions can be for the habitat selection process.

Avoidance and Ontogenesis

There is a curious problem associated with spatial active avoidance: From which point, during ontogeny (early life stage, juveniles or adult), may avoidance play an important role in preventing the population from being exposed to a stressor? The same rationale applies to recolonization and habitat selection.

Although ecotoxicological assays are preferably performed with early life stages as they are generally more sensitive, in the case of avoidance assays, this supposition might not be true because the sensory system that detects contamination can be underdeveloped in juvenile individuals and also because their moving ability can be insufficient. Therefore, we performed studies to compare the differences in the sensitivity of different post-larval stages of the estuarine whiteleg shrimp to avoid contamination. Post-larvae of 5, 10, 20, 45, 60, and 80 days were exposed to a copper gradient and responded in a different way, so that a slight trend to avoid more intensively copper gradient was observed for larvae of 10 to 45 days. From avoidance percentage values (Fig. 2), it was calculated by researchers that the copper concentration that causes an avoidance by 50 percent of the exposed population

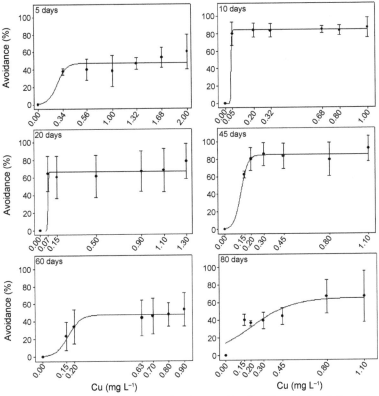

Fig. 2. Avoidance response of different post-larval stages (days) of the whiteleg shrimp *Litopenaeus vannamei* exposed to a linear copper gradient in a free-choice, non-forced exposure system.

Table 1. Values (in mg L^{-1}) of AC$_{50}$ (avoidance concentration for 50 percent of the exposed population) and confidence intervals obtained for different post-larval stages (days) of the whiteleg shrimp *Litopenaeus vannamei* exposed to a linear copper gradient.

Post-larval stages (days)	AC$_{50}$	Confidence intervals
5	1.41	0.89–5.09
10	< 0.05	nc
20	< 0.08	nc
45	< 0.15	nc
60	0.86	0.66–1.32
80	0.40	0.27–0.62

nc: not calculated.

(AC$_{50}$) and the values obtained are presented in Table 1. Our results showed that the AC for larvae of five days was the highest one, indicating less sensitivity of this stage to detect and avoid copper. Larvae of 60 and 80 days presented AC$_{50}$ almost similar values; however, the lowest AC$_{50}$ values were recorded for larval stages of 10, 20 and 45 days (lower than the lowest concentration used). Although the mechanism that can determine these differences was not studied, we hypothesized that younger larvae (5 days) were less sensitive because they were not physiologically able to detect as efficiently the copper gradient as did the larvae of 10, 20 and 45 days. On the other hand, the lower sensitivity of the larvae of 60 and 80 days could be attributed to the higher larval development that provides higher physiological tolerance to copper contamination. Our results indicated that, contrary to the assumption used in traditional toxicity assays, younger organisms are not necessarily more sensitive and, therefore, more advisable to be used in avoidance assays. The age of the organisms to be chosen should be carefully assessed using stages to ensure that they are neither unable to detect contamination or to move away from it, nor that they are less sensitive to the contamination.

Avoidance and Moribundity

At what concentrations does avoidance cease to play an appreciable role in preventing toxicity? Many studies have treated avoidance as the loss of ability to move, focusing on how contaminants impair or provoke changes in the patterns of movements linked to swimming: velocity, direction and distance. In the non-forced exposure system, this loss of ability to move is used to characterize a status of moribundity that slows down or prevents spatial active avoidance. Moribundity, therefore, can occur when the concentration of the contaminant is so high that severe toxic physiological effects take place before the reaction to avoid it from being triggered (e.g., contaminants can act as disruptors of the nervous system impairing the ability of the organism to detect and escape from contamination). For instance, moribundity was observed in the above-mentioned avoidance assay with fries of the estuarine cobia fish, at concentrations of copper of 1.6 and 1.8 µg L^{-1} (Araújo et al. 2016c). Also, stream macroinvertebrates of the genus *Anomalocosmoecus* exposed to a soluble fraction of crude oil (Araújo et al. 2014e), cladocerans and copepods exposed to the insecticide endosulfan and to metals (Gutierrez et al. 2012) and tadpoles exposed to copper (Araújo et al. 2014b) showed signs of moribundity at the highest tested concentrations. Different contaminants have also been shown to impair the swimming ability of tadpoles and, therefore, indirectly affected the avoidance ability (Wojtaszek et al. 2004, Chen et al. 2007, Denoël et al. 2013). In some cases, impairment on the neuro-muscular function has been attributed as the cause of decreasing or loss of the organism's ability to escape (Chen et al. 2007). In those cases, spatial avoidance is not possible, as mobility of organisms was impaired, therefore, traditional forced exposure plays a crucial role in assessing the potential toxicity at the individual level.

Final Considerations: Incorporating Avoidance, Recolonization and Habitat Selection into ERA Schemes

As discussed above, contaminants can trigger avoidance by organisms from a disturbed habitat. On the other hand, when a habitat starts to present signs of recovery, a displacement of the organisms towards the recovering habitat is expected. This hypothesis has recently been attested through an experiment with fish, in which the process of recolonization of a habitat recovering from an acid mine drainage contamination event could be predicted in non-forced exposure assays (Araújo et al. 2018). We observed that the acid mine drainage dilutions that elicited avoidance of 50, 20 or 80 percent of the exposed population were the same, allowing a recolonization of 50, 80 or 20 percent, respectively, of that same population. These first results point out the importance of including avoidance and recolonization responses in ERA, as complementary responses elucidating the role of contaminants in regulating the patterns of species distribution.

In assessing environmental disturbances caused by stressors, the non-forced exposure approach is a novel tool allowing a change in the paradigm related to the forced exposure scenarios traditionally and widely used in ecotoxicological studies. Although the forced exposure designs lack ecological relevance in many circumstances, it has its strengths, permitting researchers to establish precise and accurate cause-effect relationships between contaminant concentrations and a multitude of organism responses. Therefore, the non-forced exposure approach should be used as a complementary (not substitutive) tool through which potential effects on the spatial displacement of the organisms can be assessed.

Besides integrating the biological effects at the habitat level, avoidance, recolonization and habitat selection assays can help predicting secondary effects expected to occur as a result of the contamination-driven preferential spatial species distribution. The immediate effect to be expected is overpopulation in the preferred habitat and a population reduction or even extinction in the disturbed habitat. From this spatial rearrangement, changes in inter- and intra-specific relationships (e.g., competition and predation) may occur (Fleeger et al. 2003). Additionally, at mid-or long-term, disruption in some biogeochemical processes at the ecosystem level can be anticipated if the evaded population is not replaced by another functionally redundant one (De Laender et al. 2008). Due to the habitat fragmentation caused by "chemical barriers" and the isolation of the inhabitable habitats, the flux of individuals and, consequently, of genes may also be reduced, which leads to the viability decrease of the small-sized populations (Ribeiro and Lopes 2013).

As discussed in the present chapter, contaminants are not, however, the only factor determinant for the habitat selection by organisms. Many other abiotic and biotic factors other than contamination present in disturbed habitats may play an important role on the decision to move from or stay on in a given habitat. Therefore, once the balance between aversive and attractive parameters is made, if organisms decide to inhabit (prefer) a disturbed habitat rather than avoiding it, toxic effects at the individual level are expected to occur. In this case, forced exposure approaches will become relevant.

The present chapter intended to discuss the importance of adding the non-forced exposure approach to ERA, highlighting the ecological relevance of avoidance and recolonization responses (once organisms are able to detect and avoid environmental disturbers), the biological effects at the ecosystem level (loss of individuals and population downsizing in disturbed habitats) and how determining the contamination for habitat selection in complex multi-parameter field scenarios (balance between aversive and attractive factors) is. The inclusion of the non-forced exposure approach into ERA should be encouraged because it will bring the disciplines of ecology and ecotoxicology closer.

Acknowledgments

CVM Araújo is grateful to Spanish Ministry of Economy and Competitiveness for the Juan de la Cierva contract (IJCI-2014-19318). This study was also partially funded by the European Fund for Economic and Regional Development (FEDER) through the Program Operational Factors of Competitiveness (COMPETE) and National Funds though the Portuguese Foundation of Science and Technology (postdoctoral fellowship to M. Moreira-Santos—SFRH/BPD/99800/2014, contract IT057-18_7285).

References

Araújo, C.V.M., Blasco, J. and Moreno-Garrido, I. 2012. Measuring the avoidance behaviour shown by the snail *Hydrobia ulvae* exposed to sediment with a known contamination gradient. Ecotoxicology 21(3): 750–758.

Araújo, C.V.M., Moreira-Santos, M., Sousa, J.P., Ochoa-Herrera, V., Encalada, A.C. and Ribeiro, R. 2014a. Active avoidance from a crude oil soluble fraction by an Andean paramo copepod. Ecotoxicology 23(7): 1254–1259.

Araújo, C.V.M., Shinn, C., Moreira-Santos, M., Lopes, I., Espíndola, E.L.G. and Ribeiro, R. 2014b. Copper-driven avoidance and mortality in temperate and tropical tadpoles. Aquat. Toxicol. 146: 70–75.

Araújo, C.V.M., Shinn, C., Vasconcelos, A.M., Ribeiro, R. and Espíndola, E.L.G. 2014c. Preference and avoidance responses by tadpoles: the fungicide pyrimethanil as a habitat disturber. Ecotoxicology 23(5): 851–860.

Araújo, C.V.M., Shinn, C., Mendes, L.B., Delello-Schneider, D., Sanchez, A.L. and Espíndola, E.L.G. 2014d. Avoidance response of *Danio rerio* to a fungicide in a linear contamination gradient. Sci. Total Environ. 484: 36–42.

Araújo, C.V.M., Moreira-Santos, M., Sousa, J.P., Ochoa-Herrera, V., Encalada, A.C. and Ribeiro, R. 2014e. Contaminant as habitat disturbers: PAH-driven drift by Andean paramo stream insects. Ecotoxicol. Environ. Saf. 108: 89–94.

Araújo, C.V.M., Moreira-Santos, M. and Ribeiro, R. 2016a. Active and passive spatial avoidance by aquatic organisms from environmental stressors: A complementary perspective and a critical review. Environ. Int. 92-93: 405–415.

Araújo, C.V.M., Martinez-Haro, M., Pais-Costa, A.J., Marques, J.C. and Ribeiro, R. 2016b. Patchy sediment contamination scenario and the habitat selection by an estuarine mudsnail. Ecotoxicology 25(2): 412–418.

Araújo, C.V.M., Cedeño-Macías, L.A., Vera-Vera, V.C., Salvatierra, D., Rodríguez, E.N.V., Zambrano, U. and Kuri, S. 2016c. Predicting the effects of copper on local population decline of 2 marine organisms, cobia fish and whiteleg shrimp, based on avoidance response. Environ. Toxicol. Chem. 35(2): 405–410.

Araújo, C.V.M., Rodríguez, E.N.V., Salvatierra, D., Cedeño-Macias, L.A., Vera-Vera, V.C., Moreira-Santos, M. and Ribeiro, R. 2016d. Attractiveness of food and avoidance from contamination as conflicting stimuli to habitat selection by fish. Chemosphere 163: 177–183.

Araújo, C.V.M., Salvatierra, D., Vera-Vera, V.C., Cedeño-Macias, L.A., Benavides, K., Macías-Mayorga, D., Moreira-Santos, M. and Ribeiro, R. 2017. Avoidance and recolonization responses of the gastropod *Olivella semistriata* exposed to coastal sediments. pp. 239–254. *In*: Araújo, C.V.M. and Shinn, C. (eds.). Ecotoxicology in Latin America. Nova Science Publishers, USA.

Araújo, C.V.M., Moreira-Santos, M. and Ribeiro, R. 2018. Stressor-driven emigration and recolonisation patterns in disturbed habitats. Sci. Total Environ. 643: 884–889.

Ares, J. 2003. Time and space issues in ecotoxicology: population models, landscape pattern analysis, and long-range environmental chemistry. Environ. Toxicol. Chem. 22: 945–957.

Åtland, Å. and Barlaup, B.T. 1995. Avoidance of toxic mixing zones by atlantic salmon and brown trout. Environ. Pollut. 90: 203–208.

Cardoso, P.G., Sousa, E., Matos, P., Henriques, B., Pereira, E., Duarte, A.C. and Pardal, M.A. 2013. Impact of mercury contamination on the population dynamics of *Peringia ulvae* (Gastropoda): implications on metal transfer through the trophic web. Estuar. Coast. Shelf Sci. 129: 189–197.

Chapman, P.M. 1995. Ecotoxicology and pollution—key issues. Mar. Pollut. Bull. 31: 167–177.

Chen, T.-H., Gross, J.A. and Karasov, W.H. 2007. Adverse effects of chronic copper exposure in larval northern leopard frogs (*Rana pipiens*). Environ. Toxicol. Chem. 26: 1470–1475.

De Laender, F., De Schamphelaere, K.A.C., Vanrolleghem, P.A. and Janssen, C.R. 2008. Is ecosystem structure the target of concern in ecological effect assessments? Wat. Res. 42: 2395–2402.

De Lange, H.J., Sala, S., Vighi, M. and Faber, J.H. 2010. Ecological vulnerability in risk assessment —A review and perspectives. Sci. Total Environ. 408: 3871–3879.

Denoël, M., Libon, S., Kestemont, P., Brasseur, C., Focant, J.-F. and De Pauw, E. 2013. Effects of a sublethal pesticide exposure on locomotor behavior: a video-tracking analysis in larval amphibians. Chemosphere 90: 945–951.

Engström-Öst, J., Karjalainen, M. and Viitasalo, M. 2006. Feeding and refuge use by small fish in the presence of cyanobacteria blooms. Environ. Biol. Fish. 76: 109–117.

Fleeger, J.W., Carman, K.R. and Nisbet, R.M. 2003. Indirect effects of contaminants in aquatic ecosystems. Sci. Total Environ. 317: 207–233.

Gerhardt, A. 2007. Aquatic behavioral ecotoxicology—prospects and limitations. Hum. Ecol. Risk Assess. 13: 481–491.

Gutierrez, M.F., Paggi, J.C. and Gagneten, A.M. 2012. Microcrustaceans escape behavioras an early bioindicator of copper, chromium and endosulfan toxicity. Ecotoxicology 21: 428–438.

Hellou, J., Cook, A., Lalonde, B., Walker, P., Dunphy, K. and MacLeod, S. 2009. Escape and survival of *Corophium volutator* and *Ilyanassa obsoleta* exposed to freshwater and chlorothalonil. J. Environ. Sci. Health Part A 44: 778–790.

Hellou, J. 2010. Behavioural ecotoxicology, an "early warning" signal to assess environmental quality. Environ. Sci. Pollut. Res. 18: 1–11.

Kravitz, M.J., Lamberson, J.O., Ferraro, S.P., Swartz, R.C., Boese, B.L. and Specht, D.T. 1999. Avoidance response of the estuarine amphipod Eohaustorius estuarius to polycyclic aromatic hydrocarbon-contaminated, field-collected sediments. Environ. Toxicol. Chem. 18: 1232–1235.

Lefcort, H., Abbott, D.P., Cleary, D.A., Howell, E., Keller, N.C. and Smith, M.M. 2004. Aquatic snails from mining sites have evolved to detect and avoid heavy metals. Arch. Environ. Contam. Toxicol. 46: 478–484.

Lopes, I., Baird, D.J. and Ribeiro, R. 2004. Avoidance of copper contamination by field populations of *Daphnia longispina*. Environ. Toxicol. Chem. 23: 1702–1708.

Luoma, S.N. 1996. The developing framework of marine ecotoxicology: Pollutants as a variable in marine ecosystems? J. Exp. Mar. Biol. Ecol. 200: 29–55.

McKinley, A.C., Miskiewicz, A., Taylor, M.D. and Johnston, E.L. 2011. Strong links between metal contamination, habitat modification and estuarine larval fish distributions. Environ. Pollut. 159: 1499–1509.

Moreira-Santos, M., Donato, C., Lopes, I. and Ribeiro, R. 2008. Avoidance tests with small fish: determination of the median avoidance concentration and of the lowest-observed-effect gradient. Environ. Toxicol. Chem. 27: 1576–1582.

Newman, M.C. and Unger, M.A. 2001. Population Ecotoxicology. Lewis Publishers, USA.

Pedder, S.C.J. and Maly, E.J. 1986. The avoidance response of groups of juvenile brook trout. Aquat. Toxicol. 8: 111–119.

Ribeiro, R. and Lopes, I. 2013. Contaminant driven genetic erosion and associated hypotheses on alleles loss, reduced population growth rate and increased susceptibility to future stressors: an essay. Ecotoxicology 22: 889–899.

Richardson, J., Williams, E.K. and Hickey, C.W. 2001. Avoidance behaviour of freshwater fish and shrimp exposed to ammonia and low dissolved oxygen separately and in combination. New Zeal. J. Mar. Freshwat. Res. 35: 625–633.

Rosa, R., Moreira-Santos, M., Lopes, I., Picado, A., Mendonça, E. and Ribeiro, R. 2008. Development and sensitivity of a 12-h laboratory test with *Daphnia magna* Straus based on avoidance of pulp mill effluents. Bull. Environ. Contam. Toxicol. 81: 464–469.

Rosa, R., Materatski, P., Moreira-Santos, M., Sousa, J.P. and Ribeiro, R. 2012. A scaled-up system to evaluate zooplankton spatial avoidance and population immediate decline concentration. Environ. Toxicol. Chem. 31: 1301–1305.

Saunders, R.L. and Sprague, J.B. 1967. Effect of copper–zinc mining pollution on a spawning migration of Atlantic salmon. Wat. Res. 1: 419–432.

Schmitt-Jansen, M., Veit, U., Dudel, G. and Altenburger, R. 2008. An ecological perspective in aquatic ecotoxicology: Approaches and challenges. Basic Appl. Ecol. 9: 337–345.

Sherman, B.H. 2000. Marine ecosystem health as an expression of morbidity, mortality and disease events. Mar. Pollut. Bull. 41: 232–254.

Spromberg, J.A., John, B.M. and Landis, W.G. 1998. Metapopulation dynamics: Indirect effects and multiple distinct outcomes in ecological risk assessment. Environ. Toxicol. Chem. 7: 1640–1649.

Swartz, R.C., Deben, W.A., Sercu, K.A. and Lamberson, J.O. 1982. Sediment toxicity and distribution of amphipods in Commencement bay, Washington, USA. Mar. Pollut. Bull. 13: 359–364.

Thorstad, E.B., Forseth, T., Aasestad, I., Økland, F. and Johnsen, B.O. 2005. *In situ* avoidance response of adult Atlantic salmon to waste from the wood pulp industry. Wat. Air Soil Pollut. 165: 187–194.

Tierney, K.B., Sekela, M.A., Cobbler, C.E., Xhabija, B., Gledhill, M., Ananvoranich, S. and Zielinski, B.S. 2011. Evidence for behavioral preference toward environmental concentrations of urban-use herbicides in a model adult fish. Environ. Toxicol. Chem. 30: 2046–2054.

Tierney, K.B. 2016. Chemical avoidance responses of fishes. Aquat. Toxicol. 174: 228–241.

Van den Brink, P.J. 2008. Ecological risk assessment: from book-keeping to chemical stress ecology. Environ. Sci. Technol. 42: 8999–9004.

Walker, C.H., Hopkin, S.P., Sibly, R.M. and Peakall, D.B. 2001. Principles of ecotoxicology. Taylor & Francis, Boca Raton, FL, USA.

Ward, D.J., Simpson, S.L. and Jolley, D.F. 2013. Avoidance of contaminated sediments by an amphipod (*Melita plumulosa*), a harpacticoid copepod (*Nitocra spinipes*), and a snail (*Phallomedusa solida*). Environ. Toxicol. Chem. 32: 644–652.

Wojtaszek, B.F., Staznik, B., Chartrand, D.T., Stephenson, G.R. and Thompson, D.G. 2004. Effects of Vision® herbicide on mortality, avoidance response, and growth of amphibian larvae in two forest wetlands. Environ. Toxicol. Chem. 23: 832–842.

Woodward, D.C., Hansen, J.A., Bergman, H.L., Little, E.E. and DeLonay, A.J. 1995. Brown trout avoidance of metals in water characteristic of the Clark Fork River, Montana. Can. J. Fish. Aquae. Sci. 52: 2031–2037.

Toxicity of Rare Earth Elements to Marine Organisms

2

Pedro Luís Borralho Aboim de Brito

INTRODUCTION

Rare earth elements (REE) comprise the series of Lanthanides[1] (atomic numbers 57–71), scandium (atomic number 21) and yttrium (atomic number 39). Lanthanides are elements of Group IIIA of the periodic table, having similar physical and chemical properties due to their electronic configurations (IUPAC 2005). In this chapter, only lanthanides and yttrium were considered as rare earth elements due to the similarities between these elements, since the lighter element (scandium) has a relatively small ionic radius, resulting in a different chemistry from the previous elements.

Rare earths elements were first discovered in 1788, in Ytterby, Sweden, in a black mineral called Ytterbite. Later, in 1794, Professor Gadolin, from the University of Åbo (Turku), Finland, studied the same mineral and found for the first time a new kind of "earth" which he called "rare earths" (Greinacher 1981). With the development in 1885 of the lanthanum and cerium gas mantles (Welsbach mantles), for the illumination of factories and streets, the exploitation of REE mineral resources began, and has been increasing significantly until today. The major mining sites of REE's mineral ore to feed Welsbach blanket manufacturing were in northern Europe (mainly Scandinavia) and in the United States (North and South Carolina), and later in India, Brazil, and China (Inner Mongolia). Over the past 50 years, the two largest REE mining sites were Mountain Pass, California, USA, and Bayan Obo,

IPMA - Av. Alfredo Magalhães Ramalho, 6, 1495-165 Lisboa.
Email: pbrito@ipma.pt

[1] According to IUPAC (IUPAC 2005), the Lanthanides series comprise the elements lanthanum (La), cerium (Ce), praseodymium (Pr), neodymium (Nd), promethium (Pm), samarium (Sm), europium (Gd), terbium (Tb), dysprosium (Dy), holmium (Ho), erbium (er), thulium (Tm), ytterbium (yb) and lutetium (Lu).

Baotou, Inner Mongolia, China (Klinger 2015). With the crisis in 2010, as a result of the decline in REE's export quotas from China, the REE technology-dependent countries had to find other sources of raw material not only through land mining (and more recently in the deep ocean), but also through the recycling of end-of-life products (Eggert 2011).

New REE mining sites around the world have emerged in recent years, from Strange Lake, in Canada (Gysi and Williams-Jones 2013), to Mount Weld (Lynas Corporation Ltd 2017), in Western Australia, through the Kvanefjeld (Greenland Minerals and Energy Ltd 2017), in Greenland, to the seabed of Japan (Takaya et al. 2018), among others.

Due to the unique properties of REE, these elements have become valuable for the most different industrial applications, being used extensively in cutting edge technology, but also in the traditional industries such as metallurgy, petroleum and textiles. The use of REE in the most diversified applications has resulted in an increase in environmental concentrations and in exposure to these elements (see Gwenzi et al. 2018 for a review).

The aquatic environments act as REE sinks, through the absorption of these elements to suspended particulate material that eventually sediment in the bottom and resuspends because of bioturbation and hydrodynamics effects. The nature of particles, on the other hand, may promote more or less adsorption/desorption reactions in the medium. Mineral phases with high surface area, such as Fe-Mn oxyhydroxides and aluminosilicates, act as efficient REE scavengers (e.g., Johannesson and Zhou 1999, Quinn et al. 2004).

The availability of REE in these environments generally depends on several processes occurring in mobile fractions such as sorption-desorption reactions between the water column and the sediment, co-precipitation with particulate matter and colloids or the complexation with inorganic and/or organic ligands (Migaszewski and Gałuszka 2015, Quinn et al. 2004), as well as the nature and physic-chemical properties of these elements and their compounds that influence their behaviour.

REE Toxicity

Over the past three decades, most toxicity studies have relied on heavy metals (e.g., cadmium, mercury, chromium, lead, nickel, etc.) due to the high anthropogenic concentrations found in the environment. With the development of new technologies and materials, the need for new toxicological studies has significantly increased. Among other elements, REE (also classified as Technology-Critical Elements or TCE),[2] are part of a vast group of "new elements" that need in-depth studies not only in material engineering or medicine, but mainly in ecology and toxicity.

In one of the earlier REE toxicity studies, these elements were considered only slightly toxic (Haley 1965). Although there are some studies on REE toxicity today,

---

[2] These elements are termed "critical" because of their high demand versus their limited supply, and overall scarcity in economically relevant concentrations.

the information is still very scarce, and has mainly focused on classic laboratory tests using rats or fruit flies, and some terrestrial plants, namely those species of economic interest due to their use in human or animal nutrition (see Rim et al. 2013 for a literature review).

Rare earth elements toxicity is, like other chemical elements, generally influenced by the characteristics of organisms such as age, size or maturation, the form of exposure and the concentrations of each element. Dose-response relationships of REE are generally biphasic, with stimulatory or beneficial effects at low concentrations, and inhibitory or toxic effects at high concentrations (Pagano et al. 2015). Several studies have shown that low REE concentrations promote growth of both aquatic and terrestrial organisms, some of which have been used since 1990 as micronutrients in fertilizers (Zhang and Shan 2001 and references hereafter) and more recently as livestock feed additives (He and Rambeck 2000, Xun et al. 2014).

As mentioned above, REE toxicity studies have focused particularly on species of economic interest or those directly related to public health. However, some studies have been published in the last two decades involving more marine species, from bacteria to vertebrates, thus covering a greater biodiversity (González et al. 2014 and references hereafter).

Bacteria

Most studies on REE in bacteria report work on soil species (e.g., Ozaki et al. 2006, Tsuruta 2007) or investigation of alternative industrial methods for extraction and purification of rare earth oxides (REO) (e.g., Bonificio and Clarke 2016, and references hereafter, Park et al. 2017), with few studies on freshwater bacteria (Rodea-Palomares et al. 2011) and even less on marine species (González et al. 2015, Kurvet et al. 2017).

In a study evaluating the ecotoxicity of cerium (Ce), gadolinium (Gd) and lutetium (Lu) on Gram-negative bacteria *Vibrio fischeri* (González et al. 2015), the reduction of bacterial bioluminescence was observed after 30 minutes of exposure. The half-maximal effective concentration (EC_{50}) nominal value for Lu was 3,200 mg.L^{-1}, whereas for Ce and Gd these values were higher than 6,400 mg.L^{-1}. The highest inhibition was found for Lu, at all concentrations tested, with most significant differences in concentrations higher than 800 mg.L^{-1}, with toxicity following the order Lu > Gd > Ce.

Kurvet et al. (2017) studied the effects of five REE (lanthanum, cerium, praseodymium, neodymium and gadolinium) and nine doped REO on the same bacteria. The authors observed that all REO produced reactive oxygen species (ROS), while all soluble REE were toxic to the bacteria under study, with half-effective concentration, EC_{50} 3.5–21 mg.L^{-1} and minimal bactericidal concentration, MBC 6.3–63 mg.L^{-1}. However, no REO toxicity was found (EC_{50} > 500 mg.L^{-1}; MBC > 500 mg.L^{-1}), except for La$_2$NiO$_4$ (MBC 25 mg.L^{-1}), whereas the use of toxic metals (such as Ni) as REO dopants can significantly decrease their environmental safety. According to the kinetic acute bioluminescence inhibition assay results, the toxicity of REE was triggered by disturbing the integrity of the cell membrane. However, these authors state that REE and REO do not appear to have harmful effects on the bacteria since they are currently produced in moderate amounts and form insoluble salts and/or oxides in the environment.

Algae and Other Plants

Most of REE's ecotoxicity studies in algae refer to freshwater microalgae such as *Chlorella* sp. and *Raphidocelis subcapitata* (Balusamy et al. 2015, Fujiwara et al. 2008, González et al. 2015, Hao et al. 1997, Joonas et al. 2017).

Fujiwara et al. (2008) evaluated the toxicity of certain lanthanides (La and Eu) in *Chlorella kesseleri*, estimating an IC_{50} of 43,475 mg.L^{-1} for La and 43,612 mg.L^{-1} for Eu. Trials with *Chlorella* sp., exposed to high concentrations of La_2O_3 (1,000 mg.L^{-1}) nanoparticles, did not show significant toxic effects after 72 hours exposure, but enhanced growth rate and biomass production (Balusamy et al. 2015).

Joonas et al. (2017) in a study on the effects of various REO on the green algae *Raphidocelis subcapitata* observed that this species were not viable at REE concentrations above 1 mg.L^{-1}. In the growth inhibition assays (72 hours) for the elements Ce, Gd, La and Pr the EC_{50} values ranged from 1.2 to 1.4 mg.L^{-1}, whereas for the REO these values were between 1 and 98 mg.L^{-1}. The authors stated that growth inhibition could be the result of sequestration of nutrients from the growth medium, while the adverse effects of REO would be due in part to the entrapment of algae between the particle agglomerates. As a conclusion it is stated that there is presumably no acute risk for unicellular aquatic algae, since the production rates of those REO particles are negligible compared to other forms of REE.

Tai et al. (2010) investigated the toxicity of lanthanides in the single-celled marine algae *Skeletonema costatum* and observed that each element produced similar effects in this species. High concentrations of lanthanides ($\approx$ 4.0 mg.L^{-1}) resulted in a 50 percent reduction in algae growth compared to the controls after a period of 96 h of exposure. These authors found that a multielement solution with equivalent concentrations of each, lanthanide had the same inhibitory effect on cells as each individual element with the same total concentration, suggesting that this species might not be able to sufficiently differentiate between practically chemically identical elements.

In the study of exposure of the green seaweed *Ulva lactuca* to Ce^{3+}, Eu^{3+}, Gd^{3+} and Yb^{3+}, Stanley and Byrne (1990) observed that the behaviour of REE was strongly dependent on the complexation of the solution. The affinity of this marine algae for these elements varied according to the concentrations of the carbonate ion present in the solution, showing the order $Eu^{3+} > Gd^{3+} > Ce^{3+} > Yb^{3+}$.

Similarly, REE studies in aquatic plants are mostly found in freshwater species such as *Lemna minor*, *Hydrilla verticillate* or *Potamogeton pectinatus*, with few studies focused on the toxicity of lanthanides (Paola et al. 2007, Wang et al. 2007). Paola et al. (2007) found a growth of *Lemna minor* roots apexes and yellowing of leaves when exposed for 5 days to REE nitrate solutions (La, Ce, Pr and Nd). These authors also verified that the protein content decreased, while an increase of ascorbate peroxidase, dehydroascorbate reductase and ascorbate free radical reductase activities was observed. An increase in total ascorbate and glutathione content was also observed. These results led the authors to consider that *Lemna minor* could be a useful tool to study the biological effects of REE in aquatic environments.

Wang et al. (2007) observed that *Hydrilla verticillate*, when exposed to concentrations of La and Ce higher than 10 μM (1,389.05 and 1,401.16 μg.L^{-1}, respectively), presented oxidative damage evidenced by increased lipid peroxidation and decreased levels of chlorophyll and protein.

Animals

In a study to measure lanthanum acute and chronic toxicity to *Daphnia carinata*, Barry and Meehan (2000) found that, in the medium with the lowest carbonated hardness (tap water), the 48-h EC$_{50}$ of La was 43 μg.L^{-1}, contrasting with the value of 1,180 μg.L^{-1} for harder water (ASTM standard). In a diluted seawater medium (DW), the 48-h EC$_{50}$ was similar to the one of the medium with the lowest hardness (48 μg.L^{-1}) but a significant precipitation of La was observed. The chronic toxicity of *Daphnia* was measured in the DW and ASTM media. The authors observed that mortality was a more sensitive end-point than growth or reproduction in both chronic experiments, with 100 percent mortality at concentrations ≥ 80 μg.L^{-1} by day six of the experiment using DW media, but no effect on survival growth or reproduction at lower concentrations. In the ASTM media, a significant mortality to Daphnia at concentrations ≥ 39 μg.L^{-1}, while no effect of the on growth of surviving daphnids at concentrations ≥ 57 μg.L^{-1} was observed. However, the authors found that second brood clutch sizes were significantly increased at 30, 39 and 57 μg.L^{-1} compared with controls. Lanthanum also caused delayed maturation in *Daphnia*.

Oral et al. (2010), in a study to evaluate the toxicity of La^{4+} and Ce^{3+} in sea urchin *Paracentrotus lividus* embryos and sperm, observed a mortality of 100 percent in reared embryos in 1.40 μg.L^{-1} Ce^{3+}, whereas identical concentrations of La^{4+} reported 100% percent developmental defects, without causing embryonic mortality. In the embryos exposed to Ce^{3+}, but not exposed to La^{4+}, the authors observed a significant concentration-related mitotoxic effect at concentrations between 14 and 420 ng.L^{-1}. Both elements induced a decrease in the success of sperm fertilization at the highest concentration (1.40 μg.L^{-1}), while offspring exposed to Ce^{3+} showed a significant increase in developmental defects, but not observed in individuals exposed to La^{4+}.

In another study with four species of sea urchins (*Paracentrotus lividus*, *Arbacia lixula*, *Heliocidaris tuberculata* and *Centrostephanus rodgersii*), it was observed that exposure to Gd resulted in inhibition or alteration of skeleton growth in larvae (Martino et al. 2016). These authors also observed a great variability in sensitivity to Gd, with the EC$_{50}$ varying between 8.8 and 21 μg.L^{-1}. Martino et al. (2017) also found that Gd is able to affect different aspects of sea urchin *Paracentrotus lividus* development such as morphogenesis, biomineralization and stress response through autophagy. In a more recent study, Martino et al. (2018) found a general delay of the Mediterranean *Paracentrotus lividus* and of the Australian *Heliocidaris tuberculata* embryos development at 24 h post-fertilization, and a strong inhibition of skeleton growth at 48 h, while total Gd and Ca content in the larvae showed a time and concentration-dependent increase in Gd, in parallel with a reduction in Ca.

Most REE toxicity studies in fish refer to freshwater species (e.g., Cui et al. 2012, Figueiredo et al. 2018, Hongyan et al. 2002).

Hongyan et al. (2002) investigated physiological and biochemical disturbances in the liver of *Carassius auratus* exposed to different concentrations of ytterbium in solution. The results showed that glutamate-pyruvate transaminase (GPT) activity in liver was stimulated at 0.05 mg.L^{-1} and inhibited at higher Yb^{3+} concentrations, whereas the activity of the antioxidant enzyme superoxide dismutase (SOD) was stimulated at Yb^{3+} > 0.05 mg.L^{-1}, while catalase (CAT) activity was strongly inhibited after 40 days of exposure. These authors also observed that detoxifying enzymes glutathione S-transferase (GST) and glutathione peroxidase (GSH-Px) were stimulated at 0.05 mg.L^{-1} and inhibited at 0.1 mg.L^{-1} after 40 days of exposure. Among the parameters determined, CAT in goldfish liver was most sensitive to Yb^{3+}, indicating that CAT might be considered a potential tool for Yb^{3+} biomonitoring exposure in aquatic ecosystems.

Cui et al. (2012) studied the effects of La and Yb on the morphological and functional development of zebrafish embryos. The results showed that La^{3+} and Yb^{3+} delayed the development of zebrafish embryos and larvae, reducing survival and hatchability rates. These authors also found a concentration-dependent tail malformation and a more severe acute toxicity associated with ytterbium than with lanthanum.

Exposure of the European eel *Anguilla anguilla* to environmentally relevant lanthanum concentrations (120 ng.L^{-1}) for 7 days revealed a significant increase in acetylcholinesterase (AchE) activity, suggesting that La^{3+} could inhibit acetylcholine binding (Figueiredo et al. 2018). A decrease in lipid peroxidation was also observed in this study, which may indicate an important role of this element in the physical and functional activities with a free radical scavenger. These authors also observed a significant inhibition of catalase activity, indicating that the availability of La^{3+} could induce physiological impairment.

Conclusions

Further ecotoxicity studies of REE in marine organisms are necessary to better understand the effects and mechanisms of these elements at different trophic levels. As discussed in this chapter, most published studies refer to terrestrial or freshwater organisms. Data for marine organisms is still very scarce. However, the information obtained in freshwater species is generally protective because these organisms tend to be more sensitive than marine organisms. These differences in sensitivity may be related to interspecies variations rather than to the speciation of REE in the different exposure matrices.

In terms of spatial distribution, most REE derives from continental contributions and are retained in coastal transition zones, such as estuaries, with sediments being the main sinks of these elements. In this way, and from the point of view of REE ecotoxicity, these transition areas should be the focus for future studies, aiming to broaden the diversity of species studied.

Acknowledgments

The author thanks Prof. Isabel Caçador and Dr. Bernardo Duarte for the opportunity to collaborate in this chapter.

References

Balusamy, B., Taştan, B.E., Ergen, S.F., Uyar, T. and Tekinay, T. 2015. Toxicity of lanthanum oxide (La$_2$O$_3$) nanoparticles in aquatic environments. Environ. Sci. Process. Impacts 17: 1265–1270. doi:10.1039/C5EM00035A.

Barry, M.J. and Meehan, B.J. 2000. The acute and chronic toxicity of lanthanum to Daphnia carinata. Chemosphere 41: 1669–1674. doi:10.1016/S0045-6535(00)00091-6.

Bonificio, W.D. and Clarke, D.R. 2016. Rare-earth separation using bacteria. Environ. Sci. Technol. Lett. 3: 180–184. doi:10.1021/acs.estlett.6b00064.

Cui, J., Zhang, Z., Bai, W., Zhang, L., He, X., Ma, Y., Liu, Y. and Chai, Z. 2012. Effects of rare earth elements La and Yb on the morphological and functional development of zebrafish embryos. J. Environ. Sci. 24: 209–213. doi:10.1016/S1001-0742(11)60755-9.

Eggert, R.G. 2011. Minerals go critical. Nat. Chem. 3: 688.

Figueiredo, C., Grilo, T.F., Lopes, C., Brito, P., Diniz, M., Caetano, M., Rosa, R. and Raimundo, J. 2018. Accumulation, elimination and neuro-oxidative damage under lanthanum exposure in glass eels (Anguilla anguilla). Chemosphere 206. doi:10.1016/j.chemosphere.2018.05.029.

Fujiwara, K., Matsumoto, Y., Kawakami, H., Aoki, M. and Tuzuki, M. 2008. Evaluation of metal toxicity in Chlorella kessleri from the perspective of the periodic table. Bull. Chem. Soc. Jpn. 81: 478–488. doi:10.1246/bcsj.81.478.

González, V., Vignati, D.A.L., Leyval, C. and Giamberini, L. 2014. Environmental fate and ecotoxicity of lanthanides: Are they a uniform group beyond chemistry? Environ. Int. 71: 148–157. doi:10.1016/j.envint.2014.06.019.

González, V., Vignati, D.A.L., Pons, M.N., Montarges-Pelletier, E., Bojic, C. and Giamberini, L. 2015. Lanthanide ecotoxicity: First attempt to measure environmental risk for aquatic organisms. Environ. Pollut. 199. doi:10.1016/j.envpol.2015.01.020.

Greenland Minerals and Energy Ltd. 2017. 2017 Annual Report—Materials for an energy efficient future.

Greinacher, E. 1981. History of Rare Earth Applications, Rare Earth Market Today, Industrial Applications of Rare Earth Elements. doi:10.1021/bk-1981-0164.ch001.

Gwenzi, W., Mangori, L., Danha, C., Chaukura, N., Dunjana, N. and Sanganyado, E. 2018. Sources, behaviour, and environmental and human health risks of high-technology rare earth elements as emerging contaminants. Sci. Total Environ. doi:10.1016/j.scitotenv.2018.04.235.

Gysi, A.P. and Williams-Jones, A.E. 2013. Hydrothermal mobilization of pegmatite-hosted REE and Zr at Strange Lake, Canada: A reaction path model. Geochim. Cosmochim. Acta 122: 324–352. doi:10.1016/j.gca.2013.08.031.

Haley, T.J. 1965. Pharmacology and toxicology of the rare earth elements. J. Pharm. Sci. 54: 663–670. doi:10.1002/jps.2600540502.

Hao, S., Xiaorong, W., Liansheng, W., Lemei, D., Zhong, L. and Yijun, C. 1997. Bioconcentration of rare earth elements lanthanum, gadolinium and yttrium in algae (Chlorella Vulgarize Beijerinck): Influence of chemical species. Chemosphere 34: 1753–1760. doi:10.1016/S0045-6535(97)00031-3.

He, M.L. and Rambeck, W.A. 2000. Rare earth elements—A new generation of growth promoters for pigs? Arch. Anim. Nutr. doi:10.1080/17450390009381956.

Hongyan, G., Liang, C., Xiaorong, W. and Ying, C. 2002. Physiological responses of Carassius auratus to ytterbium exposure. Ecotoxicol. Environ. Saf. 53: 312–316. doi:10.1006/eesa.2002.2223.

IUPAC. 2005. Nomenclature of inorganic chemistry—IUPAC recommendations 2005. Chem. Int.—Newsmag. IUPAC 27. doi:10.1515/ci.2005.27.6.25.

Johannesson, K.H. and Zhou, X. 1999. Origin of middle rare earth element enrichments in acid waters of a Canadian High Arctic lake. Geochim. Cosmochim. Acta 63: 153–165.

Joonas, E., Aruoja, V., Olli, K., Syvertsen-Wiig, G., Vija, H. and Kahru, A. 2017. Potency of (doped) rare earth oxide particles and their constituent metals to inhibit algal growth and induce direct toxic effects. Sci. Total Environ. 593: 478–486. doi:10.1016/j.scitotenv.2017.03.184.

Klinger, J.M. 2015. A historical geography of rare earth elements: From discovery to the atomic age. Extr. Ind. Soc. 2: 572–580. doi:10.1016/j.exis.2015.05.006.

Kurvet, I., Juganson, K., Vija, H., Sihtmäe, M., Blinova, I., Syvertsen-Wiig, G. and Kahru, A. 2017. Toxicity of nine (doped) rare earth metal oxides and respective individual metals to aquatic microorganisms Vibrio fischeri and Tetrahymena thermophila. Materials (Basel) 10: 754. doi:10.3390/ma10070754.

Lynas Corporation Ltd. 2017. 2017 Annual Report.

Martino, C., Bonaventura, R., Byrne, M., Roccheri, M. and Matranga, V. 2016. Effects of exposure to gadolinium on the development of geographically and phylogenetically distant sea urchins species. Mar. Environ. Res. doi:10.1016/j.marenvres.2016.06.001.

Martino, C., Chiarelli, R., Bosco, L. and Roccheri, M.C. 2017. Induction of skeletal abnormalities and autophagy in Paracentrotus lividus sea urchin embryos exposed to gadolinium. Mar. Environ. Res. 130: 12–20. doi:10.1016/j.marenvres.2017.07.007.

Martino, C., Costa, C., Roccheri, M.C., Koop, D., Scudiero, R. and Byrne, M. 2018. Gadolinium perturbs expression of skeletogenic genes, calcium uptake and larval development in phylogenetically distant sea urchin species. Aquat. Toxicol. 194: 57–66. doi:10.1016/j.aquatox.2017.11.004.

Migaszewski, Z.M. and Gałuszka, A. 2015. The characteristics, occurrence, and geochemical behavior of rare earth elements in the environment: A review. Crit. Rev. Environ. Sci. Technol. 45: 429–471. doi:10.1080/10643389.2013.866622.

Oral, R., Bustamante, P., Warnau, M., D'Ambra, A., Guida, M. and Pagano, G. 2010. Cytogenetic and developmental toxicity of cerium and lanthanum to sea urchin embryos. Chemosphere 81: 194–8. doi:10.1016/j.chemosphere.2010.06.057.

Ozaki, T., Suzuki, Y., Nankawa, T., Yoshida, T., Ohnuki, T., Kimura, T. and Francis, A.J. 2006. Interactions of rare earth elements with bacteria and organic ligands. J. Alloys Compd. 408-412: 1334–1338. doi:10.1016/j.jallcom.2005.04.142.

Pagano, G., Guida, M., Tommasi, F. and Oral, R. 2015. Health effects and toxicity mechanisms of rare earth elements-knowledge gaps and research prospects. Ecotoxicol. Environ. Saf. 115: 40–48. doi:10.1016/j.ecoenv.2015.01.030.

Paola, I.M., Paciolla, C., D'Aquino, L., Morgana, M. and Tommasi, F. 2007. Effect of rare earth elements on growth and antioxidant metabolism in Lemna minor L. Caryologia 60: 125–128. doi:10.1080/00087114.2007.10589559.

Park, D.M., Brewer, A., Reed, D.W., Lammers, L.N. and Jiao, Y. 2017. Recovery of rare earth elements from low-grade feedstock leachates using engineered bacteria. Environ. Sci. Technol. 51: 13471–13480. doi:10.1021/acs.est.7b02414.

Quinn, K.A., Byrne, R.H. and Schijf, J. 2004. Comparative scavenging of yttrium and the rare earth elements in seawater: Competitive influences of solution and surface chemistry. Aquat. Geochemistry 10: 59–80. doi:10.1023/B:AQUA.0000038959.03886.60.

Rim, K.T., Koo, K.H. and Park, J.S. 2013. Toxicological evaluations of rare earths and their health impacts to workers: a literature review. Saf. Health Work 4: 12–26. doi:10.5491/SHAW.2013.4.1.12.

Rodea-Palomares, I., Boltes, K., Fernández-Piñas, F., Leganés, F., García-Calvo, E., Santiago, J. and Rosal, R. 2011. Physicochemical characterization and ecotoxicological assessment of CeO_2 nanoparticles using two aquatic microorganisms. Toxicol. Sci. 119: 135–145. doi:10.1093/toxsci/kfq311.

Stanley, J.K. and Byrne, R.H. 1990. The influence of solution chemistry on REE uptake by Ulva lactuca L. in seawater. Geochim. Cosmochim. Acta 54: 1587–1595. doi:10.1016/0016-7037(90)90393-Y.

Tai, P., Zhao, Q., Su, D., Li, P. and Stagnitti, F. 2010. Biological toxicity of lanthanide elements on algae. Chemosphere 80: 1031–1035. doi:10.1016/j.chemosphere.2010.05.030.

Takaya, Y., Yasukawa, K., Kawasaki, T., Fujinaga, K., Ohta, J., Usui, Y., Nakamura, K., Kimura, J.-I., Chang, Q., Hamada, M., Dodbiba, G., Nozaki, T., Iijima, K., Morisawa, T., Kuwahara, T., Ishida, Y., Ichimura, T., Kitazume, M., Fujita, T. and Kato, Y. 2018. The tremendous potential of deep-sea mud as a source of rare-earth elements, Scientific Reports. doi:10.1038/s41598-018-23948-5.

Tsuruta, T. 2007. Accumulation of rare earth elements in various microorganisms. J. Rare Earths 25: 526–532. doi:10.1016/S1002-0721(07)60556-0.

Wang, X., Shi, G.X., Xu, Q.S., Xu, B.J. and Zhao, J. 2007. Lanthanum- and cerium-induced oxidative stress in submerged Hydrilla verticillata plants. Russ. J. Plant Physiol. 54: 693–697. doi:10.1134/S1021443707050184.

Xun, W., Shi, L., Hou, G., Zhou, H., Yue, W., Zhang, C. and Ren, Y. 2014. Effect of rare earth elements on feed digestibility, rumen fermentation, and purine derivatives in sheep. Ital. J. Anim. Sci. 13: 3205. doi:10.4081/ijas.2014.3205.

Zhang, S. and Shan, X. 2001. Speciation of rare earth elements in soil and accumulation by wheat with rare earth fertilizer application. Environ. Pollut. 112: 395–405. doi:10.1016/S0269-7491(00)00143-3.

Chemical Contaminants in a Changing Ocean

3

Ana Luísa Maulvault,[1,*] *Patrícia Anacleto,*[1]
António Marques,[1] *Rui Rosa*[2] *and Mário Diniz*[3]

INTRODUCTION

The human footprint on the planet has dramatically increased since the Industrial Revolution due to the world population's constant growth, excessive use of natural resources and massive production of pollutants. This has contributed to one of the greatest environmental concerns of our time: climate change. Over the last decade, greenhouse gas (GHG; e.g., CO_2, CH_4, N_2O) emissions have reached unprecedented values, unequivocally contributing to a warmer atmosphere and ocean (Solomon et al. 2007). Apart from increased average surface seawater temperatures (up to +4°C, according to the Intergovernmental Panel for Climate Change [IPCC] for the worst-case predicted scenario; IPCC 2014), some of the most notorious climate changes affecting marine ecosystems include alterations of wind and precipitation patterns/ intensity, diminished snow cover and rising sea level with consequent changes in seawater salinity, ocean acidification (–0.4 pH units) due to the disturbance of the carbon cycle and reduced levels of dissolved oxygen (IPCC 2014). Depending on the region, each effect can occur alone or in combination with other effects, representing additional challenges to the resilience of marine ecosystems. Moreover, one climate change effect may act in synergism or potentiate the occurrence of another. Regardless of whether acting alone or combined, climate change effects will certainly have an impact on marine biota, affecting their fitness, metabolism, reproduction, recruitment and distribution, among other ecological features (Rosa et al. 2014, 2016), raising

[1] Division of Aquaculture and Seafood Upgrading. Portuguese Institute for the Sea and Atmosphere, I.P. (IPMA), Rua Alfredo, Magalhães Ramalho, 6, 1495-006 Lisboa Portugal.
[2] MARE – Marine and Environmental Sciences Centre, Laboratório Marítimo da Guia, Faculdade de Ciências da Universidade de Lisboa, Av. Nossa Senhora do Cabo, 939, 2750-374 Cascais, Portugal.
[3] UCIBIO, REQUIMTE Chemistry Department, Centre of Fine Chemistry and Biotechnology, Faculty of Sciences and Technology, Nova University of Lisbon (CQFB-FCT/UNL), 2829-516 Caparica, Portugal. Emails: panacleto@ipma.pt; amarques@ipma.pt; rarosa@fc.ul.pt; mesd@fct.unl.pt
* Corresponding author: aluisa@ipma.pt

the need for species to adapt to the new prevailing environmental conditions or, in extreme cases, leading to extinction.

In marine ecosystems, particularly those more vulnerable to anthropogenic impacts, such as estuaries and coastal areas, biota are exposed to several stressors, as they are surrounded by an array of chemical contaminants and depend on food availability and fluctuations of environmental conditions. Such stressors can further induce physiological stress, emphasise contaminants' toxicity or even determine species' success in a changing climate (Marques et al. 2015, Jager 2016, Maulvault et al. 2016, 2017). As will be discussed later in this chapter, chemical contaminants' availability in marine sediments/water column and toxicity to biota are strongly influenced by environmental drivers, such as temperature, pH, upwelling and stratification events (e.g., Noyes et al. 2009, Marques et al. 2010). By altering species' physiological status and, at the same time, exacerbating many forms of water pollution, deleterious impacts on marine organisms can be expected if the climate continues changing as forecasted, including changes in contaminants' uptake, retention and detoxification rates (e.g., Maulvault et al. 2016, 2017).

Given the current lack of empirical data, with most available information being based on mechanistic approaches, the interaction between environmental conditions and pollution is still unclear. Thus, there is an urgent need to further explore this research topic to better forecast the ecological consequences of climate change. In this way, the main aim of this chapter is to provide an overview of the expected challenges that marine species will face in tomorrow's ocean, based on mechanistic toxicological models as well as on findings from the most recent laboratory and field ecotoxicological studies. Despite a wide variety of environmental factors, acting alone or combined with other stressors, which also deserve attention, the impacts of ocean warming, acidification, changes of salinity regimes and hypoxia on contaminants' toxicity will be primarily addressed here, given the prevalence of these variables in the current state of the art with respect to climate change's effects on marine ecosystems.

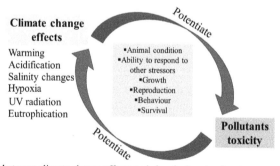

Fig. 1. Interaction between climate change effects and chemical contaminants.

Chemical Contaminants in Tomorrow's Ocean

Contaminants' fate and bioavailability

Organic compounds (OCs)

This group of chemical compounds includes pesticides, polychlorinated biphenyl (PCBs), flame retardants, dioxins and polycyclic aromatic hydrocarbons (PAHs), among others. Temperature is one of the most relevant factors influencing their distribution, half-life in biological compartments, volatilisation and re-emission (Gouin and Wania 2007, Teran et al. 2012), altering OC's partitioning into the different phases (solid, liquid and gas) (Macdonald et al. 2002). Increased temperatures can likely enhance OCs' volatilisation and, consequently, the exchanges between the ocean and atmosphere, which can result in slight reductions in contaminant exposure by marine biota (Armitage et al. 2011, Nadal et al. 2015). Depending on the geographic area, increased temperatures and precipitation can directly influence salinity levels in aquatic systems. Given their lower solubility in saltwater, OCs' bioavailability and toxicity can be enhanced at higher salinities (Noyes et al. 2009). Conversely, the opposite effects in OCs' bioavailability can be expected in regions of the planet where precipitation and snow/ice melting will increase, resulting in the high input of freshwater into the ocean.

Toxic metals (e.g., Hg, Pb, Cd, and As)

Solubility and speciation of toxic metals are largely dependent on seawater temperature and pH (Hoffmann et al. 2012). Thus, environmental variations associated with climate change will certainly have preponderant effects on metals' availability and concentration in the marine environment by altering their behaviour and transfer from sediments into the water column and vice versa, as well as their toxicity (Marques et al. 2010). Additionally, since metal inputs into aquatic systems are strongly linked to climate events, such as snow melting and precipitation, alterations in elemental profiles, distribution and concentrations due to climate change are expected (Marques et al. 2010, Hoffmann et al. 2012). Due to metals' ability to precipitate, bind, or be released from sediments is largely determined by environmental characteristics, such as pH, cation exchange capacity, organic matter content, redox conditions and chloride content, varying salinities can likely affect metal mobility in intertidal sediments, bioavailability and toxicity to biota, as demonstrated in previous studies (Du Laing et al. 2002, 2008, Hatje et al. 2003). Low dissolved oxygen levels (hypoxia) can also facilitate the release of metals from sediments to the water column (Schiedek et al. 2007). Apart from changes in bioavailability and mobility, increased seawater temperatures can also have a prevalent role in elements' speciation and, consequently, their toxicity to biota (Booth and Zeller 2005, Maulvault et al. 2016).

Pharmaceuticals and personal care products (PPCPs)

These compounds are currently among the least studied groups of contaminants, especially within a climate change context. Despite being relatively new pollutants, for which the available data are still limited, recent studies suggest that environmental conditions play a key role in the chemical behaviour, degradation and metabolisation of such pollutants, emphasising the need to consider and further assess the potential effects of climate change when investigating the toxicological risks of PPCPs (e.g., Azzouz and Ballesteros 2013). Since most substances are extremely sensitive to light, heat and surrounding pH conditions (Moreno et al. 2009, Welankiwar et al. 2013, Gul et al. 2015), the expected warming, increased UV radiation due to a depleted ozone layer and acidification can likely exacerbate PPCPs' degradation in the aquatic environment, depending on the degree of stability of each compound, and enhance their toxicity to biota (Macdonald et al. 2005, Schiedek 2007, Azzouz and Ballesteros 2013).

Contaminants' Carry Over and Toxicity

When assessing the effects of climate change from an ecotoxicological point of view, two aspects should be taken into account: (i) bioaccumulation and (ii) toxicity. Adapting to one set of environmental stressors implies great physiologic efforts and, therefore, increased susceptibility to other existing stressors. For those reasons, the interactions between climate change and chemical contaminants can be looked at from two different angles. On one hand, by affecting biota's metabolism, climate changes can result in altered bioaccumulation/detoxification mechanisms and toxicological responses to contaminant exposure (e.g., Noyes et al. 2009, Marques et al. 2010, Maulvault et al. 2016, 2017). On the other hand, organisms already living on the edge of their physiological capacities due to constant inputs of pollutants will certainly exhibit lower tolerance to environmental changes and will struggle to adapt to the new prevailing conditions (Noyes et al. 2009, Marques et al. 2010, Maulvault et al. 2016). The following section will present an overview of the available data, which were gathered from recent field and laboratory studies and reports concerning the ecotoxicological responses of marine species subjected to the combination of climate change effects and pollution.

Ocean warming

Out of the environmental variables affected by climate change, to date, temperature is one of the best documented parameters from a marine and ecological point of view (e.g., Madeira et al. 2012, 2013, Anacleto et al. 2014, Maulvault et al. 2017). Since most marine species are ectothermic, temperature is a crucial variable to their physiological functioning. Although many organisms have evolved to cope with daily or seasonal temperature variations, when multiple environmental stressors take place concurrently, species' resilience to temperature peak events or drastic season changes may be surpassed, thus compromising their survival (e.g., Madeira et al. 2012).

Stressors

Warming:
* Changes in OCs' fate, distribution and half–life in biological compartments
* Enhanced OCs volatilisation and re-emission
* *(In combination with increased precipitation)* Higher OCs degradation – lower volatilization!
* Changes in metal availability, transfer from sediments to water column and vice-versa , and speciation
* Increased PPCPs degradation into more toxic compounds

Acidification:
* Changes in OCs half–life in biological compartments
* Changes in metal availability, transfer from sediments to water column and vice-versa , and speciation
* Increased PPCPs degradation into more toxic compounds

Salinity changes:
* Enhanced OCs' bioavailability at higher salinities, but opposite effects in areas with reduced salinity
* Changes in metal solubility, mobility in intertidal sediments, bioavailability and toxicity
* Increased complexation of metals (and lower toxicity) at higher salinities, but opposite effects in areas with reduced salinity

Hypoxia:
* Facilitated release of metals from sediments to the water column

Responses

• Reduced thermal tolerance or reduced ability to cope with contaminants exposure
* Higher contaminant bioaccumulation/ elimination
* Changes in animal growth efficiency
* Increased contaminant biotransformation into more toxic compounds
* Increased ROS formation
* Increased susceptibility to diseases

• Increased contaminant toxicity (metals and ionic compounds)
* Enzymatic impairment
* Depression of the immune system
* Increased susceptibility to diseases
* Increased ROS formation
* Enhanced neurotoxicity of contaminants
* Behavioural alterations

* Lower toxicity of metals at higher salinities, but the opposite at lower salinities
* Increased toxicity of OCs at higher salinities, but the opposite at lower salinities
* Changes in metals' internal transport through the ionic channels
* Changes in PPCPs toxicity
* Overall increased contaminant toxicity in bivalves at lower salinity, particularly due to their valve closing reflex

* Metabolic depression
* Enhanced contaminant bioaccumulation
* Diminished contaminant detoxification
* Increased ROS formation
* Depression of the immune system

Fig. 2. Summary of climate change effects and observed ecotoxicological responses of marine species exposed to climate change and contaminants.

Metabolic changes induced by thermal stress are among the most studied physiological responses in marine biota (e.g., Neuheimer et al. 2011, Holt and Jørgensen 2015, Madeira et al. 2016). In general, organisms subjected to warmer temperatures exhibit enhanced metabolism, accompanied by increased ventilation and feeding rates in response to higher metabolic demands. Such changes can translate into higher contaminant bioaccumulation (contaminants dissolved in the water column, i.e., via respiration, or present in feeds or natural preys, i.e., via ingestion) and elimination rates (i.e., contaminant metabolisation and excretion) (e.g., Maulvault et al. 2016, Sampaio et al. 2016). Following the current lack of empirical data on this subject, and within the framework of the European project ECsafeSEAFOOD (reference no. 311820), the bioaccumulation of several emerging contaminants was evaluated in marine bivalves (*Mytilus galloprovincialis* and *Ruditapes philippinarum*) exposed to different climate change scenarios (warming and acidification). Thus, results gathered in this pilot laboratory study showed that bioaccumulation patterns are largely dependent on the behaviour and chemical properties of each compound, as warming induced higher bioaccumulation of some contaminants (e.g., sotalol, carbamazepine, triclosan, and TBBPA), but also lower bioaccumulation of others (e.g., PFOA and PFOS) (Maulvault, unpublished data). Furthermore, this study revealed that the interaction of warming and acidification can either exacerbate contaminants' bioaccumulation (e.g., sulfamethoxazole and TBBPA) or impair it (e.g., venlafaxine, citalopram, methylparaben and iAs; Maulvault et al. 2018; Serra-Compte et al. 2018).

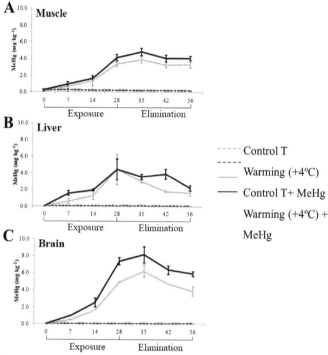

Fig. 3. MeHg concentrations in muscle, liver and brain of *D. labrax* exposed to warming and dietary MeHg during 56 days of experiment (28 days of exposure followed by 28 days of elimination) (adapted from Maulvault et al. 2016).

To adequately assess the relation between warming and contaminants' bioaccumulation/elimination patterns, animal growth efficiency is a parameter that deserves careful consideration when interpreting data, because (i) an enhanced metabolism also implies high energetic costs to biota, often leading to lower animal condition and fitness (Johnston and Dunn 1987, Stauber et al. 2016), (ii) animal feeding rates cannot be seen as directly proportional to their metabolic demands, since the amount of ingested food is largely determined not only by prey/feed availability but also by predator behaviour (Dijkstra et al. 2013). In this context, bioenergetics models become a crucial tool, although currently poorly developed, towards a more accurate and consistent estimation of contaminants' bioaccumulation rates in climate change scenarios (Dijkstra et al. 2013).

In addition to changes in contaminant bioaccumulation patterns, alterations in animals' metabolism and condition induced by warming can also exacerbate compound toxicity, lowering the ability of an organism to successfully respond to contaminants and/or to detoxify them (Sampaio et al. 2016, Maulvault et al. 2017) or increasing their biotransformation into more toxic compounds (Stauber et al. 2016). Such trends were observed in two recent studies performed on juvenile seabass (*Dicentrachus labrax*) exposed to warming and MeHg acting alone or in combination (Maulvault et al. 2016, 2017). In the former study, animals subjected to warmer temperatures (22°C versus 18°C) exhibited higher MeHg bioaccumulation, but also lower elimination of

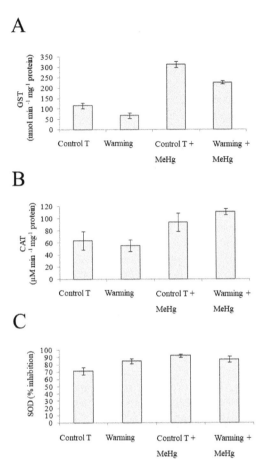

Fig. 4. GST, CAT and SOD activities in the liver of *D. labrax* after 28 days of exposure to warming and/or dietary MeHg (adapted from Maulvault et al. 2017).

this compound in the liver, that is, the primary organ responsible for contaminant metabolisation and subsequent transport to other organs or hepato-biliary excretion (Maulvault et al. 2016). As for the latter work, contaminated specimens subjected to increased seawater temperatures revealed, overall, depressed enzymatic activity, for instance, in terms of glutathione S-transferase (GST), which is a major second phase detoxification enzyme, and catalase (CAT) and superoxide dismutase (SOD), which play key roles in cell defence against the formation of reactive oxygen species (ROS) induced by stress (Maulvault et al. 2017).

Ocean acidification

At first glance, the effects of ocean acidification seem to be more noticeable and mostly dramatic in calcified organisms, to which the development and growth are intrinsically dependent on the equilibrium of the calcium carbonate cycle. For this reason, over the years, great research efforts have been channelled towards the assessment of potential ecological threats of ocean acidification to marine invertebrates, where most of the studies have been focused on coral reefs, followed by bivalves and crustaceans (e.g., Kleypas et al. 2006, Hoegh-Guldberg et al. 2007, Cohen and Holcomb 2009). Nevertheless, although fish species are known to be relatively tolerant to pH variations, since they are capable of adjusting their internal pH according to the surrounding levels (Fabry et al. 2008), recent studies have revealed that hypercapnia can lead to body malformations, as well as changes in buoyancy and loss of spatial orientation (Gutowska et al. 2010, Pimentel et al. 2014).

As previously discussed, contaminants' chemical properties are largely influenced by environmental conditions, with metals and other ionic compounds being particularly affected by the surrounding seawater pH levels. Such is the case of the PPCP triclosan (TCS), which becomes increasingly protonated and loses its negative charge as pH decreases, which, on the other hand, translates into enhanced compound availability and toxicity to biota (Orvos et al. 2002, Rowett et al. 2016). Indeed, the recent study of Rowett et al. (2016), using the freshwater amphipod *Gammarus pulex* as a model organism, evidenced increased TCS toxicity under lower pH levels. As argued by these authors, such a trend is explained by the fact that lipid membranes are generally impermeable to ionised molecular forms, and that TCS requires a pH value around 8.0 pH units to become ionised (i.e., in its less toxic form) (Lyndall et al. 2010, Rowett et al. 2016). Similarly, two recent studies using marine bivalves (i.e., *Scrobicularia plana*, Freitas et al. 2016, *Ruditapes philippinarum*, Munari et al. 2016) also reported synergistic interactions between the surrounding pH level and the PPCPs carbamazepine and diclofenac, with specimens exhibiting higher mortality and oxidative stress when exposed to the combination of acidification plus PPCPs than those exposed to each stressor separately.

When addressing the toxicological impacts of acidification, it is also worthwhile to consider its effects on biota's behaviour. Such effects are mostly attributed to the disruption of the ionic balance in proton-based neurotransmitter receptors, such as the $GABA_A$ neurotransmitter, which can likely translate into increased animal anxiety and boldness (Hamilton et al. 2014, Munday et al. 2014, Sampaio et al. 2016, Lai et al. 2017). Furthermore, recent studies have evidenced that, given the high conservation of the nervous system throughout evolution within vertebrates, water pollutants that are neurotoxic to humans (e.g., MeHg) or were designed to modulate specific human behaviours (i.e., psycho-active drugs, such as antidepressants, anxiolytics and anticonvulsants) may also promote similar responses in fish (Valenti et al. 2012, Brooks 2014, Fong and Ford 2014, Sampaio et al. 2016). Despite the recent findings consistently suggesting that ocean acidification can potentially promote neurotoxicological aspects of these contaminants (Bisesi et al. 2014, Sampaio et al. 2016, Maulvault et al. 2017), further research on this topic should be carried out to accurately explore different fish behavioural cues, neurological functioning and the ecotoxicological impacts of climate change.

Salinity

Salinity constitutes a critical environmental variable vis-à-vis marine biota, both to stenohaline organisms that exhibit a narrow tolerance range (thereby requiring stable salinity regimes) as well as euryhaline species (since osmoregulatory mechanisms involved in the maintenance of the internal homeostasis are energy-demanding) (Sampaio and Bianchini 2002). Hence, because salinity variations lead to metabolic changes and great energy costs, bioenergetic models can provide accurate estimations of the physiological and ecological impacts of salinity fluctuations (e.g., Normant et al. 2012, Chen et al. 2017). Moreover, the use of such models becomes further relevant when chemical contaminants are also involved in order to determine compound toxicokinetics under osmotic stress, adding to the equation important variables such as feed intake, feed efficiency, growth, excretion and ventilation (e.g., Chen et al. 2017).

Interactions between salinity and chemical contaminants are complex, since salinity by itself is able to largely influence marine species' physiological status and affect compound speciation and toxicity (Marques et al. 2010). In this sense, several studies, not necessarily focusing on climate change effects, have been conducted to assess the effects of different salinity levels on metal bioaccumulation and toxicity (e.g., Yung et al. 2015, Piazza et al. 2016, Zhou et al. 2017). By becoming increasingly more complexed, metals tend to be less toxic as salinity increases (e.g., Modassir 2000, Yung et al. 2015, Piazza et al. 2016, Zhou et al. 2017). Such was the case of Zn oxide nanoparticles' toxicity to the diatom *Thalassiosira pseudonana* (Yung et al. 2015). In this study, increased compound aggregation and decreased ion release were observed at higher salinity, thus leading to lower bioavailability and toxicity associated to Zn^{+2} ions (Yung et al. 2015). Similarly, early life stages of the barnacle *Amphibalanus amphitrite* exposed to Cd also revealed increased animal mortality, immobilisation, and changes in swimming speed at lower salinity ranges (Piazza et al. 2016). In opposition to this trend, an earlier study using mangrove clams (*Polymesoda erosa*) showed lower Hg toxicity at lower salinity levels, which could be related to enhanced detoxification (i.e., increased ventilation rates, metabolisation, and excretion) of this contaminant, as well as changes in this element's speciation and bioavailability (Modassir 2000). It is also noteworthy that salinity changes may additionally affect the internal transport of metals through ionic channels. Zhang and Wang (2007) observed in the seabream *Acanthopagrus schlegeli* that the calcium channel was actively involved in the uptake of Cd and Zn (transcellular uptake) at low salinity ranges, but not at high salinity levels.

As opposed to metals, and despite the limited information available, overall, persistent organic pollutants' (POPs) toxicity seems to be enhanced at elevated salinities (Noyes et al. 2009). For instance, Song and Brown (1998) observed increased mortality in brine shrimp *Artemia* sp. exposed to the organophosphate pesticide dimethoate when salinity increased three to four times in relation to the iso-osmotic salinity value of this species. In an earlier study focused on the impacts of severe droughts due to climate change, physiological and histological alterations observed in the gills of the fish species *Sarotherodon melanotheron* after short-term waterborne DDT contamination were potentiated by fluctuating salinities (Riou et al. 2012).

As for PPCPs, distinct and controversial ecotoxicological responses have been observed when marine biota was subjected to different salinity regimes. For instance, similarly to metals, sulfapyridine, sulfamethoxazole, sulfadimethoxine and trimethoprim showed to be less toxic to the green microalgae *Chlorella vulgaris* at higher salinities (Borecka et al. 2016). In this study, the authors argued that such toxicity decrease could be associated with the reduced permeability of algal cell walls to PPCPs, as higher salinities promoted elevated concentrations of inorganic monovalent cations, which, in turn, showed the ability to bind to algal surfaces' countercharges (hydroxyl functional groups, Borecka et al. 2016). Following this pattern, González-Ortegón et al. (2013) observed reduced growth (about 18 percent) in larvae of the marine shrimp *Palaemon serratus* exposed to clotrimazole and low salinity in comparison to specimens subjected to reference salinity levels.

Finally, and particularly to marine bivalve species, increased contaminant toxicity at lower salinity can also be associated to the fact that hypohaline conditions promote valve closing. As argued by Correia et al. (2015) in their study on *R. philippinarum* exposed to paracetamol, by having such a defense strategy (i.e., valve closing reflex under stress), bivalves can prevent the absorption of more contaminants from the external media, but the drawback is that they are unable to excrete metabolised forms of the contaminants already bioaccumulated.

Hypoxia

Hypoxic and anoxic zones in the ocean can occur as a result of (i) increased eutrophication (nutrients and pollutants), particularly in coastal areas, which leads to biomass increase and, consequently, to enhanced respiration rates (Nixon and Buckley 2002, Keeling et al. 2010), or (ii) increased seawater stratification due to warming and/or salinity changes (Diaz and Rosenberg 2008). So, when exploring potential physiological and ecotoxicological responses associated to hypoxia, it is of utmost importance to assess interdependence relations between hypoxia and other abiotic factors such as temperature, salinity, nutrients and suspended organic particles. In this sense, the interdependent impacts of hypoxia and increased temperature are the most widely described in the literature, suggesting that oxygen availability plays a key role in marine organisms' thermal tolerance limits (e.g., Anttila et al. 2013, Artigaud et al. 2014, Pédron et al. 2017). Interactive effects between acidification and hypoxia were observed in the physiological status and energy budget of *M. coruscus*, clearly evidencing further metabolic depression when both stressors were combined (Sui et al. 2016). Rosa et al. (2013) also observed that during cuttlefish (*Sepia officinalis*) embryogenesis, the already harsh conditions inside the egg capsules can be further exacerbated by hypoxia acting in synergism with warming and acidification, likely leading to premature hatching and smaller post-hatching animal size, and, consequently, compromising the survival of this species and the successful development of posterior ontogenetic stages.

As for the effects of hypoxia from a toxicological perspective, the information available to date is extremely limited. However, the few available studies, which were mostly performed on freshwater species, suggest that hypoxia potentially enhances contaminants' toxicity, mostly because it can significantly impair biota's metabolism and detoxification mechanisms (Marques et al. 2010). Mustafa et al. (2012) observed enhanced oxidative DNA damage, histopathological alterations in different tissues and changes in feed conversion efficiency and growth rates in

European carp (*Cyprinus carpio*) exposed to hypoxia and dietary Cu. Wang et al. (2014) reported interactive effects of hypoxia and titanium dioxide nanoparticles (nano-TiO$_2$) in the immune system of the mussel *Perna viridis*, with increased ROS formation, and lower total hemocyte counts in specimens subjected to low dissolved oxygen levels. Conversely, recent studies have also evidenced that pollutants can, on the other hand, hamper biota's ability to cope with the physiological stress induced by hypoxic events (e.g., Negreiros et al. 2011, Gorokhova et al. 2013). Importantly, a recent study regarding the interactions between hypoxia and Cu exposure suggested that hypoxia may act in opposite ways during the embryonic development of three-spined stickleback (*Gasterosteus aculeatus*), that is, having a protective effect on metal toxicity before embryo hatches, but exacerbating Cu toxicity after hatching (Fitzgerald et al. 2017). The authors further argued that this pattern can occur in teleost species, as these toxicological interactions were also observed in zebrafish embryos (Fitzgerald et al. 2016).

Conclusions and Perspectives

Although many possible interactions between climate change and chemical contaminants have been pointed out, to date, little research has been conducted addressing such interactions in either laboratory or field conditions, and even fewer studies address marine species. Thus, the current lack of empirical data clearly suggests that greater efforts are needed to understand how multiple environmental stressors interact with each other. This will enable researchers to develop broader management and efficient mitigation strategies in tomorrow's ocean.

Future research should take into consideration regional trends (i.e., abiotic factors, pollution levels and types of contaminants) when addressing potential ecotoxicological responses to climate change, as the alterations of prevailing environmental conditions will certainly not affect marine ecosystems in the same way across the planet. Furthermore, because environmental stressors will unlikely occur in isolation, or all at once, different combinations of contaminant mixtures and abiotic conditions (exploring less pronounced to more severe climate change scenarios) should be investigated to gain a broader view on ecotoxicological impacts of climate change. Finally, apart from empirical toxicological data, and given the multiple physiological and metabolic alterations involved, bioenergetic models should be further explored in climate change contexts, as they provide accurate insights on biota's welfare and are resourceful tools to understand contaminants' toxicokinetics in a changing climate.

Acknowledgements

The authors would like to thank the Fundação para a Ciência e Tecnologia (FCT) through the strategic projects UID/MAR/04292/2013 granted to MARE and UID/Multi/04378/2013 granted to UCIBIO, the contracts of AM and RR in the framework of the IF program, as well as, the PhD Grant of ALM (SFRH/BD/103569/2014) and post-PhD Grant of PA (SFRH/BPD/100728/2014).

References

Anacleto, P., Maulvault, A.L., Lopes, V.M., Repolho, T., Diniz, M., Nunes, M.L., Marques, A. and Rosa, R. 2014. Physiological ecology does not explain invasive success: an integrative and comparative analysis of biological responses in native and alien invasive clams in an ocean warming context. Part A Comp. Biochem. Physio. 175: 28–37.

Anttila, K., Dhillon, R.S., Boulding, E.G., Farrell, A.P., Glebe, B.D., Elliott, J.A.K., Wolters, W.R. and Schulte, P.M. 2013. Variation in temperature tolerance among families of Atlantic salmon (*Salmo salar*) is associated with hypoxia tolerance, ventricle size and myoglobin level. J. Exp. Bio. 216: 1183–1190.

Armitage, J.M., Quinn, C.L. and Wania, F. 2011. Global climate change and contaminants: an overview of opportunities and priorities for modeling the potential implications for long-term human exposure to organic compounds in the arctic. J. Environ. Monit. 13: 1532–1546.

Artigaud, S., Lacroix, C., Pichereau, V. and Flye-Sainte-Marie, J. 2014. Respiratory response to combined heat and hypoxia in the marine bivalves *Pecten maximus* and *Mytilus* spp. Part A Comp. Biochem. Physio. 175: 135–140.

Azzouz, A. and Ballesteros, E. 2013. Influence of seasonal climate differences on the pharmaceutical, hormone and personal care product removal efficiency of a drinking water treatment plant. Chemosphere 93(9): 2046–2054.

Bisesi Jr, J.H., Bridges, W. and Klaine, S.J. 2014. Reprint of: Effects of the antidepressant venlafaxine on fish brain serotonin and predation behavior. Aqua. Tox. 151: 88–96.

Booth, S. and Zeller, D. 2005. Mercury, food webs, and marine mammals: Implications of diet and climate change for human health. Environ. Health Persp. 113(5): 521–526.

Borecka, M., Białk-Bielińska, A., Haliński, L.P., Pazdro, K., Stepnowski, P. and Stolte, S. 2016. The influence of salinity on the toxicity of selected sulfonamides and trimethoprim towards the green algae *Chlorella vulgaris*. J. Hazard Mat. 308: 179–186.

Brooks, B.W. 2014. Fish on prozac: Ten years later. Aqua. Tox. 151: 61–67.

Chen, W.-Q., Wang, W.-X. and Tan, Q.-G. 2017. Revealing the complex effects of salinity on copper toxicity in an estuarine clam *Potamocorbula laevis* with a toxicokinetic-toxicodynamic model. Environ. Poll. 222: 323–330.

Cohen, A.L. and Holcomb, M. 2009. Why corals care about ocean acidification: uncovering the mechanism. Oceanography 22: 118–127.

Correia, B., Freitas, R., Figueira, E., Soares, A.M.V.M. and Nunes, B. 2015. Oxidative effects of the pharmaceutical drug paracetamol on the edible clam *Ruditapes philippinarum* under different salinities. Comp. Biochem. Physio. Part C: Tox. Pharma. 179: 116–124.

Dale, B., Edwards, M. and Reid, P.C. 2006. Climate change and harmful algae blooms. pp. 367–378. *In:* Granéli, E. and Turner, J.T. (eds.). Ecological Studies. Springer-Verlag. Heidelberg, Berlin.

Danovaro, R., Corinaldesi, C., Dell'anno, A., Fuhrman, J.A., Middelburg, J.J., Noble, R.T. and Suttle, C.A. 2011. Marine viruses and global climate change. FEMS Microbio. Rev. 35(6): 993–1034.

De Silva, S.S. and Soto, D. 2009. Climate change and aquaculture: potential impacts, adaptation and mitigation. pp. 151–212. *In*: Cochrane, K., De Young, C., Soto, D. and Bahri, T. (eds.). Climate Change Implications for Fisheries and Aquaculture: Overview of Current Scientific Knowledge. FAO Fisheries and Aquaculture Technical Paper No. 530. Rome, FAO.

Diaz, R.J. and Rosenberg, R. 2008. Spreading dead zones and consequences for marine ecosystems. Science 321: 926–929.

Dijkstra, J.A., Buckman, K.L., Ward, D., Evans, D.W., Dionne, M. and Chen, C.Y. 2013. Experimental and natural warming elevates mercury concentrations in estuarine fish. PLoS One 8(3): e58401.

Du Laing, G., Bogaert, N., Tack, F.M.J., Verloo, M.G. and Hendrickx, F. 2002. Heavy metal contents (Cd, Cu, Zn) in spiders (*Pirata Piraticus*) living in intertidal sediments of the river Scheldt estuary (Belgium) as affected by substrate characteristics. STOTEN 289 (1-3): 71–81.

Du Laing, G., De Vos, R., Vandecasteele, B., Lesage, E., Tack, F.M.G. and Verloo, M.G. 2008. Effect of salinity on heavy metal mobility and availability in intertidal sediments of the Scheldt estuary. Est. Coast. She. Scie. 77: 589–602.

Fabry, V.J., Seibel, B.A., Feely, R.A. and Orr, J.C. 2008. Impacts of ocean acidification on marine fauna and ecosystem processes. ICES J. Mar. Scie. 65: 414–432.

Fitzgerald, J.A., Jameson, H.M., Dewar Fowler, V.H., Bond, G.L., Bickley, L.K., Uren Webster, T.M., Bury, N.R., Wilson, R.J. and Santos, E.M. 2016. Hypoxia suppressed copper toxicity during early development in zebrafish embryos in a process mediated by the activation of the HIF signalling pathway. Environ. Sci. Tech. 50: 4502–4512.

Fitzgerald, J.A., Katsiadaki, I. and Santos, E.M. 2017. Contrasting effects of hypoxia on copper toxicity during development in the three-spined stickleback (*Gasterosteus aculeatus*). Environ. Poll. 222: 433–443.

Fong, P.P. and Ford, A.T. 2014. The biological effects of antidepressants on the molluscs and crustaceans: A review. Aqua. Tox. 151: 4–13.

Freitas, R., Almeida, A., Calisto, V., Velez, C., Moreira, A., Schneider, R.J., Esteves, V.I., Wronad, F.J., Figueira, E. and Soares, A.M.V.M. 2016. The impacts of pharmaceutical drugs under ocean acidification: New data on single and combined long-term effects of carbamazepine on *Scrobicularia plana*. STOTEN 541: 977–985.

Giraudo, M., Douville, M., Letcher, R.J. and Houde, M. 2017. Effects of food-borne exposure of juvenile rainbow trout (*Oncorhynchus mykiss*) to emerging brominated flame retardants 1,2-bis(2,4,6-tribromophenoxy)ethane and 2-ethylhexyl-2,3,4,5-tetrabromobenzoate. Aqua. Tox. 186: 40–49.

González-Ortegón, E., Blasco, J., Le Vay, L. and Giménez, L. 2103. A multiple stressor approach to study the toxicity and sub-lethal effects of pharmaceutical compounds on the larval development of a marine invertebrate. J. Hazard Mat. 263: 233–238.

Gorokhova, E., Löf, M., Reutgard, M., Lindström, M. and Sundelin, B. 2013. Exposure to contaminants exacerbates oxidative stress in amphipod *Monoporeia affinis* subjected to fluctuating hypoxia. Aqua. Tox. 127: 46–53.

Gouin, T. and Wania, F. 2007. Time trends of arctic contamination in relation to emission history and chemical persistence and partitioning properties. Environ. Scie. Tech. 41: 5986–5992.

Gul, W., Basheer, S., Karim, F. and Ayub, S. 2015. Effect of acid, base, temperature and UV light on amlodipine besylate. Int. J. Adv. Res. Chem. Scie. 2: 21–24.

Gutowska, M.A., Melzner, F., Pörtner, H.O. and Meier, S. 2010. Cuttlebone calcification increases during exposure to elevated seawater pCO_2 in the cephalopod *Sepia officinalis*. Mar. Bio. 157: 1653–1663.

Hamilton, T.J., Holcombe, A. and Tresguerres M. 2014. CO_2-induced ocean acidification increases anxiety in rockfish via alteration of GABAA receptor functioning. Proc. Royal Soc. B: Bio. Scie. 281(1775): 20132509.

Hatje, V., Payne, T.E., Hill, D.M., McOrist, G., Birch, G.F. and Szymczak, R. 2003. Kinetics of trace element uptake and release by particles in estuarine waters: effects of pH, salinity, and particle loading. Environ. Int. 29(5): 619–629.

Hoegh-Guldberg, O., Mumby, P.J., Hooten, A.J., Steneck, R.S., Greenfield, P., Gomez, E., Harvell, C.D., Sale, P.F., Edwards, A.J., Caldeira, K., Knowlton, N., Eakin, C.M., Iglesias-Prieto, R., Muthiga, N., Bradbury, R.H., Dubi, A., Hatziolos M.E. 2007. Coral reefs under rapid climate change and ocean acidification. Science 318: 1737–1742.

Hoffmann, L.J., Breitbarth, E., Boyd, P.W. and Hunter, K.A. 2012. Influence of ocean warming and acidification on trace metal biogeochemistry. Mar. Eco. Prog. Ser. 470: 191–205.

Holt, R.E. and Jørgensen, C. 2015. Climate change in fish: effects of respiratory constraints on optimal life history and behavior. Bio. Lett. 11: 20141032.

Honorato, T.B.M., Boni, R., Silva, P.M. and Marques-Santos, L.F. 2017. Effects of salinity on the immune system cells of the tropical sea urchin *Echinometra lucunter*. J. Exp. Mar. Bio. Eco. 486: 22–31.

Hwang, J., Suh, S.-S., Chang, M., Park, S.Y., Ryu, T.K., Lee, S. and Lee, T.-K. 2014. Effects of triclosan on reproductive parameters and embryonic development of sea urchin, *Strongylocentrotus nudus*. Ecotox. Environ. Saf. 100: 148–152.

Imsland, A.K., Gústavsson, A., Gunnarsson, S., Foss, A., Árnason, J., Arnarson, I., Jónsson, A.F., Smáradóttir, H. and Thorarensen, H. 2008. Effects of reduced salinities on growth, feed conversion efficiency and blood physiology of juvenile Atlantic halibut (*Hippoglossus hippoglossus* L.). Aquaculture 274: 254–259.

IPCC. 2014. Climate change 2014: Impacts, adaptation, and vulnerability. Part A: global and sectoral aspects. Contribution of working group II to the fifth assessment report of the intergovernmental panel on climate change. pp. 1132. *In*: Field, C.B., Barros, V.R., Dokken, D.J., Mach, K.J., Mastrandrea, M.D., Bilir, T.E., Chatterjee, M., Ebi, K.L., Estrada, Y.O., Genova, R.C., Girma, B., Kissel, E.S., Levy, A.N., MacCracken, S., Mastrandrea, P.R. and White, L.L. (eds.). Cambridge University Press, Cambridge, United Kingdom and New York, NY, USA.

Jager, T. 2016. Dynamic modeling for uptake and effects of chemicals. pp. 77–81. *In*: Blasco, J., Chapman, P.M., Campana, O. and Hampel, M. (eds.). Marine Ecotoxicology: Current Knowledge and Future Issues. Academic Press, Elsevier, Oxford, UK.

Jantzen, C.E., Annunziato, K.M. and Cooper, K.R. 2016. Behavioral, morphometric, and gene expression effects in adult zebrafish (*Danio rerio*) embryonically exposed to PFOA, PFOS, and PFNA. Aqua. Tox. 180: 123–130.

Johnston, I.A. and Dunn, J. 1987. Temperature acclimation and metabolism in ectotherms with particular reference to teleost fish. Symp. Soc. Exp. Bio. 41: 67–93.

Keeling, R.F., Kortzinger, A. and Gruber, N. 2010. Ocean deoxygenation in a warming world. Annual Rev. Mar. Scie. 2: 199–229.

Kleypas, J.A., Feely, R.A., Fabry, V.J., Langdon, C., Sabine, C.L. and Robbins, L.L. 2006. Impacts of ocean acidification on coral reefs and other marine calcifiers: A guide for future research. Contribution No. 2897 from NOAA/Pacific Marine Environmental Laboratory. 87pp.

Lai, F., Fagernes, C.E., Jutfelt, F. and Nilsson, G.E. 2017. Expression of genes involved in brain GABAergic neurotransmission in three-spined stickleback exposed to near-future CO_2. Cons. Physio. 5(1): cox004.

Li, J.Y., Tang, J.Y.M., Jin, L. and Escher, B.I. 2013. Understanding bioavailability and toxicity of sediment-associated contaminants by combining passive sampling with *in vitro* bioassays in an urban river catchment. Environ. Tox. Chem. 32: 2888–2896.

Lyndall, J., Fuchsman, P., Bock, M., Barber, T., Lauren, D., Leigh, K., Perruchon, E. and Capdevielle, M. 2010. Probabilistic risk evaluation for triclosan insurface water, sediment, and aquatic biota tissues. Integr. Environ. Assess. Manag. 6: 419–440.

MacDonald, R., MacKay, D. and Hickie, B. 2002. A contaminant amplification in the environment. Environ Scie. Techn. 36: 456–462.

Macdonald, R.W., Harner, T. and Fyfe, J. 2005. Recent climate change in the arctic and its impact on contaminant pathways and interpretation of temporal trend data. STOTEN 342: 5–86.

Madeira, D., Narciso, L., Cabral, H. and Vinagre, C. 2012. Thermal tolerance and potential impacts of climate change on coastal and estuarine organisms. J. Sea Res. 70: 32–41.

Madeira, D., Narciso, L., Cabral, H.N., Vinagre, C. and Diniz, M.S. 2013. Influence of temperature in thermal and oxidative stress responses in estuarine fish. Comp. Biochem. Physio. Part A: Mol. Integ. Physio. 166: 237–243.

Madeira, D., Araújo, J.E., Vitorino, R., Capelo, J.L., Vinagre, C. and Diniz, M.S. 2016. Ocean warming alters cellular metabolism and induces mortality in fish early life stages: A proteomic approach. Environ. Res. 148: 164–176.

Marcogliese, D.J. 2008. The impact of climate change on the parasites and infectious diseases of aquatic animals. Rev. Scie. Tech. 27(2): 467–84.

Marques, A., Nunes, M.L., Moore, S.K. and Strom, M.S. 2010. Climate change and seafood safety: Human health implications. Food Res. Int. 43(7): 1766–1779.

Marques, A., Diogène, J. and Rodriguez-Mozaz, S. 2015. Non-regulated environmental contaminants in seafood: Contributions of the ECsafeSEAFOOD EU project. Environ. Res. Part B 143: 1–2.

Maulvault, A.L., Custódio, C., Anacleto, P., Repolho, T., Pousão, P., Nunes, M.L., Diniz, M., Rosa, R. and Marques, A. 2016. Bioaccumulation and elimination of mercury in juvenile seabass (*Dicentrarchus labrax*) in a warmer environment. Environ. Res. 149: 77–85.

Maulvault, A.L., Barbosa, V., Alves, R., Custódio, A., Anacleto, A., Repolho, T., Pousão Ferreira, P., Rosa, R., Marques, A. and Diniz, M. 2017. Ecophysiological responses of juvenile seabass (*Dicentrarchus labrax*) exposed to increased temperature and dietary methylmercury. STOTEN 586: 551–558.

Maulvault, A.L., Camacho, C., Barbosa, V., Alves, R., Anacleto, P., Fogaça, F., Kwadijk, C., Kotterman, M., Cunha, S.C., Fernandes, J.O., Rasmussen, R.R., Sloth, J.J., Aznar-Alemany, O., Eljarrat, E., Barceló, D. and Marques, A. 2018. Assessing the effects of seawater temperature and pH on the bioaccumulation of emerging chemical contaminants in marine bivalves. Environ. Res. 161: 236–247.

Modassir, Y. 2000. Effect of salinity on the toxicity of mercury in mangrove Clam, *Polymesoda erosa* (Lightfoot 1786). Asian Fish. Scie. 13: 335–341.

Moreno, A.H., Da Silva, M.F.C. and Salgado, H.R.N. 2009. Stability study of azithromycin in ophthalmic preparations. Braz. J. Pharma. Scie. 45: 2.

Munari, M., Chemello, G., Finos, L., Ingrosso, G., Giani, M. and Marin, M.G. 2016. Coping with seawater acidification and the emerging contaminant diclofenac at the larval stage: A tale from the clam *Ruditapes philippinarum*. Chemosphere 160: 293–302.

Munday, P.L., Cheal, A.J., Dixson, D.L., Rummer, J.L. and Fabricius, K.E. 2014. Behavioural impairment in reef fishes caused by ocean acidification at CO_2 seeps. Nat. Clim. Chan. 4: 487–492.

Mustafa, S.A., Davies, S.J. and Jha, A.N. 2012. Determination of hypoxia and dietary copper mediated sub-lethal toxicity in carp, *Cyprinus carpio*, at different levels of biological organization. Chemosphere 87: 413–422.

Nadal, M., Marquès, M., Mari, M. and Domingo, J.L. 2015. Climate change and environmental concentrations of POPs: A review. Part A, Environ. Res. 143: 177–185.

Negreiros, L.A., Silva, B.F., Paulino, M.G., Fernandes, M.N. and Chippari-Gomes, A.R. 2011. Effects of hypoxia and petroleum on the genotoxic and morphological parameters of *Hippocampus reidi*. Part C, Comp. Biochem. Physio. 153: 408–414.

Neuheimer, A.B., Thresher, R.E., Lyle, J.M. and Semmens, J.M. 2011. Tolerance limit for fish growth exceeded by warming waters. Nat. Clim. Chan. 1: 110–113.

Nie, H., Liu, L., Huo, Z., Chen, P., Ding, J., Yang, F. and Yan, X. 2017. The HSP70 gene expression responses to thermal and salinity stress in wild and cultivated Manila clam *Ruditapes philippinarum*. Aquaculture 470: 149–156.

Nixon, S.W. and Buckley, B.A. 2002. A strikingly rich zone—nutrient enrichment and secondary production in coastal marine ecosystems. Estuaries 25: 782–796.

Normant, M., Król, M. and Jakubowska, M. 2012. Effect of salinity on the physiology and bioenergetics of adult Chinese mitten crabs *Eriocheir sinensis*. J. Exp. Mar. Bio. Eco. 416-417: 215–220.

Noyes, P.D., McElwee, M.K., Miller, H.D., Clark, B.W., Van Tiem, L.A., Walcott, K.C., Erwin, K.N. and Levin, E.D. 2009. The toxicology of climate change: environmental contaminants in a warming world. Environ. Int. 35: 971–986.

Nugegoda, D. and Kibria, G. 2016. Effects of environmental chemicals on fish thyroid function: Implications for fisheries and aquaculture in Australia. Gen. Com. Endocr. 244: 40–53.

Orvos, D.R.,Versteeg, D.J., Inauen, J., Capdevielle, M. and Rothenstein, A. 2002. Aquatic toxicity of triclosan. Environ. Tox. Chem. 21: 1338–1349.

Pédron, N., Artigaud, S., Infante, J.-L.Z., Le Bayon, N., Charrier, G., Pichereau, V. and Laroche, J. 2017. Proteomic responses of European flounder to temperature and hypoxia as interacting stressors: Differential sensitivities of populations. STOTEN 586: 890–899.

Piazza, V., Gambardella, C., Canepa, S., Costa, E., Faimali, M. and Garaventa, F. 2016. Temperature and salinity effects on cadmium toxicity on lethal and sublethal responses of *Amphibalanus amphitrite* nauplii. Ecotox. Environ. Saf. 123: 8–17.

Pimentel, M., Faleiro, F., Dionísio, G., Repolho, R., Pousão-Ferreira, P., Machado, J. and Rosa, R. 2014. Defective skeletogenesis and oversized otoliths in fish early stages in a changing ocean. J. Exp. Bio. 217: 2062–2070.

Riou, V., Ndiaye, A., Budzinski, H., Dugué, R., Le Ménach, K., Combes, Y., Bossus, M., Durand, J.D., Charmantier, G. and Lorin-Nebel, C. 2012. Impact of environmental DDT concentrations on gill adaptation to increased salinity in the tilapia *Sarotherodon melanotheron*. Comp. Biochem. Physio. Part C: Pharma. Tox. Endocr. 156: 7–16.

Rodd, A.L., Messier, N.J., Vaslet, C.A. and Kane, A.B. 2017. A 3D fish liver model for aquatic toxicology: Morphological changes and Cyp1a induction in PLHC-1 microtissues after repeated benzo(a)pyrene exposures. Aquatic Toxicology 186: 134–144.

Rosa, R., Trubenbach, K., Repolho, T., Pimentel, M., Faleiro, F., Boavida-Portugal, J., Baptista, M., Lopes, V.M., Dionisio, G., Leal, M.C., Calado, R. and Portner, H.O. 2013. Lower hypoxia thresholds of cuttlefish early life stages living in a warm acidified ocean. Proc. Royal Soc. London, Series B: Bio. Scie. 280: 20131695.

Rosa, R., Trübenbach, K., Pimentel, M.S., Boavida-Portugal, J., Faleiro, F., Baptista, M., Calado, R., Pörtner, H.O. and Repolho, T. 2014. Differential impacts of ocean acidification and warming on winter and summer progeny of a coastal squid (*Loligo vulgaris*). J. Exp. Bio. 217: 518–525.

Rosa R., Paula, J.R., Sampaio, E., Pimentel, M., Lopes, A.R., Baptista, M., Guerreiro, M., Santos, C., Campos, D., Almeida-Val, V.M.F., Calado, R., Diniz, M. and Repolho, T. 2016. Neurooxidative damage and aerobic potential loss of sharks under elevated CO_2 and warming. Mar. Bio. 163: 119.

Rosa, R., Rummer, J.L. and Munday, P.L. 2017. Biological responses of sharks to ocean acidification. Bio. Lett. 13.

Rowett, C.J., Hutchinson, T.H. and Comber, S.D.W. 2016. The impact of natural and anthropogenic dissolved organic carbon (DOC), and pH on the toxicity of triclosan to the crustacean *Gammarus pulex* (L.). STOTEN 565: 222–231.

Sampaio, E., Maulvault, A.L., Lopes, V.M., Paula, J.R., Barbosa, V., Alves, R., PousãoFerreira, P., Repolho, T., Marques, A. and Rosa, R. 2016. Habitat selection disruption and lateralization impairment of cryptic flatfish in a warm, acid, and contaminated ocean. Mar. Bio. 163–217.

Sampaio, E., Lopes, A.R., Francisco, S., Paula, J.R., Pimentel, M., Maulvault, A.L., Repolho, T., Grilo, T.F., Pousão-Ferreira, P., Marques, A. and Rosa, R. 2017. Ocean acidification dampens warming and contamination effects on the physiological stress response of a commercially important fish. Biogeoscie. Discuss. doi:10.5194/bg-2017-147.

Sampaio, L.A. and Bianchini, A. 2002. Salinity effects on osmoregulation and growth of the euryhaline flounder *Paralichthys orbignyanus*. J. Exp. Mar. Bio. Eco. 269: 187–196.

Schiedek, D., Sundelin, B., Readman, J.W. and Macdonald, R.W. 2007. Interactions between climate change and contaminants. Mar. Poll. Bull. 54: 1845–1856.

Serra-Compte, A., Maulvault, A.L., Camacho, C., Álvarez-Muñoz, D., Barceló, D., Rodríguez-Mozaz, S. and Marques, A. 2018. Effects of water warming and acidification on bioconcentration, metabolization and depuration of pharmaceuticals and endocrine disrupting compounds in marine mussels (Mytilus galloprovincialis). Environ. Poll. 236: 824–834.

Soares, P., Andrade, A., Santos, T., Silva, S., Silva, J., Santos, A., Souza, E. Magliano da Cunha, F., Wanderley Teixeira, V., Sales Cadena, M.R., Bezerra de Sá, F., Bezerra de Carvalho, Jr., L. and Gonçalves Cadena, P. 2016. Acute and chronic toxicity of the benzoylurea pesticide, lufenuron, in the fish, *Colossoma macropomum*. Chemosphere 161: 412–421.

Solomon, S., Qin, D., Manning, M., Alley, R.B., Berntsen, T., Bindoff, N.L., Chen, Z., Chidthaisong, A., Gregory, J.M., Hegerl, G.C., Heimann, M., Hewitson, B., Hoskins, B.J., Joos, F., Jouzel, J., Kattsov, V., Lohmann, U., Matsuno, T., Molina, M., Nicholls, N., Overpeck, J., Raga, G., Ramaswamy, V., Ren, J., Rusticucci, M., Somerville, R., Stocker, T.F., Whetton, P., Wood, R.A. and Wratt, D. 2007. Technical summary. *In:* Solomon, S., Qin, D., Manning, M., Chen, Z., Marquis, M., Averyt, K.B., Tignor, M. and Miller, H.L. (eds.). Climate Change 2007: The Physical Science Basis. Contribution of Working Group Ito the Fourth Assessment Report of the Intergovernmental Panel on Climate Change Cambridge University Press, Cambridge, United Kingdom and New York, NY, USA.

Song, M.Y. and Brown, J.J. 1998. Osmotic effects as a factor modifying insecticide toxicity on Aedes and Artemia, Ecotox. Environ. Safe. 41: 195–202.

Stauber, J.L., Chariton, A. and Apte, S. 2016. Global change. pp. 294–303. *In*: Blasco, J., Chapman, P.M., Campana, O. and Hampel, M. (eds.). Marine Ecotoxicology: Current Knowledge and Future Issues. Academic Press, Elsevier, Oxford, UK.

Sui, Y., Kong, H., Huang, X., Dupont, S., Hu, M., Storch, D., Portner, H.-O., Lu, W. and Wang, Y. 2016. Combined effects of short-term exposure to elevated CO_2 and decreased O_2 on the physiology and energy budget of the thick shell mussel *Mytilus coruscus*. Chemosphere 155: 207–216.

Teran, T., Lamon, L. and Marcomini, A. 2012. Climate change effects on POPs' environmental behaviour: a scientific perspective for future regulatory actions. Atm. Poll. Res. 3: 466–476.

Topal, A., Alak, G., Ozkaraca, M., Yeltekin, A.C., Comakl, S., Acıl, G., Kokturk, M. and Atamanalp, M. 2017. Neurotoxic responses in brain tissues of rainbow trout exposed to imidacloprid pesticide: Assessment of 8-hydroxy-2-deoxyguanosine activity, oxidative stress and acetylcholinesterase activity. Chemosphere 175: 186–191.

Urbina, M.A. and Glover, C.N. 2015. Effect of salinity on osmoregulation, metabolism and nitrogen excretion in the amphidromous fish, inanga (*Galaxias maculatus*). J. Exp. Mar. Bio. Eco. 473: 7–15.

Valenti, T.W., Gould, G.G., Berninger, J.P., Connors, K.A., Keele, N.B., Prosser, K.N. and Brooks, B.W. 2012. Human therapeutic plasma levels of the selective serotonin reuptake inhibitor (SSRI) sertraline decrease serotonin reuptake transporter binding and shelter-seeking behavior in adult male fathead minnows. Environ. Scie. Tech. 46(4): 2427–2435.

Wang, Y., Hu, M., Li, Q., Li, J., Lin, D. and Lu, W. 2014. Immune toxicity of TiO_2 under hypoxia in the green-lipped mussel *Perna viridis* based on flow cytometric analysis of hemocyte parameters. STOTEN 470-471: 791–799.

Welankiwar, A., Saudagar, S., Kumar, J. and Barabde, A. 2013. Photostability testing of pharmaceutical products. Int. Res. J. Pharma. 4: 11–15.

Yung, M.M.N., Wong, S.W.Y., Kwok, K.W.H., Liu, F.Z., Leung, Y.H., Chan, W.T., Li, X.Y., Djurišić, A.B. and Leung, K.M.Y. 2015. Salinity-dependent toxicities of zinc oxide nanoparticles to the marine diatom Thalassiosira pseudonana. Aqua. Tox. 165: 31–40.

Zenker, A., Cicero, M.R., Prestinaci, F., Bottoni, P. and Carere, M. 2014. Bioaccumulation and biomagnification potential of pharmaceuticals with a focus to the aquatic environment. Journal of Environmental Management 133: 378–387.

Zhang, L. and Wang, W.-X. 2007. Waterborne cadmium and zinc uptake in a euryhaline teleost *Acanthopagrus schlegeli* acclimated to different salinities. Aqua. Tox. 84: 173–181.

Zhou, Y., Zhang, W., Guo, Z. and Zhang, L. 2017. Effects of salinity and copper co-exposure on copper bioaccumulation in marine rabbitfish *Siganus oramin*. Chemosphere 168: 491–500.

4 Effects of Harmful Algal Bloom Toxins on Marine Organisms

Lopes, V.M.,[1,2,*] *Costa, P.R.*[2,3] *and Rosa, R.*[1]

INTRODUCTION

Phytoplanktonic communities are vital to marine ecosystems. These communities constitute the basis of marine food webs throughout the planet, providing food for filter-feeding organisms, such as bivalves and planktivorous fish and also a number of vertebrate and invertebrate larval stages. Algal blooms are natural occurrences, defined as the sudden overgrowth of microscopic algae under optimal environmental conditions, reaching up to millions of cells per litre (Hallegraeff 1993). These blooms are typically beneficial for the ecosystem, increasing feeding opportunities for countless organisms. However, if toxin-producing microalgae undergo this sudden overgrowth, it can lead to harmful algal blooms (HABs). Despite the fact that approximately 2 percent of microalgae species produce toxins (Hallegraeff 2014, Smayda 1997), HABs can significantly impact marine communities.

In the marine realm, the majority of HAB-toxins are produced by dinoflagellates and diatoms (Table 1). Biochemically, phycotoxins are secondary metabolites that can have a wide range of effects. They can act on the nervous system (brevetoxins), which can induce permanent short-term memory loss (domoic acid) or cause sensorimotor impairment, leading to death (paralytic shellfish toxins) and act on the digestive tract, inducing gastrointestinal distress (diarrhetic shellfish toxins). During the last decades, several new toxins and new toxin derivatives, such as gymnodimines, azaspiracids, pterotoxins, pinnatoxins and hydroxybenzoate saxitoxin, okadaic and domoic acid

[1] MARE – Marine Environmental Sciences Centre, Laboratório Marítimo da Guia, Faculdade de Ciências da Universidade de Lisboa, Portugal.
[2] IPMA – Portuguese Institute for the Sea and Atmosphere, Avenida de Brasília, 1449-006 Lisboa, Portugal.
[3] CCMAR – University of Algarve, Campus of Gambelas, 8005-139 Faro, Portugal.
* Corresponding author: vmlopes@fc.ul.pt

Table 1. The most common toxins produced by marine phytoplankton.

Toxin	Toxic phytoplankton species	Mode of action	Toxin family
Saxitoxins	*Alexandrium* sp., *Gymnodinium catenatum*, *Pyrodinium bahamense*	Inhibition of voltage-gated sodium channels in neural cells	Paralytic shellfish toxins
Domoic acid	*Pseudo-nitzschia* spp., *Amphora coffaeaiformis*, *Nitzschia* sp.,	Binding to glutamate receptors in neural cells causing constant influx of Ca^{2+}	Amnesic shellfish toxins
Brevetoxins	*Karenia brevis, Karenia* sp., *Chatonella* cf. *verrucosa, C. antiqua, C. marina, Fibrocapsa japonica, Heterosigma akashiwo*	Binding to voltage-sensitive sodium channels causing membrane depolarization	Neurotoxic shellfish toxins
Okadaic acid and dinophysistoxins	*Dinophysis* sp., *Prorocentrum* sp.	Inhibition of activity of protein phosphatase 1 and 2	Diarrhetic shellfish toxins

analogues have been described, mostly due to scientific and technological advances (Cruz et al. 2006, Miles et al. 2000, Negri et al. 2003, Satake et al. 1998, Takada et al. 2000, Zaman et al. 1997). In addition, changes on global climate conditions and anthropogenic pressures have been conducting several tropical and subtropical endemic HAB-toxins, namely ciguatoxins, palytoxins and brevetoxins to expand their geographical range into temperate waters (Botana et al. 2015, Villareal et al. 2007).

There is a great body of available information regarding the effects of these toxins in marine organisms, although, the information is much dispersed. Therefore, and with recent increases in HAB frequency and intensity, this chapter aims to update and summarize the available information on the effects of different phycotoxins in marine organisms.

Routes of Toxin Exposure

Toxin transfer can be foodborne or waterborne, that is via food web transfer or through exposure to toxins dissolved in the water after their excretion or cell release (Fig. 1). The most likely pathway of toxin transfer is when toxin-producing species bloom, thus achieving massive concentrations in the water column. However, there are many potential toxin vectors (Fig. 2), depending mostly on the ecology of the toxin producer (pelagic or epibenthic) and the organism's likelihood of exposure to the toxin.

If an organism is exposed to a sudden bloom of toxin-producing microalgae, the toxin concentrations will certainly trigger immediate physiological and behavioural alterations and ultimately cause the death of the organism. In addition, the continuous exposure to low HAB-toxin concentrations can lead to chronic effects.

Here, the pathways of exposure will be divided into direct and indirect contact with the toxin-producer. Through ingestion of toxic phytoplanktonic cells by filter-feeding organisms, such as bivalve molluscs, zooplankton and planktivorous fish, the toxins present inside the cell can accumulate in the predator's viscera. This can create

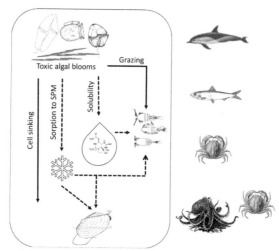

Fig. 1. Foodborne and waterborne exposure to HAB-toxins. Solid lines illustrate the well documented route of dietary exposure; dashed lines illustrate the less studied routes of dissolved toxins exposure.

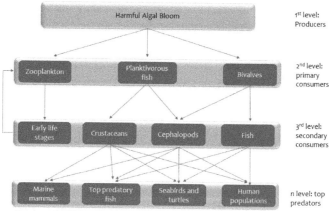

Fig. 2. Schematic view of HAB-toxins food web transfer.

a chain of vectors throughout the food web, potentially eliciting adverse effects in marine communities. Depending on the vector, these toxins can be transferred to humans and cause a variety of shellfish poisonings, due to the ingestion of contaminated shellfish, such as Amnesic Shellfish Poisoning (ASP), Paralytic Shellfish Poisoning (PSP), Neurotoxic Shellfish Poisoning (NSP), Diarrhetic Shellfish Poisoning (DSP) and other syndromes (Table 1).

Some microalgae species produce exotoxins, or exudates, that are released into the water column, causing other organisms to come inadvertently in contact with these compounds. Similarly, when the bloom becomes senescent, the cells lyse and release the toxins to the surrounding environment (Lefebvre et al. 2008), opening another possible pathway to the organisms' direct contact with the toxins. Lastly, there are other HAB-species which segregate the toxin on the outer surface of their cells, potentially inducing damage upon contact. These toxins' effects on marine organisms will be discussed in detail in the following sections.

HAB-toxin Effects on Marine Organisms

Paralytic shellfish toxins

Paralytic shellfish toxins (PSTs) are one example of phycotoxins produced by dinoflagellates, and one of the most abundant and toxic phycotoxins in oceans worldwide. Dinoflagellates from three genera, *Alexandrium, Gymnodinium* and *Pyrodinium*, produce saxitoxin or a suite of over 50 derivatives (Anderson et al. 2012), which the most frequent can be divided according to their chemical structure into carbamoyl, decarbamoyl and sulfamate toxins. These compounds block the conduction of electrical impulses in neural cells through the inhibition of voltage-gated sodium channels on these cell's membranes. This leads to membrane hyperpolarization and results in paralysis in muscle cells as determined in laboratory animals (Kao and Nishiyama 1965, Ritchie and Rogart 1977). PSTs have shown to elicit a wide range of effects on marine organisms, from sublethal and recoverable effects to events of mass mortality in fish, marine mammals and seabirds (Tables 2a and 2b).

PSTs producers inhabit the pelagic realm, the same habitat as many planktonic species. Therefore, plankton species come in contact with PSTs through contact with PST-producing cells or their exudates. It has been shown that many planktonic organisms can be affected by these toxins. In some cases, PSTs inhibit the growth of diatoms, haptophytes and heliozoans, and induce disruptions in swimming behaviour leading to death in ciliates (Table 2a). Diatoms and haptophytes had reduced growth rates after being placed in water previously conditioned by the PSTs producer *A. lusitanicum*, presumably releasing the watersoluble toxins into the culture medium (Blanco and Campos 1988).

Regarding the effects of PSTs on planktonic grazers, there are several studies indicating that some species can selectively avoid ingestion of toxic dinoflagellates, while others do not (Turner and Tester 1997). The latter group can present different effects, with some species being less affected by the toxins, namely *Euterpina acutifrons*, and other species, such as *Acartia grani*, presenting high mortality rates (Costa et al. 2008, 2012).

Direct exposure of bivalve molluscs to PST-producers has been shown to elicit negative effects, as summarized in Table 2a. Exposure to dinoflagellate cells increased shell valve closure in many bivalve species (e.g., *Crassostrea virginica*, *Mytilus edulis*), leading to decreased filtration rates, potentially impacting the animal's normal feeding behaviour. *A. minutum* and purified STX exposure in *C. gigas* resulted in decreased phagocytic activity and ROS production in oyster hemocytes (Mello et al. 2013), leading to higher susceptibility of contracting an infection. Also, the presence of the toxic dinoflagellates decreased byssus production in *M. edulis* and *Geukensia demissa*. However, byssus production in mussels (*M. edulis*) that have been previously exposed to these toxins was less affected.

Table 2a. Documented cases of marine invertebrates exposed to and affected by paralytic shellfish toxins (PSTs).

	Target species	PST source	Route of exposure	Levels of exposure	Tissues analysed	Effects	Observations	References
Phytoplankton	*Isochrysis galbana*	*A. lusitanicum*	Exposure in pre-conditioned seawater with *A. lusitanicum*	-	-	Reduced growth rates	Seawater previously containing *A. lusitanicum* cultures	(Blanco and Campos 1988)
	Pavlova lutheri	*A. lusitanicum*	Exposure in pre-conditioned seawater with *A. lusitanicum*	-	-	Reduced growth rates	-	(Blanco and Campos 1988)
	Skeletonema costatum	*A. lusitanicum*	Exposure in pre-conditioned seawater with *A. lusitanicum*	-	-	Reduced growth rates	-	(Blanco and Campos 1988)
Ciliates	*Favella ehrenbergii*	*A. tamarense*	Exposure to *A. tamarense* cells	4×10^6 cells L^{-1}	-	Disruption of swimming patterns, immobilization, swelling and death	-	(Hansen 1989)
		A. ostenfeldii	Exposure to *A. ostenfeldii* cells	Higher than 2×10^6 cells L^{-1}	-	Backward swimming, swelling and death	-	(Hansen et al. 1992)
Crustaceans	*Acartia clausi*	*A. lusitanicum*	Exposure to *A. lusitanicum* cells	Up to 1600 μg C L^{-1}		Egg production limited	-	(Dutz 1998)

Copepod	Prey	Exposure	Concentration	Culture	Effect		Reference
A. hudsonica	A. minutum	Exposure to A. minutum cells	Up to 0.7 µg C ml⁻¹	A. minutum cultures	Increased feeding rates with increasing dinoflagellate densities, lower hatching success and nauplii production	-	(Frangopulos et al. 2000)
	Alexandrium spp.	Exposure to Alexandrium spp. cells	Up to 1 × 10³ cells L⁻¹	-	Non-selectively fed on Alexandrium spp.	-	(Teegarden et al. 2001)
	A. fundyense	Exposure to A. fundyense cells	-	-	Naïve populations had decreased somatic growth, size at maturity, egg production and survival	-	(Colin and Dam 2004)
	Protogonyaulax tamarensis (now A. tamarense)	Exposure to A. tamarense cells	25 × 10⁴–30 x 10⁴ cells L⁻¹	-	Lower activity, reduced feeding rates	-	(Ives 1987)
A. tonsa	Alexandrium spp.	Exposure to Alexandrium spp. cells	Up to 25 pg STX eq cell⁻¹	Dinoflagellate cultures	Avoided feeding on toxic Alexandrium spp.	-	(Teegarden 1999)
Calamus helgolandicus	G. tamarensis (now A. tamarense)	Exposure to A. tamarense cells	-	-	Reduced fecundity	-	(Gill and Harris 1987)

Table 2a contd. ...

...Table 2a contd.

Target species	PST source	Route of exposure	Levels of exposure	Tissues analysed	Effects	Observations	References
Calanus finmarchicus	A. excavatum	Exposure to A. excavatum cells	45 µg STX kg^{-1}	Copepods	Avoided feeding on toxic A. excavatum	-	(Turriff et al. 1995)
Centropages hamatus	Alexandrium spp.	Exposure to Alexandrium spp. cells	Up to 31 pg STX eq cell^{-1}	Dinoflagellate cultures	Avoided feeding on toxic Alexandrium spp.	-	(Teegarden 1999)
Eurytemora herdmani	Alexandrium spp.	Exposure to Alexandrium spp. cells	Up to 31 pg STX eq cell^{-1}	Dinoflagellate cultures	Non-selectively fed on Alexandrium spp.	-	(Teegarden 1999)
Euterpina acutifrons	A. minutum and G. catenatum	Exposure to A. minutum and G. catenatum cells	Up to 10^6 cells L^{-1} (A. minutum) and 17.5 × 10^4 cells L^{-1} (G. catenatum)	-	Reduced naupliar activity at low cell densities, immobility at higher cell densities	-	(Bagøein et al. 1996)
Palaemonetes pugio	Gonyaulax monilata (now A. monilatum)	Exposure to A. monilatum cells	Up to 1.2 × 10^6 cells L^{-1}	-	High mortality rates following molting	-	(Sievers 1969)
Pseudocalamus sp.	P. tamarensis (now A. tamarense)	Exposure to A. tamarense cells	Up to 6 × 10^5 cells ml^{-1}	-	Lower activity, reduced feeding rates	-	(Ives 1987)
Nauplii of Semibalanus balanoides	Alexandrium spp.	Exposure to Alexandrium spp. cells	Up to 1 × 10^3 cells L^{-1}	-	Avoided feeding on toxic Alexandrium spp.	-	(Teegarden et al. 2001)
Temora longicornis	G. tamarensis (now A. tamarense)	Exposure to A. tamarense cells	-	-	Reduced fecundity	-	(Gill and Harris 1987)

Annelids	Neanthes succinea	Gonyaulax monilata (now A. monilatum)	Exposure to A. monilatum cells	Up to 1.2 × 10^6 cells L^{-1}	-	High mortality rates following spawning	-	(Sievers 1969)
	Polydora websteri	Gonyaulax monilata (now A. monilatum)	Exposure to A. monilatum cells	Up to 1.2 × 10^6 cells L^{-1}	-	High mortality rates following spawning	-	(Sievers 1969)
Molluscs	Larvae of Argopecten irradians concentricus	A. tamarense	Exposure to A. tamarense cells	Up to 11 pg STX eq cell^{-1}	A. tamarense cultures	Activity and growth inhibition, lower attachment rates and reduced climbing rates	-	(Yan et al. 2003)
	A. ventricosus	G. catenatum	Exposure to G. catenatum cells	Up to 4 × 10^6 cells L^{-1}	-	Lower feeding activity, paralysis, increase in hemocytes, epithelial melanisation, increase in pseudofaeces production	Paralysis was found to be reversible	(Escobedo-Lozano et al. 2012)
	Larvae of Chlamys farreri	A. tamarense	Exposure to A. tamarense cells	Up to 5.9 × 10^9 g STX L^{-1}	A. tamarense cultures	High mortality rates, lower hatching rates	-	(Yan et al. 2001)
	Crassostrea gigas	G. washingtonensis (now A. catenella)	Exposure to A. catenella cells	5 × 10^4 cells L^{-1}	-	Reduced pumping activity, increased valve activity	-	(Dupuy and Sparks1967)
		A. minutum	Exposure to A. minutum cells	Up to 12 × 10^4 cells L^{-1}	-	Decreased valve activity, clearance and filtration rates	-	(Lassus et al. 1999)

Table 2a contd. ...

...Table 2a contd.

Target species	PST source	Route of exposure	Levels of exposure	Tissues analysed	Effects	Observations	References
	A. tamarense and A. minutum	Exposure to dinoflagellate cells	Up to 12 × 10⁸ cells L⁻¹	-	Reduced clearance rates		(Laabir and Gentien 1999)
	A. minutum	Exposure to A. minutum cells	5 × 10⁶ cells L⁻¹	-	Mono and diacylglycerols reduced in digestive gland, inflammation of gastrointestinal tract, modified spermatozoa and mitochondria	-	(Haberkorn et al. 2010)
	A. minutum	Exposure to A. minutum cells	15 × 10⁶ cells L⁻¹	-	Altered circadian rhythm	-	(Tran et al. 2015)
	A. minutum	Exposure to A. minutum cells and purified STX	2 × 10⁴ cells L⁻¹ and 0.05 µg STX L⁻¹	-	Reduced hemocyte phagocytic activity, decreased ROS production	Results obtained in both treatments	(Mello et al. 2013)
Larvae of C. gigas	A. tamarense	Exposure to A. tamarense cells	Up to 10⁸ cells⁻¹	-	High mortality rates	-	(Matsuyama et al. 2001)
	A. taylori	Exposure to A. taylori cells	Up to 10⁸ cells⁻¹	-	High mortality rates	-	(Matsuyama et al. 2001)
C. virginica	G. monilata (now A. monilatum)	Exposure to A. monilatum cells	Up to 1.2 × 10⁶ cells L⁻¹	-	Inhibition of byssus production and shell closure	-	(Sievers 1969)
Brachiodontes recurvus	G. monilata (now A. monilatum)	Exposure to A. monilatum cells	Up to 1.2 × 10⁶ cells L⁻¹	-	Inhibition of byssus production and shell closure	-	(Sievers 1969)

Species	Dinoflagellate	Treatment	Concentration	Tissue	Effect		Reference
Geukensia demissa	*P. tamarensis* (now *A. tamarense*)	Exposure to *A. tamarense* cells	$2.5–5.5 \times 10^5$ cells L^{-1}	-	Inhibited cardiac activity	-	(Gainey and Shumway 1988)
	A. tamarense	Exposure to *A. tamarense* cells	10^5 cells L^{-1}	-	Reduced clearance rates	-	(Lesser and Shumway 1993)
	P. tamarensis (now *A. tamarense*)	Exposure to *A. tamarense* cells	5×10^5 cells L^{-1}	-	Reduced clearance rates, increased mucus production	-	(Shumway and Cucci 1987)
	P. tamarensis (now *A. tamarense*)	Exposure to *A. tamarense* cells	10^6 cells L^{-1}	-	Inhibition of byssus production	-	(Shumway et al. 1987)
Mercenaria mercenaria	*A. tamarense*	Exposure to *A. tamarense* cells	10^5 cells L^{-1}	-	Reduced clearance rates	-	(Lesser and Shumway 1993)
Mya arenaria	*P. tamarensis* (now *A. tamarense*)	Exposure to *A. tamarense* cells	$2.5-5.5 \times 10^5$ cells L^{-1}	-	Inhibited cardiac activity	-	(Gainey and Shumway 1988)
	A. tamarense	Exposure to *A. tamarense* cells	Up to 77×10^4 µg STX eq kg^{-1} in viscera	Viscera and other tissues	Naïve populations had higher toxicity and mortality, reduced clearance rates, oxygen consumption rates and burrowing capacity	-	(MacQuarrie and Bricelj 2008)
	A. tamarense	Exposure to *A. tamarense* cells	10^5 cells L^{-1}	-	Reduced clearance rates	-	(Lesser and Shumway 1993)

Table 2a contd. ...

...*Table 2a contd.*

Target species	PST source	Route of exposure	Levels of exposure	Tissues analysed	Effects	Observations	References
	P. tamarensis (now *A. tamarense*)	Exposure to *A. tamarense* cells	5×10^5 cells L^{-1}	-	Reduced clearance rates	-	(Shumway and Cucci 1987)
	A. excavata and *A. tamarense*	Exposure to *Alexandrium* spp. cells	Up to 30.4×10^4 µg STX eq kg^{-1}	Soft tissue	Burrowing incapacity	-	(Bricelj et al. 1996)
Larvae and post-larvae *M. arenaria*	*A. tamarense*	Exposure to *A. tamarense* cells	Up to 64-69 pg STX eq cell^{-1}	Soft tissue	Larvae not affected, post larvae exhibited high mortalities, paralysis, burrowing incapacity	Naïve populations were more severely affected	(Bricelj et al. 2010)
Mytilus edulis	*P. tamarensis* (now *A. tamarense*)	Exposure to *A. tamarense* cells	2.5–5.5×10^5 cells L^{-1}	-	Inhibited cardiac activity	Transient inhibition, long term inhibition and long-term excitation	(Gainey and Shumway 1988)
	Dissolved STX	Intramuscular injection	3330 µg STX kg^{-1}	Digestive glands	Higher GST activity	-	(Gubbins et al. 2001)
	A. tamarense	Exposure to *A. tamarense* cells	10^5 cells L^{-1}	-	Reduced clearance rates	-	(Lesser and Shumway 1993)
	P. tamarensis (now *A. tamarense*)	Exposure to *A. tamarense* cells	5×10^5 cells L^{-1}	-	Increased mucus production	-	(Shumway and Cucci 1987)
	P. tamarensis (now *A. tamarense*)	Exposure to *A. tamarense* cells	10^6 cells L^{-1}	-	Inhibition of byssus production	-	(Shumway et al. 1987)

M. chilensis	*A. catenella*	Exposure to *A. catenella* cells	Up to 7240 µg STX eq kg^{-1}	Soft tissue	Clearance rates, ingestion of organic matter, and absorption efficiency decreased at the start of the experiment.	Effects reversible after three days	(Navarro and Contreras 2010)
Nodipecten subnodosus	*G. catenatum*	Intramuscular injection of GTX 2/3	Up to 0.18 µg STX	-	Paralysis, mantle retraction, inhibition of shell closure, hemocyte reduction		(Estrada et al. 2010)
N. subnodosus	*G. catenatum*	Exposure to *G. catenatum* cells	2 × 10^6 cells individual^{-1}	-	Production of pseudofaeces, partial shell closure, increase in melanisation, increased activity of GPx and lipid peroxidation in gills, decrease in SOD activity in gills	-	(Estrada et al. 2007)
Ostrea edulis	*P. tamarensis* (now *A. tamarense*)	Exposure to *A. tamarense* cells	2.5–5.5 × 10^5 cells L^{-1}	-	Decreased in heart rate	Effects reversible	(Gainey and Shumway 1988)
Perna canaliculus	*A. tamarense*	Exposure to *A. tamarense* cells	12950 µg STX eq kg^{-1}	Soft tissue	No mortalities, normal byssus production and oxygen consumption rates	Oxygen consumption rate increased after 1 h of exposure but normalized after 24 h recovery	(Marsden and Shumway 1992)

Table 2a contd. ....

...Table 2a contd.

Target species	PST source	Route of exposure	Levels of exposure	Tissues analysed	Effects	Observations	References
P. viridis	A. tamarense	Exposure to A. tamarense cells	10^5 cells L^{-1}	-	Reduced clearance rates	-	(Lesser and Shumway 1993)
Placopecten magellanicus	P. tamarensis (now A. tamarense)	Exposure to A. tamarense cells	5×10^5 cells L^{-1}	-	Increased clearance rates	-	(Shumway and Cucci 1987)
Ruditapes phillipinarum	A. tamarense	Exposure to A. tamarense cells	Up to 170 µg STX eq kg^{-1}	Soft tissue	Decreased absorption efficiency and reduced scope for growth with higher toxin concentration, decreased clearance and growth rates	-	(Li et al. 2002)
	A. tamarense	Exposure to A. tamarense cells	2×10^5 cells L^{-1}	-	Hepatic GPx, gill LPO were positively correlated with PSP concentrations, GST presented negative correlation	-	(Choi et al. 2006)
Spisula solidissima	A. tamarense	Exposure to A. tamarense cells	10^5 cells L^{-1}	-	Reduced clearance rates	-	(Lesser and Shumway 1993)

Table 2b. Documented cases of marine vertebrates exposed to and affected by paralytic shellfish toxins (PSTs). IP – Intraperitoneal; IC – Intracoelomic.

	Target species	PST source	Route of exposure	Levels of exposure	Tissues analysed	Effects	Observations	References
Fish	Chanos chanos fingerlings	A. minutum	Dinoflagellate cells and extracts	-	-	Damage to the gills (hyperplasia, edema and necrosis)	Exposure to different cell densities	(Chen and Chou 2001)
	Clupea harengus harengus	PSTs extracted from G. excavata (now A. tamarense)	Oral and IP (intraperitoneal injection)	Up to 5240 µg STX eq kg^{-1}	Viscera	Irregular swimming behaviour, loss of balance, shallow and arrhythmic breathing	Death occurred after 20–60 min of exposure in both routes	(White 1981)
	Larval Clupea harengus harengus	P. tamarensis (now A. tamarense)	Dinoflagellate cells and extracts, via prey	Exposure up to 15×10^5 cells L^{-1}	-	Paralysis and erratic swimming, increased mortality	Heart stopped 20 min after complete immobilization	(Gosselin et al. 1989)
	Larval Clupea harengus pallasi	Dissolved STX	Uptake from surrounding seawater	4000 µg STX eq kg^{-1}/day for 7 days	-	Reduced spontaneous and touch-activated swimming	Effects reversible	(Lefebvre et al. 2005)
	Cyprinodon variegatus	G. monilata (now A. monilatum)	Exposure to A. monilatum cells	Up to 1.2×10^6 cells L^{-1}	-	100 percent mortality rate	-	(Sievers1969)
	Larval C. variegatus	A. fundyense	Prey (copepod Coullana canadensis)	9–12 µg STX eq kg^{-1}	Whole body	Death after consuming 6–12 contaminated copepods	Reduced prey capture and predator avoidance	(Samson et al. 2008)
	Diplodus sargus	PSTs extracted from G. catenatum	Intracoelomic injection (IC)	15.2 µg STXeq kg^{-1}	Liver	Increased GST activity and erythrocyte nuclear abnormalities	-	(Costa et al. 2012)

Table 2b contd. ...

Table 2b contd. ...

Target species	PST source	Route of exposure	Levels of exposure	Tissues analysed	Effects	Observations	References
Fundulus heteroclitus	Dissolved STX	Uptake from surrounding seawater	0, 75, and 150 ppb, nominal concentrations	-	Decreased c-Fos expression, paralysis, decreased activity, floating	-	(Saliermo et al. 2006)
Larval *Fundulus heteroclitus*	*A. fundyense*	Prey (copepod *Coullana canadensis*) and *A. fundyense*	17–25 μg STX eq kg^{-1}	Whole body	Death after consuming 6–12 copepods and following direct ingestion of dinoflagellate cells	Reduction in prey capture, swimming performance	(Samson et al. 2008)
Larval *Mallotus villosus*	*P. tamarensis* (now *A. tamarense*)	Exposure to *A. tamarense* cells	Up to 15×10^5 cells L^{-1}	-	Paralysis and erratic swimming, increased mortality	Heart stopped 20 min after complete immobilization	(Gosselin et al. 1989)
Pollachius virens	PSTs extracted from *G. excavata* (now *A. tamarense*)	Oral and IP	Up to 3430 μg STX eq kg^{-1}	Viscera	Irregular swimming behaviour, loss of balance, shallow and arrhythmic breathing	Death occurred after 20–60 min of exposure in both routes	(White 1981)
Pseudopleuronectes americanus	PSTs extracted from *G. excavata* (now *A. tamarense*)	Oral and IP	Up to 8370 μg STX eq kg^{-1}	Viscera	Irregular swimming behaviour, loss of balance, shallow and arrhythmic breathing	Death occurred after 20–60 min of exposure in both routes	(White 1981)
Recently-settled *P. americanus*	*A. fundyense*	Prey (copepod *Coullana canadensis*)	41–58 μg STX eq kg^{-1}	Whole body	Death after consuming 6–12 contaminated copepods	Reduced swimming abilities	(Samson et al. 2008)

Species	Toxin source	Route	Concentration	Tissue	Effects	Outcome	Reference
Onchorhynchus mykiss	Dissolved STX	IP injection	1.752 µg kg⁻¹	-	Elevated levels of cortisol, decreased attack latency times	-	(Bakke et al. 2010)
Salmo salar	Dissolved STX	IP injection	10 µg STX kg⁻¹	-	Loss of balance, decreased respiration rate, muscle bursts and twitching	-	(Bakke et al. 2010)
	PSTs extracted from *G. excavata* (now *A. tamarense*)	Oral and IP	Up to 9410 µg STX eq kg⁻¹	Viscera	Irregular swimming behaviour, loss of balance, shallow and arrhythmic breathing	Death occurred after 20–60 min of exposure in both routes	(White 1981)
Seabirds							
Uria aalge californica, Gavia arctica pacifica, Melanitta fusca deglandi, Lunda cirrhata, Puffinus griseus, Fulmarus glacialis, Diomedea nigripes	*Gonyaulax catenella* (now *A. catenella*)	Prey (small fish and crustaceans)	-	-	Death	-	(Mckernan and Scheffer 1942)
Larus argentatus	*G. catenella* (now *A. catenella*)	Prey (small fish and crustaceans)	-	-	Death	-	(Mckernan and Scheffer 1942)
	A. tamarense	Likely *Ammodytes hexapterus*	1100 (intestine) and 480 (brain) µg STX eq kg⁻¹	Intestine and brain of dead birds	Death	-	(Levasseur et al. 1996)
	G. excavata (now *A. tamarense*)	Likely *A. hexapterus*	970 µg STX eq kg⁻¹ in *A. hexapterus*	Prey	Death	-	(Nisbet 1983)

Table 2b contd....

Table 2b contd. ...

Target species	PST source	Route of exposure	Levels of exposure	Tissues analysed	Effects	Observations	References
L. dominicus	*G. catenella* (now *A. catenella*), *G. grindleyi* (now *Protoceratium reticulatum*)	Prey (mussels)	-	-	Death	Some bird populations decreased drastically following dinoflagellate outbreak	(Shumway et al. 2003)
L. hartlaubii	*G. catenella* (now *A. catenella*), *G. grindleyi* (now *Protoceratium reticulatum*)	Prey (mussels)	-	-	Death	Some bird populations decreased drastically following dinoflagellate outbreak	(Shumway et al. 2003)
L. occidentalis	*G. catenella* (now *A. catenella*)	Prey (small fish and crustaceans)	-	-	Death	-	(Mckernan and Scheffer 1942)
Haematopus moquini	*G. catenella* (now *A. catenella*), *G. grindleyi* (now *Protoceratium reticulatum*)	Prey (mussels)	-	-	Death	Some bird populations decreased drastically following dinoflagellate outbreak	(Shumway et al. 2003)
L. atricila	*G. excavata* (now *A. tamarense*)	Likely *A. hexapterus*	970 µg STX eq kg^{-1} in *A. hexapterus*	-	Death	-	(Nisbet 1983)

	Species	Causative organism	Vector/Prey	Toxin concentration	Tissue	Effects	Notes	Reference
	Phalacrocorax aristotelis	G. tamarensis (now A. tamarense)	-	Not determined in birds. Up to 6000 MU (mouse units) 100 g⁻¹ in mussel meat	-	Intestinal hemorrhaging; loss of balance and stagger, vomiting, death	Drastic population decline	(Armstrong et al. 1978)
	Anas rubripes	G. tamarensis (now A. tamarense)	Prey (mussels)	20000–40000 µg STX eq kg⁻¹	-	Death	Other bird species affected nearby	(Shumway et al. 2003)
	Sterna hirundo, S. paradisaea, S. dougallii	G. excavata (now A. tamarense)	Likely A. hexapterus	970 µg STX eq kg⁻¹ in A. hexapterus	-	Death	-	(Nisbet 1983)
Marine Mammals	Monachus monachus	A. minutum	-	3030–3390 µg STX eq kg⁻¹	Liver	Death	Analysis on stranded animals	(Costas and Lopez-Rodas 1998)
		G. catenatum or Pyrodinium bahamense var. compressa	Prey	6–900 µg dcSTX kg⁻¹	Liver, kidney, muscle, brain	Death	PST producer uncertain, analysis on stranded animals	(Reyero et al. 2000)
		A. minutum, G. catenatum, D. acuta	Prey	2–280 µg dcSTX kg⁻¹ liver 8–64 µg dcSTX kg⁻¹ brain	Liver, kidney, muscle, brain	Lethargy, sensorimotor impairments, paralysis, death	PST producer uncertain	(Hernández et al. 1998)
	Megaptera novaeangliae	-	Horse mackerel (Scomber scombrus)	-	-	Death	PST producer uncertain, analysis on stranded animals	(Geraci et al. 1989)

Similar experiments performed on greenshell mussels (*Perna canaliculus*) showed that mussels presented oxygen consumption and clearance rates similar to the control group after 24 h exposure to *A. tamarense* (Marsden and Shumway 1992).

Exposure of *M. chilensis* to *A. catenella* for 21 days (Navarro and Contreras 2010) resulted in lowered clearance rates, organic matter intake and absorption efficiency at the start of the experiment, followed by an increase to levels similar to the ones presented in the control group. The STX uptake steadily increased throughout the experiment, similarly to the mussels' excretion rates. Oxygen consumption rates seemed unaffected by the ingestion of this toxic species, revealing that this species may possess defence mechanisms that allow them to feed safely on this dinoflagellate.

On the other hand, some scallop and clam species presented negative effects when exposed to this toxin. The scallop *Nodipecten subnodosus* presented paralysis of the adductor muscle while maintaining the digestive tract functioning after receiving intramuscular injections of GTX 2/3. Also, the hemocyte number decreased and they presented mantle retraction and their shells remained open up to 40 days after the exposure (Estrada et al. 2010). In a different study on the same species (Estrada et al. 2007), exposure to *G. catenatum* cells resulted in the increased production of mucus, pseudo-faeces, melanisation and hemocyte aggregation in gill tissue. Biochemically, an antioxidative stress response to the toxin was shown in gill tissue, this tissue being the first to come in contact with the toxin. There was an increase in glutathione peroxidase (GPx) activity and lipid peroxidation, along with a decrease in superoxide dismutase activity, indicating oxidative and cellular damage. Similar results were obtained when feeding *G. catenatum* cultures to scallops (*Argopecten ventricosus*) (Escobedo-Lozano et al. 2012). The scallops presented paralysis of the adductor muscle, lower feeding activity, increased pseudo-faeces production, increased number of hemocytes in gill, mantle and adductor muscle tissue and epithelial melanisation in gill and mantle tissue. These results indicate that scallops have efficient mechanisms that protect them against lethal effects from external toxicants. Studies conducted in *Ruditapes phillipinarum* feeding on *A. tamarense* cultures for 6 (Li et al. 2002) and 15 days (Choi et al. 2006) showed reduced scope for growth, decreased absorption efficiency, clearance and growth rates after six days of exposure, and increased activity of GPx in hepatopancreas and gill lipid peroxidation with increasing toxin burden after 15 days of exposure.

PST effects on bivalve early stages are comparatively less understood. Bricelj et al. (2010) addressed this issue by exposing larvae, post-larvae and juveniles of *Mya arenaria* to *A. tamarense*. They showed that larvae were not significantly affected by the dinoflagellate due to the fact that the cells were too large for prey capture; thus, the larvae did not accumulate the toxin or displayed any intoxication symptoms. On the other hand, the post-larvae presented decreased burrowing capacity, here used as proxy of sensitivity to PSTs. Also, the post-larvae had increased mortality rates, especially in populations that are not usually exposed to *A. tamarense* blooms, while juvenile were less susceptible.

PST effects on bivalves are species-specific and seem to differ geographically within the same species and life stages. Species that are usually exposed to toxic dinoflagellate blooms seem to be more resistant, and appear to have developed defence mechanisms that allow them to cope with high PST levels, unlike other species in areas less affected by blooms.

These studies reveal that bivalves are not immune to the effects of PST contamination, and there are some species with higher sensitivity to these toxins. This may pose additional concerns over the ecosystem's health and elicit negative economic impacts, since some of these species are commercially farmed in shellfish aquacultures, and blooms may occur in farmed areas.

PSTs have long been associated with fish kills. Fish can be directly exposed to the toxins, as is the case of planktivorous fish such as sardines, herring and anchovies, or indirectly through feeding on vectors, affecting many levels of the marine food web, from groupers and hake to sturgeons and artificially fed fish, such as farmed salmon.

Only a few events have been directly linked to PST contamination, since these events are unpredictable and sporadic, many times leading to inconclusive data. For a complete list of fish kills associated with PSTs refer to Costa (2016, Table 1).

When studying PST's effect on fish, it is procedurally simpler to inject the toxin intracoelomically (IC), in order to closely control the given concentration. Standard STX is the toxin most commonly administered. However, despite the benefits, these methods are less ecologically relevant, since the toxins do not enter directly in the coelom, and STX is but a fraction of the toxins produced by the dinoflagellate species. Nevertheless, these studies provide windows into the symptoms presented by fish and insight into the effects of these neurotoxins.

STX's effects on killifish (*Fundulus heteroclitus*) were quantified regarding the expression of c-Fos protein (Salierno et al. 2006), responsible for regulating neural cells' survival, and is associated with long term memory (Sadananda and Bischof 2002). It was shown that the expression of this protein decreased, and the fish presented behavioural alterations including paralysis, lethargy and loss of balance. STX most likely affects the neural pathways responsible for swimming. In Atlantic salmon (*Salmo salar*), it was shown that STX crosses the blood-brain barrier and that sublethal doses of this toxin affect the activity of brain subregions in the central nervous system (CNS), possibly affecting the organism's cognitive abilities (Bakke and Horsberg 2007). Intracoelomic injections of STX in white seabream (*Diplodus sargus*) resulted in an increase of glutathione-S-transferase (GST) activity, an enzyme responsible for removing xenobiotics, among many other roles. STX also induced DNA damage (chromosome breaks or loss) and increased erythrocyte nuclear abnormalities (Costa et al. 2012).

In order to simulate bloom conditions, milkfish (*Chanus chanus*) fingerlings were exposed to STX extract and *A. minutum* cells in increasing concentrations and cell density, respectively (Chen and Chou 2001). After 24 h, the fish presented oedema, hyperplasia and necrosis in gill lamellae. The exposure also resulted in increased mortality rates in the treatments with higher cell density and STX concentrations, due to increasing oxygen demand following gill damage. Similar results reporting gill damage following fish exposure to PST-producing dinoflagellate cells were found in salmon and trout (Mortensen 1985). White (1981) reported high mortality rates after 20 to 60 minutes in Atlantic herring (*Clupea harengus harengus*), American pollock (*Pollachius virens*), winter flounder (*Pleuronectes americanus*), Atlantic salmon (*S. salar*) and cod (*Gadus morhua*) when dosed intraperitonially (IP) or orally with toxin extracts from *A. tamarense* cultures. Prior to death, the fish presented loss of balance, immobilization and arrhythmic breathing, consistent with the symptoms described here in adult fish.

In early stages of development, some fish species present different ecologies than the adults, starting out as planktonic larvae, and thus occupying the same niche as the pelagic dinoflagellate PST-producers. Also, earlier stages of development are likely more vulnerable to the effects of these toxins as they possess higher mass-specific metabolic rates and they lack fully developed detoxification systems (Vasconcelos et al. 2010).

Overall, when fish early stages are exposed to bloom simulations in experimental conditions, it resulted in extremely high mortality rates, nearing the totality of the experimental population, besides the sublethal effects displayed by the young (Table 2b). Fish early stages can be exposed to the toxin through feeding on zooplanktonic vectors, such as copepods, or through direct exposure to the dissolved toxins. Recently settled flounders (*P. americanus*), sheepshead minnow (*Cyprinodon variegatus*) and mummichog larvae (*F. heteroclitus*) were fed with contaminated copepods, acting as vectors of *A. fundyense* (Samson et al. 2008). After consuming 6–12 contaminated copepods, the fish died. In this study, the fish were also fed with fewer copepods, resulting in a variety of effects, such as reduced swimming abilities, prey capture success, predator avoidance and overall activity (Table 2b).

Gosselin et al. (1989) exposed capelin and herring larvae to three different treatments to ascertain the effects of PSTs through different routes of exposure, recurring to both direct exposure through feeding the larvae with *A. tamarense* cells in increasing densities, and placing toxin extracts in the experimental tanks. Indirect exposure was achieved by feeding the larvae with contaminated microzooplankton. Capelin and herring larvae fed on *A. tamarense* swam erratically, lost motility and sank to the bottom paralysed, dying after 20 minutes of exposure, contrary to the lack of effect when exposed directly to the toxin dissolved. Feeding these larvae with contaminated zooplankton elicited similar results as feeding directly on the dinoflagellate, resulting in paralysis and high mortality rates. However, exposure of herring larvae to dissolved STX resulted in a reversible dose-dependent suite of sensorimotor impairments (Lefebvre et al. 2005), such as spontaneous swimming and tactile response inhibition. Also, it was shown that older larvae were more susceptible to the dissolved toxin, likely due to the degree of gill and body maturation leading to higher toxin uptake.

Monk seal populations have been greatly impacted by PST outbreaks. In the late 1990s in Cape Blanc Peninsula over 100 monk seals died following PST intoxication. Tissue analysis revealed PSTs in brain tissue, suggesting that these toxins were present in the seal's nervous system (Costas and Lopez-Rodas 1998, Reyero et al. 2000). The cause has been attributed to PSTs since there were high levels of these toxins in many fish species that the seals prey upon (Reyero et al. 2000). Dying organisms presented many behavioural alterations, lethargy, paralysis and sensorimotor discoordination (Hernández et al. 1998).

Earlier, in 1987, over a dozen humpback whales washed ashore dead along Nantucket Sound. The cause of the stranding was ascertained by analysing fish and whale tissues. It was determined that one of the fish species analysed, Atlantic mackerel (*Scomber scombrus*), presented high levels of STX and stomachal content analysis revealed that the whales were previously feeding on this species. It was worth noting that the time-lapse between the onset of the first symptoms and death (approximately 90 minutes, Geraci et al. 1989), suggesting a very quick process, characteristic of severe STX intoxication.

Seabird deaths due to HAB-toxins have been comparatively overlooked. However, there have been countless events where many seabird species died following the ingestion of contaminated fish and shellfish (Table 2b). Shumway et al. (2003) extensively reviewed all the registered seabird deaths that were linked with HAB-toxins, including PSTs. PSTs were reported to cause loss of motor coordination and paralysis, resulting in the bird's inability to feed and thus, causing death by starvation. Female terns presented an inability to lay eggs due to sublethal onset of paralysis, resulting in the egg breaking inside the body and causing fatal haemorrhages. Other species presented severe inflammation of the gastro-intestinal tract and haemorrhages in the intestines and brain.

Understanding the effects of PSTs, which are produced worldwide, is of vital importance, since, as reviewed here, the range of possible consequences is very wide. In some cases, the toxins directly affect the species, causing high mortalities, and in other cases the toxin accumulates and is transferred throughout many levels of the marine food web, causing indirect damage to the ocean's health, communities and human populations.

Amnesic shellfish toxins

Domoic acid (DA) is a potent neurotoxin produced by some species belonging to two genera of diatoms, *Pseudo-nitzschia* and *Nitzschia*. This toxin is known to cause Amnesic Shellfish Poisoning (ASP), and the attention on this toxin and its possible consequences was focused after an incident involving the death of three people in 1987 following the ingestion of mussels contaminated with DA. Afterwards, most coastal countries developed monitoring programs, regularly analysing bivalve tissue for DA and other phycotoxins in order to prevent foodborne illnesses. These monitoring programs have been successful at avoiding further human casualties. Nevertheless, there have been many events of DA intoxication in marine organisms.

DA acts in neural cells, competing for the same receptors as glutamate, an excitatory neurotransmitter. By having less affinity for these receptors, glutamate fails to bind normally, causing excessive concentrations of glutamate outside the synapses, triggering AMPA, kainate and NMDA receptors' activation, permanently opening the neural cell's membrane, leading to excessive influx of Ca^{2+} (Berman and Murray 1997). This causes membrane depolarization and subsequent degeneration of neural cells. The higher concentration of glutamatergic receptors in the hippocampus, the brain region responsible for memory acquisition and learning, is the cause for the memory loss.

In Table 3 we summarized the effects of DA in marine organisms. Bivalves are common vectors of this toxin and the effects of this toxin seem somewhat overlooked. DA seems to affect haemolymph chemistry, increase the number of hemocytes as well as increase cholinesterase activity and DNA damage. On the other

Table 3. Documented cases of marine organisms exposed to and affected by domoic acid (DA). IP – Intraperitoneal; IC – Intracoelomic.

	Target species	DA source	Route of exposure	Toxicity	Tissues analysed	Effects	Observations	References
Crustaceans	*Tigriopus californicus*	Dissolved DA	Uptake from seawater	8.62 μM	-	Death	-	(Shaw et al. 1997)
Molluscs	*C. gigas*	*P. pungens* f. *multiseries*	Exposure to diatom cells	Up to 0.86 μg DA g^{-1}	Soft tissue	Increased number and activity of hemocytes	Effects reversible	(Jones et al. 1995)
		P. pungens f. *multiseries*	Exposure to diatom cells	36.3 μg DA g^{-1}	Soft tissue	Respiratory acidosis, haemolymph hypercapnia and increased bicarbonate, low haemolymph PO$_2$ levels	Haemolymph PO$_2$ and pH returned to normal	(Jones et al. 1995)
	M. californiamus	*P. pungens* f. *multiseries*	Exposure to diatom cells	3.6 μg DA g^{-1}	Soft tissue	Increased haemolymph pH, decreased PO$_2$, decreased PCO$_2$	Effects reversible	(Jones et al. 1995)
	M. edulis	Dissolved DA	Intramuscular injection	Up to 50 μg DA g^{-1}	-	Cholinesterase activity, DNA damage and number of hemocytes increased, phagocytic activity decreased	-	(Dizer et al. 2001)
	Larvae of *P. maximus*	Dissolved DA	Uptake from seawater	5.21 pg DA individual^{-1}	Whole body	Decreased growth, shell length and survival	Exposure of 25 days	(Liu et al. 2007)
	P. maximus	Dissolved DA	Feed incorporated with DA	Up to 302.5 ng DA g^{-1}	Whole body	Decreased growth and survival	-	(Liu et al. 2008)

Fish					
O. kisutch	Dissolved DA	IC injections	Up to 34 µg DA g^{-1}	-	Circle, spiral and upside-down swimming	-	(Lefebvre et al. 2007)
O. mykiss	Dissolved DA	IP injections	0.75 mg DA kg^{-1}	-	Increased cortisol levels, decreased attack latency time	-	(Bakke et al. 2010)
Engraulis mordax	Dissolved DA	IC injections	Up to 14 µg DA g^{-1}	-	Spinning, disorientation, inability to school, death	-	(Lefebvre et al. 2001)
S. salar	Dissolved DA	IP injections	6 mg DA kg^{-1}	-	Increase in metabolic activity	-	(Bakke and Horsberg 2007)
F. heteroclitus	Dissolved DA	IP injections	5 mg DA kg^{-1}	-	Increased c-Fos activity expression in optic brain regions	-	(Salierno et al. 2006)
Sparus aurata	Dissolved DA	IP injections	Up to 9 mg DA kg^{-1}	-	Death at higher concentrations, spiral, circle and upside-down swimming	Sublethal effects (swimming anomalies) reversible after 24 h	(Nogueira et al. 2010)
Sigamus oramin	Dissolved DA	IC injections	2 x 10^3 µg DA kg^{-1}	-	Increased CYP1A activity (protein involved in xenobiotic metabolism)	-	(Wang et al. 2008)

Table 3 contd. ...

...Table 3 contd.

	Target species	DA source	Route of exposure	Toxicity	Tissues analysed	Effects	Observations	References
Sea birds	*Callithrix jacchus*	Dissolved DA	IP injection	4 mg DA kg⁻¹	-	Epileptic seizures, death	-	(Perez-Mendes et al. 2005)
	Pelecanus occidentalis	*Pseudo-nitzschia* sp.	Prey (*S. scombrus*)	37.17×10^3 µg DA kg⁻¹	Digestive tract	Death	150 birds dead in 5 days	(Sierra-Beltrán et al. 1997)
		P. australis	Prey (*E. mordax*)	27.9×10^3 µg DA kg⁻¹	Pelican stomach	Hemorrhage, tissue necrosis, death	Death occurred after ~ 60 min of exposure	(Work et al. 1993)
		P. australis	Prey (*E. mordax*)	$45–48 \times 10^3$ µg DA kg⁻¹	Pelican stomach	Death		(Fritz et al. 1992)
	Phalacrocorax penicillatus	*P. australis*	Prey (*E. mordax*)	27.9×10^3 µg DA kg⁻¹	Cormoran stomach	Hemorrhage, tissue necrosis, death	Death occurred after ~ 60 min of exposure	(Work et al. 1993)
		P. australis	Prey (*E. mordax*)	$45–48 \times 10^3$ µg DA kg⁻¹	Cormoran stomach	Death		(Fritz et al. 1992)
Marine mammals	*Balaenoptera acutorostrata*	*P. australis*	Prey (*E. mordax*)	258×10^3 µg DA g⁻¹	Whale feces	Death	-	(Fire et al. 2010)
	Callorhinus ursinus	*Pseudo-nitzschia* sp.	Likely through prey	18600×10^3 ng DA kg⁻¹	Seal's feces	Ataxia, seizures, lesions in brain and heart, death	-	(Lefebvre et al. 2010)
	Kogia breviceps	*Pseudo-nitzschia* sp.	Likely through prey	13.56×10^3 ng DA kg⁻¹	Whale feces	Death	-	(Fire et al. 2009)

Species	Causative organism	Exposure route	DA concentration	Sample type	Symptoms		Reference
K. sima	*Pseudo-nitzschia* sp.	Likely through prey	967×10^3 ng DA kg^{-1}	Whale feces	Death	-	(Fire et al. 2009)
Zalophus californianus	*Pseudo-nitzschia* sp.	Likely through prey	96.8×10^3 µg DA kg^{-1}	Sea lion's feces	Ataxia, head weaving, disorientation, seizures and death	-	(Bargu et al. 2011)
	P. australis	Likely through prey	Up to 182.01×10^3 µg DA L^{-1}	Sea lion's feces	Ataxia, head weaving, disorientation, seizures and death	-	(Scholin et al. 2000)
	Pseudo-nitzschia sp.	Maternal transfer	44×10^3 ng DA L^{-1}	Stomach contents of premature pups	Premature births, abortions, reproductive failure and brain edema	-	(Goldstein et al. 2009)
	Pseudo-nitzschia sp.	Maternal transfer	261×10^3 ng DA L^{-1}	Maternal urine	Premature births, abortions and reproductive failure	-	(Brodie et al. 2006)
	P. australis	Prey (*E. mordax*)	-	-	Seizures, hippocampal atrophy, neural necrosis	-	(Silvagni et al. 2005)
	-	-	-	-	Heart and brain lesions, death	-	(Zabka et al. 2009)
	P. australis	Prey (*E. mordax*)	136.5×10^3 µg DA kg^{-1}	Sea lion's feces	Death	-	(Lefebvre et al. 1999)

hand, exposure to this toxin decreased phagocytic activity, growth and survival rates. In some cases, the effects of exposure to the DA-producing diatoms were reversible and the organisms recovered after a short period of time (up to 24 h), emphasizing the notion that bivalves are quite resilient to DA and other toxins.

Information regarding the effects of DA in wild animals is very scarce and limited to marine mammals and seabirds. Domoic acid's effects on fish have been studied through IC injection (Table 3). This technique allows the use of known DA concentrations without dispersal throughout the organism's body. However, this method does not always allow for ecologically relevant DA concentrations to be used or for natural DA uptake and transfer between body tissues to take place naturally. Regarding the effects of DA in fish, most studies concluded that it causes abnormal swimming behaviour, including spiral, circle and upside-down swimming, and ultimately death. Other effects that can escalate DA toxicosis are inability to school in *Engraulis mordax* (Lefebvre et al. 2001), possibly making the fish easier targets for predators, disrupting the balance of the food web during diatom blooms.

Killifish (*F. heteroclitus*) IC injected with up to 9 mg DA kg^{-1} showed that c-Fos activity, a protein associated with long term memory (Sadananda and Bischof 2002), increased in several brain regions, indicating neuronal stress following exposure. Variations in c-Fos expression can lead to effects at the behavioural levels, as observed in Salierno et al. (2006), such as disorientation and loss of equilibrium.

One of the main groups affected by domoic acid are marine mammals, more specifically sea lions. There is an extensive record of sea lion deaths going back nearly two decades, when over 400 sea lions (*Zalophus caifornianus*) were found stranded or displayed neurological symptoms associated with DA intoxication, later confirmed by detecting DA in sea lions' tissues (Scholin et al. 2000). The cause of death was attributed to ingestion of contaminated anchovies, a common food source for these mammals. Behavioural tests and magnetic resonance imaging (MRI) performed on sea lions displaying intoxication symptoms revealed abnormal behaviours, such as head weaving, ataxia and severe disorientation. The MRIs showed hippocampal lesions damaging hippocampal-thalamic networks. Also, DA has been detected in the stomach contents of premature pups and shown to elicit premature births, abortions and death of pregnant sea lions (Brodie et al. 2006) due to the consumption of contaminated prey, possibly endangering this species' populations. Throughout the years, many other events of sea lion mortality have been attributed to DA intoxication (Table 3).

Studies regarding the interaction of HAB-toxins occurring simultaneously is very scarce. However, between February–April 2008, over 100 dolphins (*T. truncatus*) were found stranded along the coast of Texas, and their tissues were positive for DA, brevetoxins and okadaic acid (Fire et al. 2011), although in different proportions and only a small percentage was positive for more than one toxin. The mass stranding may be linked to an interaction of these three toxins, but without historical data and means of comparison, it may not be safe to conclude so.

DA, as mentioned above, acts on glutamatergic receptors, mainly present in organisms with developed brains, possibly explaining the discrepancy between the effects caused in vertebrates and invertebrates. Invertebrates are likely less affected by this toxin, since they mostly lack complex brains and possess effective elimination systems, as in the case of bivalve molluscs. The fact that bivalves seem to be less affected and are efficient at eliminating DA does not exclude the sublethal effects that it may cause in other organisms higher up the food web, through chronic ingestion of contaminated prey.

Brevetoxins

Brevetoxins (BTXs—PbTx1-10, BTX 1-4) are produced by dinoflagellates and raphidophytes and can cause Neurotoxic Shellfish Poisoning (NSP). These toxins are a group of complex polycyclic polyether compounds that alter the properties of membranes in excitable cells by binding to voltage sensitive sodium channels in nerve cells, leading to membrane depolarization and disrupting normal processes in nerve cells (Landsberg 2002, Lopes et al. 2013). The term "red tides" is mainly associated with blooms of *Karenia brevis* (= *Gymnodinium breve*), which proliferate in great concentrations and their pigments discolour the surrounding seawater. This dinoflagellate is mainly responsible for causing many events of mass mortality in marine organisms, with reports dating back to the 19th century (Glennan 1887). Although no human deaths have been related to brevetoxins, human populations are affected at the sublethal level, mostly through consumption of contaminated shellfish or aerosol inhalation during red tide events (Landsberg 2002). Effects of brevetoxins in marine organisms, from copepods to marine mammals, are summarized in Table 4.

Brevetoxins can have allelopathic effects on other species of phytoplankton, as shown in Prince et al. (2008), where cultures of *Asterionellopsis glacialis*, *Prorocentrum minimum* and *Skeletonema costatum* presented decreased growth rates when their individual cultures were mixed with *K. brevis* cultures and exudates.

Studies on copepods revealed that these organisms are very sensitive to *K. brevis*, by presenting accelerated heart rates, loss of motor control, suppressed swimming behaviour, lethargy, paralysis, regurgitation and decreased survival and growth (Cohen et al. 2007, Huntley et al. 1987, Huntley et al. 1986, Sykes and Huntley 1987, Turner et al. 1996). Other BTXs producers have elicited negative impacts upon feeding in copepods (Uye 1986, Uye and Takamatsu 1990).

Bivalves, as one of the main vector of phycotoxins, also accumulate high concentrations of BTX, with very few studies regarding its effect. Leverone et al. (2007) showed that when exposed to *K. brevis* cells and extracts caused reduced clearance rates in *A. irradians, C. virginica, Mercenaria mercenaria* and *P. viridis*. Recruitment of bay scallop (*A. irradians concentricus*) in North Carolina was greatly affected by *K. brevis* blooms, jeopardizing the sustainability of scallop beds in the region (Summerson and Peterson 1990).

Red tides can also affect the species' abundance and richness of the impacted areas, in some cases nearly wiping out many important benthic infaunal species (Simon and Dauer 1972), decreasing richness in fish species by 50 percent and causing a decrease in invertebrate species' abundance in general (Dupont et al. 2010).

Knowledge on the effects of BTXs in fish is very scarce beside the numerous accounts of fish kills associated with BTXs (Gunter et al. 1948, 1947, Rounsefell and Nelson 1966, Thronson and Quigg 2008). Flaherty and Landsberg (2011) reported reduced annual recruitment in *Cynoscion nebulosus, C. arenarius*, and *Sciaenops*

Table 4. Documented cases of marine organisms exposed to and affected by okadaic acid (OA) and dinophysistoxins (DSTs).

	Target species	OA source	Route of exposure	Toxicity	Tissues analysed	Effects	Observations	References
Molluscs	*M. edulis*	Dissolved OA	Mixed with algal diet	2.5 nM OA	-	Increased DNA fragmentation	-	(McCarthy et al. 2014)
	M. galloprovincialis	Dissolved OA	*In vitro* exposure	Up to 500 nM OA	-	DNA damage leading to necrosis and apoptosis in gill tissue	-	(Prego-Faraldo et al. 2015)
		Dissolved OA	Mixed with mussel feed	Up to 6.5 μg OA	-	Up-regulation of gene transcripts associated with stress response	-	(Manfrin et al. 2010)
	C. gigas	Dissolved OA	Mixed with algal diet	2.5 nM OA	-	Increased DNA fragmentation	-	(McCarthy et al. 2014)
	C. virginica	*P. minimum*	Exposure to dinoflagellate cells	3.9×10^6 cells ml^{-1}	-	Larvae presented lower growth rates, and slower development	-	(Wikfors and Smolowitz 1995)
	P. perna	*P. lima*	Exposure to dinoflagellate cells	Up to 10000 cells mussel^{-1}	-	Higher incidence of micronuclei and nuclear lesions in hemocytes	-	(Carvalho Pinto-Silva et al. 2005)
		Dissolved OA	Uptake from seawater	0.3 μg OA	-	Higher incidence of micronuclei in hemocytes	-	(Carvalho Pinto-Silva et al. 2003)

Effects of Harmful Algal Bloom Toxins on Marine Organisms 71

	R. decussatus	*P. lima* extracts and cells	Exposure to extracts and dinoflagellate cells	Up to 100 nM OA and 20000 cells ml^{-1}	-	DNA damage in gill tissue	-	(Flórez-Barrós et al. 2011)
Fish	*Dicentrarchus labrax*	*P. lima* pre-conditioned seawater	Exposure to pre-conditioned sewater with *P. lima*	Previously with 9×10^3 cells ml^{-1}	-	Reduced feeding reflexes, abnormal swimming patterns	-	(Ajuzie 2007)
		P. lima	Exposure to dinoflagellate cells	4.5×10^3 cells ml^{-1}	-	Death	-	(Ajuzie 2007)

ocellatus and Riley et al. (1989) revealed that upon hatching, larvae of *S. ocellatus* were negatively affected by BTXs, developing deformities, swimming erratically and being paralysed before dying. Seabird deaths attributed to brevetoxicosis date back to the early 70's, when thousands of lesser scaup (*Athya affinis*) died concurrently in a red tide event (Forrester et al. 1977), along with many other seabird species throughout the years (Landsberg et al. 2009). Also, a number of sea turtles strandings have been positively linked to red tide events, as the number of strandings increases during red tides (Landsberg et al. 2009).

The highly endangered Florida manatees (*Trichechus manatus latirostris*) have suffered great impacts from red tides. In 1996, over a hundred manatees died following a red tide caused by *K. brevis* (Bossart et al. 1998), likely through prolonged exposure to BTX aerosols or ingestion of contaminated seawater. Recently, Flewelling et al. (2005) showed that seagrass (*Thalassia testudinum*) accumulates high concentrations of BTXs, the main manatee food source, opening a likely pathway for manatees' BTX uptake. Since 1996, many other events of marine mammal mortality have been attributed to BTXs intoxication (reviewed in Landsberg et al. 2009). What is worth noting is the death of 107 bottlenose dolphins (*Tursiops truncatus*), following a red tide in 2002. Again, Flewelling et al. (2005), analysed dolphin tissues and undigested fish remains, and concluded that both had very high levels of BTXs, enough to cause brevetoxicosis and death. Until then, fish were not considered likely vectors of this toxin, as it caused the fish to die in a very short period of time, limiting the toxin transfer higher up the food chain. Several mass fish kills followed red tides (Landsberg 2002, 2009, Steidinger et al. 1973), drawing attention to this toxin's ichthyotoxic potential. Many studies have revealed that fish are more sensitive to BTXs dissolved in seawater than ingestion of *K. brevis* cells. It was shown that fish survive direct exposure to the dinoflagellate cells, whereas exposure to the dissolved toxin leads to death (Landsberg et al. 2009).

The effects of BTXs are still poorly understood, especially in marine invertebrates, despite the great impact red tides have on ecosystems worldwide.

Diarrhetic shellfish toxins

Diarrhetic Shellfish Toxins (DSTs) are produced by many species of *Dinophysis* and *Prorocentrum*, both cosmopolitan genera occurring worldwide. *Dinophysis* are pelagic species, whereas *Prorocentrum* are typically epibenthic. DSTs are lipophilic toxins, comprising okadaic acid (OA) and dinophysistoxins (DTX).

OA was first isolated from the marine sponge *Halichondria melanodocia*. It was later discovered that this toxin was produced by a dinoflagellate that was accumulated in the sponge through filter-feeding. OA specifically inhibits the activity of protein phosphatase 1 and 2, two of the main protein phosphatases in mammals, increasing protein phosphorylation. Gastrointestinal distress symptoms may arise from the loss of balance in membrane transport and substance secretion, resulting in loss of fluids.

The effects of OA and dinophysistoxins on marine organisms are summarized in Table 5. DSP outbreaks are not caused any human fatalities (Hallegraeff et al. 2003), and of all the other shellfish poisonings, it can be considered the mildest, with patients fully recovering after few days. However, OA has been identified as a tumour promoting compound (Suganuma et al. 1988), posing additional threats to human health and marine life alike.

OA has been documented to be actively accumulated in sponges, which increases the sponge's immune system against parasites and bacterial infections (Schröder et al. 2006, Wiens et al. 2003).

It is worth noting that, although greatly lacking supportive evidence, *P. lima* has been shown to elicit allelopathic effects when grown in culture with other microalgal species. OA and DTX-1 produced by *P. lima* inhibit the growth of the other species present, and, at lower concentrations it enhanced growth in *P. lima* cultures (Windust et al. 1996). Similarly, when exposed to OA, *Dunaliella tertiolecta* cultures decreased cell density and increased oxidative stress response. Additionally, OA inhibited the ability for electron transport in photosystem II, impairing photosynthesis (Perreault et al. 2012).

Regarding other marine organisms, OA has been found to accumulate in numerous shellfish species, the main vectors of this toxin. Blue mussels have been reported to accumulate high concentrations of OA for long periods of time (up to five months, Shumway 1990). Typically, bivalves are not directly affected by the toxins, most likely due to their rapid clearance rates and the fact that many species can convert the parent compound into less toxic derivatives (Suzuki et al. 1999). Most studies regarding OA toxicity beyond DSP symptoms have been performed in mice and human cell lines. The toxic effects of OA on marine organisms is still fairly unknown. Recently it was shown that OA induced genotoxic and cytotoxic effects on bivalves (McCarthy et al. 2014). OA has been shown to elicit the formation of micronuclei and nuclear lesions in mussel hematocytes (Carvalho Pinto-Silva et al. 2003, 2005) and up-regulation of gene transcripts associated with stress response (Manfrin et al. 2010). In clams (*R. decussatus*), OA induced higher DNA damage to clams' gills than in hemocytes when exposed to lower OA concentrations, as opposed to when they were exposed to high OA concentrations for a shorter period of time (Flórez-Barrós et al. 2011). Mussels (*M. galloprovincialis*) presented similar results, with hemocytes having less DNA damage than gill tissue when exposed to OA, and gills presenting increased DNA damage at lower OA concentrations (Prego-Faraldo et al. 2015). Gills are the first tissue to come in contact with the toxin, and the lack of response when exposed to higher concentrations suggests: (i) very efficient defence mechanism in bivalves and (ii) detoxification pathways by metabolizing OA into less toxic compounds.

While the target protein phosphatases are as sensitive to OA *in vitro* exposure in blue mussels (*M. edulis*) as they are in other organisms (Svensson and Förlin 1998), there are no records of mussel mortalities due to OA exposure. Mussels can be exposed to OA throughout the year; thus, it is proposed that they possess detoxification mechanisms. Svensson et al. (2003), found that *M. edulis* can accumulate OA in the lysosomal system, therefore, preventing cellular damage in hematocytes.

Marine turtles have been found to accumulate OA in their tissues, produced by the epibenthic *Prorocentrum* spp., likely present on the surface of the algae that the turtles consume. Coincidentally, two DSP-producing *Prorocentrum* species (*P. lima* and *P. concavum*) occur where there is high risk of fibropapillomatosis, a neoplastic

Table 5. Documented cases of marine organisms exposed to and affected by brevetoxins (BTXs).

	Target species	Brevetoxin source	Route of exposure	Toxicity	Tissues analysed	Effects	Observations	References
Phytoplankton	*Asterionellopsis glacialis*	*K. brevis* extracellular exudates	Exposure to exudates	Up to 55 ng L^{-1} PbTx-2	-	Inhibition of growth	Indicates allelopathy	(Prince et al. 2008)
	Skeletonema costatum	*K. brevis* extracellular exudates	Exposure to exudates	Up to 55 ng L^{-1} PbTx-2	-	Inhibition of growth	Indicates allelopathy	(Prince et al. 2008)
	Prorocentrum minimum	*K. brevis* extracellular exudates	Exposure to exudates	Up to 55 ng L^{-1} PbTx-2	-	Inhibition of growth	Indicates allelopathy	(Prince et al. 2008)
Crustaceans	*A. tonsa*	*K. brevis*	Exposure to dinoflagellate cells	Up to 2400 cells ml^{-1}		Decreased survival and egg production	-	(Prince et al. 2006)
		Ptychodiscus brevis (now *K. brevis*)	Exposure to dinoflagellate cells	Up to 19567 cells ml^{-1}	-	Increased feeding rates	-	(Turner and Tester 1989)
		K. brevis	Exposure to dinoflagellate cells and brevetoxins	Up to 1×10^7 cells L^{-1} and 15 µg PbTx-2 L^{-1}	-	Increased mortality at higher concentrations	-	(Cohen et al. 2007)
	C. pacificus	*P. brevis* (now *K. brevis*)	Exposure to dinoflagellate cells	0.68 ng C $cell^{-1}$		Loss of motor control, increased heart rate and lethargy	-	(Huntley et al. 1987)
		P. brevis (now *K. brevis*)	Exposure to dinoflagellate cells	0.68 ng C $cell^{-1}$	-	Avoided feeding, increased heart rate and loss of motor control	-	(Huntley et al. 1986)
		P. brevis (now *K. brevis*)	Exposure to dinoflagellate cells	Up to 1000 ng C $cell^{-1}$	-	Increased heart rates and loss of motor control	-	(Sykes and Huntley 1987)

	Centropages typicus	P. brevis (now K. brevis)	Exposure to dinoflagellate cells	Up to 19567 cells ml⁻¹	–	Decreased feeding rates	–	(Turner and Tester 1989)
		K. brevis	Exposure to dinoflagellate cells and brevetoxins	Up to 1×10^7 cells L⁻¹ and 15 µg PbTx-2 L⁻¹	–	Increased mortality at higher concentrations	–	(Cohen et al. 2007)
	Labidocera aestiva	P. brevis (now K. brevis)	Exposure to dinoflagellate cells	Up to 19567 cells ml⁻¹	–	Increased feeding rates	–	(Turner and Tester 1989)
	Oncaea venusta	P. brevis (now K. brevis)	Exposure to dinoflagellate cells	Up to 19567 cells ml⁻¹	–	Increased feeding rates	–	(Turner and Tester 1989)
	Paracalamus quasimodo	P. brevis (now K. brevis)	Exposure to dinoflagellate cells	Up to 845 cells ml⁻¹	–	Decreased feeding rates	–	(Turner and Tester 1989)
	Temora turbinata	K. brevis	Exposure to dinoflagellate cells and brevetoxins	Up to 5×10^6 cells L⁻¹ and 15 µg PbTx-2 L⁻¹	–	Suppressed swimming behaviour	–	(Cohen et al. 2007)
Molluscs	A. irradians	P. brevis (now K. brevis)	Exposure to dinoflagellate cells	–	–	Decreased recruitment	–	(Summerson and Peterson 1990)
		K. brevis	Exposure to dinoflagellate cells and extracts	Up to 22000 cells ml⁻¹	–	Reduction in clearance rates	–	(Leverone et al. 2007)
	C. virginica	K. brevis	Exposure to dinoflagellate cells and extracts	Up to 24600 cells ml⁻¹	–	Reduction in clearance rates	–	(Leverone et al. 2007)
	M. mercenaria	K. brevis	Exposure to dinoflagellate cells and extracts	Up to 23100 cells ml⁻¹	–	Reduction in clearance rates	–	(Leverone et al. 2007)

Table 5 contd. ...

...Table 5 contd.

	Target species	Brevetoxin source	Route of exposure	Toxicity	Tissues analysed	Effects	Observations	References
Fish	P. viridis	K. brevis	Exposure to dinoflagellate cells and extracts	Up to 23800 cells ml^{-1}	-	Reduction in clearance rates		(Leverone et al. 2007)
	Larval Sciaenops ocellatus	P. brevis (now K. brevis)	Exposure to dinoflagellate cells	Up to 2040 cells ml^{-1}	-	Deformities, abnormal swimming, paralysis and death	-	(Riley et al. 1989)
Sea turtles	Caretta caretta	K. brevis	Likely through prey	-	-	Death	-	(Landsberg et al. 2009)
	Chelonia mydas	K. brevis	Likely through prey	-	-	Death	-	(Landsberg et al. 2009)
	Lepidochelys kempii	K. brevis	Likely through prey	-	-	Death	-	(Landsberg et al. 2009)
Seabirds	Aythya affinis	G. breve (now Karenia brevis)	Prey (contaminated clams)	-	-	Lethargy, ataxia, death	Over 6000 specimens found dead or moribund	(Forrester et al. 1977)
		G. breve (now K. brevis)	Prey (contaminated clams)	-	-	Weakness, inability to dive, unresponsiveness, loss of reflexes, death	-	(Quick and Henderson 1974)
		G. breve (now K. brevis)	Prey (contaminated clams)	-	-	Dead or weakened		(Schreiber et al. 1975)
	Phalacrocorax auritus	K. brevis	Prey (contaminated fish)	-	-	Cerebellar ataxia, death	-	(Kreuder et al. 2002)

Species	Organism	Source	Concentration	Tissue	Effect	Impact	Reference
Sterna maxima	K. brevis	-	Up to 33.1 PbTx ng g⁻¹	Kidney, liver, testes, heart muscle, stomach, intestines, lung	Death	Analysis performed on beached specimens	(Vargo et al. 2006)
Larus atricilla	K. brevis	-	16.2 PbTx ng g⁻¹	Kidney	Death	Analysis performed on beached specimens	(Vargo et al. 2006)
Marine Mammals Trichechus manatus latirostris	K. brevis	Aerosols or ingested seawater	-	-	Death following pulmonary oedema and hemmorhage, congestion of liver and kidneys	Over 100 manatees died	(Bossart et al. 1998)
	K. brevis	Contaminated seagrass	Up to 300 ng g⁻¹	Liver	Death	~ 30 manatees died	(Flewelling et al. 2005)
	G. breve (now K. brevis)	Possibly ingestion of contaminated ascidians	-	-	Congestion and hemorrhage in brain tissue, death	~ 40 manatees died	(O'Shea et al. 1991)
Tursiops truncatus	K. brevis	Prey	Up to 613 ng g⁻¹	Feces	Death	Over 100 dolphins died	(Flewelling et al. 2005)
	P. brevis (now K. brevis)	Likely through prey	Up to 15820 ng g⁻¹	Liver	Death	Over 700 dolphins died	(Geraci 1989)
	G. breve (now K. brevis)	Prey	-	-	Death	Over 100 dolphins died	(Mase et al. 2000)
	K. brevis	Prey	Up to 2896 ng g⁻¹	Stomach contents	Death	-	(Fire et al. 2007)

disease specific to sea turtles. Therefore, OA may play an important role in this disease's etiology (Landsberg 2002, Landsberg et al. 1999).

Although not positively linked to DSTs, many seabird deaths occurred after DSP and other toxin's outbreaks (Shumway et al. 2003). The presence of DSTs is likely to decrease the organism's fitness and well-being, making them more vulnerable to other toxicants.

Despite being regarded as a less dangerous toxin, OA has been shown to cause a wide array of responses and effects in marine organisms, highlighting the potential of this toxin to affect many other organisms, including human populations chronically ingesting low doses of a tumour promoting toxin.

Future Directions and Concluding Remarks

It is predicted that many changes in the world's oceans will occur. Increasing temperature and CO_2 concentrations are but two of the many factors affecting HAB distribution, frequency and intensity. HAB ecology is complex, and it is dependent on the interaction of many factors, including ocean stratification, oceanic currents, nutrient availability and precipitation (Wells et al. 2015). Currently, there are a number of studies on the effect of climate change in HABs. However, the interactions simulated are scarce and do not allow for species adaptation and plasticity. Temperature fluctuations directly affect phytoplankton communities. Typically, with increasing temperatures, phytoplanktonic species tend to have higher growth rates until a species-specific temperature threshold is met (Wells et al. 2015). There is growing evidence that HABs are increasing in frequency and intensity throughout the globe (Dolah 2000), and further studies are needed to better understand the shifts in HAB ecology and physiology in these new conditions.

It is evident from the present work that there is a great lack of knowledge on the effects that HAB toxins have on early stages of development, a very sensitive and critical stage in an organism's life. Moreover, most toxins may be chronically accumulated and, in many cases, not elicit outward signs of toxicity. Here, we reviewed the effects that HAB toxins have on marine organisms, more specifically, the four main groups of toxins (PSP, ASP, NSP and DSP). Still, there has been growing evidence that in addition to the increasing intensity and toxicity of these blooms, new and emerging toxins, such as palytoxins, cyclic imines, tetrodotoxins and ciguatoxins, are occurring in regions previously undetected (Soliño et al. 2014). Also, the co-occurrence of emerging toxins with endemic HAB-toxins may lead to additive, synergistic or antagonistic effects on marine organisms; however, the available data is not sufficient to confirm and characterize such effects. Multidisciplinary studies are necessary to comprehensively understand the effects of exposure to multiple toxins. Thus, there is great need to reach out to policy makers and work alongside with the monitoring programs already implemented in many countries worldwide, to better understand the risk faced by organisms exposed to marine toxins in the natural environment, the consequences on an ecosystem's stability and to develop models of biotoxins kinetics useful to predict the toxic effects highlighted in this study.

Acknowledgements

This study had the support of Fundação para a Ciência e Tecnologia (FCT), through the strategic project UID/MAR/04292/2013 granted to MARE and UID/Multi/04326/2019 granted to CCMAR. The research leading to these results has received funding from the project Cigua (PTDC/CTA-AMB/30557/2017) supported by the Portuguese Foundation for Science and Technology (FCT) and FEDER. The authors would like to thank the Portuguese Foundation for Science and Technology for the "Investigador FCT" grants to RR for a project grant PTDC/BIA-BMA/28317/2017, and the Ph.D. and PRC and the Ph.D. scholarship to V.M. Lopes (SFRH/BD/97633/2013).

References

Ajuzie, C.C. 2007. Palatability and fatality of the dinoflagellate *Prorocentrum lima* to *Artemia salina*. J. Appl. Phycol. 19: 513–519. doi:10.1007/s10811-007-9164-9.

Anderson, D.M., Alpermann, T.J., Cembella, A.D., Collos, Y., Masseret, E. and Montresor, M. 2012. The globally distributed genus *Alexandrium*: multifaceted roles in marine ecosystems and impacts on human health 1410. doi:10.1016/j.hal.2011.10.012.

Armstrong, H., Coulson, J.C., Hawkey, P. and Hudson, M.J. 1978. Further mass seabird deaths from paralytic shellfish poisoning. British Birds 71: 58–68.

Bagøein, E., Miranda, A., Reguera, B. and Franco, J.M. 1996. Effects of two paralytic shellfish toxin producing dinoflagellates on the pelagic harpacticoid copepod *Euterpina acutifrons*. Mar. Biol. 126: 361–369. doi:10.1007/BF00354618.

Bakke, M.J. and Horsberg, T.E. 2007. Effects of algal-produced neurotoxins on metabolic activity in telencephalon, optic tectum and cerebellum of Atlantic salmon (*Salmo salar*). Aquat. Toxicol. 85: 96–103. doi:10.1016/j.aquatox.2007.08.003.

Bakke, M.J., Hustoft, H.K. and Horsberg, T.E. 2010. Subclinical effects of saxitoxin and domoic acid on aggressive behaviour and monoaminergic turnover in rainbow trout (*Oncorhynchus mykiss*). Aquat. Toxicol. 99: 1–9. doi:10.1016/j.aquatox.2010.03.013.

Bargu, S., Goldstein, T., Roberts, K., Li, C. and Gulland, F. 2011. *Pseudo-nitzschia* blooms, domoic acid, and related California sea lion strandings in Monterey Bay, California. Mar. Mammal Sci. 28: 237–253. doi:10.1111/j.1748-7692.2011.00480.x.

Beltrán, A.S., Palafox-Uribe, M., Grajales-Montiel, J., Cruz-Villacorta, A. and Ochoa, J.L. 1997. Sea bird mortality at Cabo San Lucas, Mexico: evidence that toxic diatom blooms are spreading. Toxicon. 35: 447–453.

Berman, F.W. and Murray, T.F. 1997. Domoic acid neurotoxicity in cultured cerebellar granule neurons is mediated predominantly by NMDA receptors that are activated as a consequence of excitatory amino acid release. J. Neurochem. 69: 693–703. doi:9231729.

Blanco, J. and Campos, M.J. 1988. The effect of water conditioned by a PSP producing dinoflagellate on the growth of four algal species used as food for invertebrates. Aquaculture 68: 289–298.

Bossart, G.D., Baden, D.G., Ewing, R.Y., Roberts, B. and Wright, S.D. 1998. Brevetoxicosis in Manatees (*Trichechus manatus latirostris*) from the 1996 Epizootic: Gross, Histologic, and Immunohistochemical Features. Toxicol. Pathol. 26: 276–282. doi:10.1177/019262339802600214.

Botana, L.M., Louzao, C. and Vilariño, N. (eds.). 2015. Climate Change and Marine and Freshwater Toxins. De Gruyter, Germany.

Bricelj, V.M., Cembella, A.D., Laby, D., Shumway, S.E. and Cucci, T.L. 1996. Comparative physiological and behavioral responses to PSP toxins in two bivalve molluscs, the softshell clam, *Mya arenaria*, and surfclam, *Spisula solidissima*. Harmful Toxic Algal Bloom. 405–408.

Bricelj, V.M., MacQuarrie, S.P., Doane, J.a.E. and Connell, L.B. 2010. Evidence of selection for resistance to paralytic shellfish toxins during the early life history of soft-shell clam, *Mya arenaria*, populations. Limnol. Oceanogr. 55: 2463–2475. doi:10.4319/lo.2010.55.6.2463.

Brodie, E.C., Gulland, F.M.D., Greig, D.J., Hunter, M., Jaakola, J., Leger, J.St., Leighfield, T.A. and Van Dolah, F.M. 2006. Domoic acid causes reproductive failure in california sea lions (*Zalophus californianus*). Mar. Mammal Sci. 22: 700–707. doi:10.1111/j.1748-7692.2006.00045.x.

Carvalho Pinto-Silva, C.R., Ferreira, J.F., Costa, R.H.R., Belli Filho, P., Creppy, E.E. and Matias, W.G. 2003. Micronucleus induction in mussels exposed to okadaic acid. Toxicon. 41: 93–97. doi:10.1016/S0041-0101(02)00214-3.

Carvalho Pinto-Silva, C.R., Creppy, E.E. and Matias, W.G. 2005. Micronucleus test in mussels *Perna perna* fed with the toxic dinoflagellate *Prorocentrum lima*. Arch. Toxicol. 79: 422–426. doi:10.1007/s00204-004-0645-1.

Chen, C.Y. and Chou, H.N. 2001. Ichthyotoxicity studies of milkfish *Chanos chanos* fingerlings exposed to a harmful dinoflagellate *Alexandrium minutum*. J. Exp. Mar. Bio. Ecol. 262: 211–219. doi:10.1016/S0022-0981(01)00291-X.

Choi, N.M.C., Yeung, L.W.Y., Siu, W.H.L., So, I.M.K., Jack, R.W., Hsieh, D.P.H., Wu, R.S.S. and Lam, P.K.S. 2006. Relationships between tissue concentrations of paralytic shellfish toxins and antioxidative responses of clams, *Ruditapes philippinarum*. Mar. Pollut. Bull. 52: 572–578. doi:10.1016/j.marpolbul.2006.01.009.

Cohen, J.H., Tester, P.A. and Forward, R.B. 2007. Sublethal effects of the toxic dinoflagellate *Karenia brevis* on marine copepod behavior. J. Plankton Res. 29: 301–315. doi:10.1093/plankt/fbm016.

Colin, S.P. and Dam, H.G. 2004. Testing for resistance of pelagic marine copepods to a toxic dinoflagellate. Evol. Ecol. 18: 355–377. doi:10.1007/s10682-004-2369-3.

Costa, P.R., Pereira, P., Guilherme, S., Barata, M., Nicolau, L., Santos, M.A., Pacheco, M. and Pousão-Ferreira, P. 2012. Biotransformation modulation and genotoxicity in white seabream upon exposure to paralytic shellfish toxins produced by *Gymnodinium catenatum*. Aquat. Toxicol. 106: 42–47.

Costa, P.R., Pereira, P., Guilherme, S., Barata, M., Nicolau, L., Santos, M.A., Pacheco, M. and Pousão-Ferreira, P. 2012. Biotransformation modulation and genotoxicity in white seabream upon exposure to paralytic shellfish toxins produced by *Gymnodinium catenatum*. Aquat. Toxicol. 106: 42–47.

Costa, P.R. 2014. Impact and effects of paralytic shellfish poisoning toxins derived from harmful algal blooms to marine fish. Fish Fish. 226–248. doi:10.1111/faf.12105.

Costa, R.M. Pereira, L.C.C. and Ferrnández, F. 2012. Deterrent effect of *Gymnodinium catenatum* Graham PSP-toxins on grazing performance of marine copepods. Harmful Algae 17: 75–82. doi:10.1016/j.hal.2012.03.002.

Costas, E. and Lopez-Rodas, V. 1998. Paralytic phycotoxins in monk seal mass mortality. Vet. Rec. 142: 643–644. doi:10.1136/vr.142.23.643.

Cruz, P.G., Daranas, A.H., Fernández, J.J., Souto, M.L. and Norte, M. 2006. DTX5c, a new OA sulphate ester derivative from cultures of *Prorocentrum belizeanum*. Toxicon. 47: 920–924. doi:10.1016/j.toxicon.2006.03.005.

Dizer, H., Fischer, B., Harabawy, A.S.A., Hennion, M.C. and Hansen, P.D. 2001. Toxicity of domoic acid in the marine mussel *Mytilus edulis*. Aquat. Toxicol. 55: 149–156.

Dolah, F.M. Van. 2000. Marine algal toxins: Origins, health effects, and their increased occurrence. Environ. Health Perspect. 108: 133–141. doi:10.1289/ehp.00108s1143.

Dupont, J.M., Hallock, P. and Jaap, W.C. 2010. Ecological impacts of the 2005 red tide on artificial reef epibenthic macroinvertebrate and fish communities in the eastern Gulf of Mexico 415: 189–200. doi:10.3354/meps08739.

Dupuy, J.L. and Sparks, A.K. 1967. *Gonyaulax washingtonensis*, its relationship to *Mytilus californianus* and *Crassostrea gigas* as a source of paralytic shellfish toxin in Sequim Bay, Washington. Proc. Natl. Shellfish. Assoc. 58: 2.

Dutz, J. 1998. Repression of fecundity in the neritic copepod *Acartia clausi* exposed to the toxic dinoflagellate *Alexandrium lusitanicum*: Relationship between feeding and egg production. Mar. Ecol. Prog. Ser. 175: 97–107. doi:10.3354/meps175097.

Escobedo-Lozano, A.Y., Estrada, N., Ascencio, F., Contreras, G. and Alonso-Rodriguez, R. 2012. Accumulation, biotransformation, histopathology and paralysis in the Pacific calico scallop *Argopecten ventricosus* by the paralyzing toxins of the dinoflagellate *Gymnodinium catenatum*. Mar. Drugs 10: 1044–1065. doi:10.3390/md10051044.

Estrada, N., de Jesús Romero, M., Campa-Córdova, A., Luna, A. and Ascencio, F. 2007. Effects of the toxic dinoflagellate, *Gymnodinium catenatum* on hydrolytic and antioxidant enzymes, in tissues of the giant lions-paw scallop *Nodipecten subnodosus*. Comp. Biochem. Physiol.—C Toxicol. Pharmacol. 146: 502–510. doi:10.1016/j.cbpc.2007.06.003.

Estrada, N., Rodríguez-Jaramillo, C., Contreras, G. and Ascencio, F. 2010. Effects of induced paralysis on hemocytes and tissues of the giant lions-paw scallop by paralyzing shellfish poison. Mar. Biol. 157: 1401–1415. doi:10.1007/s00227-010-1418-4.

Fire, S.E., Fauquier, D., Flewelling, L.J., Henry, M., Naar, J., Pierce, R. and Wells, R.S. 2007. Brevetoxin exposure in bottlenose dolphins (*Tursiops truncatus*) associated with *Karenia brevis* blooms in Sarasota Bay, Florida 827–834. doi:10.1007/s00227-007-0733-x.

Fire, S.E., Wang, Z., Leighfield, T.A., Morton, S.L., McFee, W.E., McLellan, W.A., Litaker, R.W., Tester, P.A., Hohn, A.A., Lovewell, G., Harms, C., Rotstein, D.S., Barco, S.G., Costidis, A., Sheppard, B., Bossart, G.D., Stolen, M., Durden, W.N. and Van Dolah, F.M. 2009. Domoic acid exposure in pygmy and dwarf sperm whales (*Kogia* spp.) from southeastern and mid-Atlantic U.S. waters. Harmful Algae 8: 658–664. doi:http://dx.doi.org/10.1016/j.hal.2008.12.002.

Fire, S.E., Wang, Z., Berman, M., Langlois, G.W., Morton, S.L., Sekula-Wood, E. and Benitez-Nelson, C.R. 2010. Trophic transfer of the harmful algal toxin domoic acid as a cause of death in a Minke whale (Balaenoptera acutorostrata) stranding in Southern California. Aquat. Mamm. 36: 342–350.

Fire, S.E., Wang, Z., Byrd, M., Whitehead, H.R., Paternoster, J. and Morton, S.L. 2011. Co-occurrence of multiple classes of harmful algal toxins in bottlenose dolphins (*Tursiops truncatus*) stranding during an unusual mortality event in Texas, USA. Harmful Algae 10: 330–336. doi:10.1016/j.hal.2010.12.001.

Flaherty, K.E. and Landsberg, J.H. 2011. Effects of a Persistent Red Tide (*Karenia brevis*) Bloom on Community Structure and Species-Specific Relative Abundance of Nekton in a Gulf of Mexico Estuary 417–439. doi:10.1007/s12237-010-9350-x.

Flewelling, L.J., Naar, J.P., Abbott, J.P., Baden, D.G., Barros, N.B., Bossart, G.D., Bottein, M.-Y.D., Hammond, D.G., Haubold, E.M. and Heil, C.A. 2005. Brevetoxicosis: Red tides and marine mammal mortalities. Nature 435: 755–756.

Flewelling, L.J., Naar, J.P., Abbott, J.P., Baden, D.G., Barros, N.B., Bossart, G.D., Bottein, M.-Y.D., Hammond, D.G., Haubold, E.M., Heil, C.A., Henry, M.S., Jacocks, H.M., Leighfield, T.A., Pierce, R.H., Pitchford, T.D., Rommel, S.A., Scott, P.S., Steidinger, K.A., Truby, E.W., Van Dolah, F.M. and Landsberg, J.H. 2005. Brevetoxicosis: Red tides and marine mammal mortalities. Nature 435: 755–756. doi:10.1038/nature435755a.

Flórez-Barrós, F., Prado-Alvarez, M., Méndez, J. and Fernández-Tajes, J. 2011. Evaluation of genotoxicity in gills and hemolymph of clam *Ruditapes decussatus* fed with the toxic dinoflagellate *Prorocentrum lima*. J. Toxicol. Environ. Health. A 74: 971–979. doi:10.1080/15287394.2011.582025.

Forrester, D.J., Gaskin, J.M., White, F.H., Thompson, N.P., Quick, J.A., Henderson, G.E., Woodbard, J.C. and Robertson, W.D. 1977. An epizootic of waterfowl associated with a red tide episode in florida. J. Wildl. Dis. 13: 160–167. doi:10.7589/0090-3558-13.2.160.

Frangopulos, M., Guisande, C., Maneiro, I., Riveiro, I. and Franco, J. 2000. Short-term and long-term effects of the toxic dinoflagellate *Alexandrium minutum* on the copepod *Acartia clausi*. Mar. Ecol. Prog. Ser. 203: 161–169. doi:10.3354/meps203161.

Fritz, L., Quilliam, M.A. and Jeffrey, L.C.W. 1992. An outbreak of domoic acid poisoning attributed to the pennate diatom *Pseudonitzschia australis*. J. Phycol. 28: 439–442.

Gainey, L.F. and Shumway, S.E. 1988. Physiological effects of *Protogonyaulax tamarensis* on cardiac activity in bivalve molluscs. Comp. Biochem. Physiol. Part C, Comp. 91: 159–164. doi:10.1016/0742-8413(88)90182-X.

Geraci, J.R. 1989. Clinical investigation of the 1987–88 mass mortality of bottlenose dolphins along the U.S. central and south Atlantic coast. Natl. Mar. Fish. Serv. U.S. Navy, Mar. Mammal Commmission 1–63.

Geraci, J.R., Anderson, D.M., Timperi, R.J., St. Aubin, D.J., Early, G.A., Prescott, J.H. and Mayo, C.A. 1989. Humpback Whales (*Megaptera novaeangliae*) fatally poisoned by dinoflagellate Toxin. Can. J. Fish. Aquat. Sci. doi:10.1139/f89-238.

Gill, C.W. and Harris, R.P. 1987. Behavioral responses of the copepods *Calanus helgolandicus* and *Temora longicornis* to dinoflagellate diets. J. Mar. Biol. Assoc. United Kingdom 67: 785–801. doi:10.1017/S0025315400057039.

Glennan, A.H. 1887. Fish killed by poisonous water. Bull. United States Fish Comm. 6: 10–11. doi:10.1126/science.11.265.149.

Goldstein, T., Zabka, T.S., DeLong, R.L., Wheeler, E.A., Ylitalo, G., Bargu, S., Silver, M., Leighfield, T., Van Dolah, F., Langlois, G., Sidor, I., Dunn, J.L. and Gulland, F.M.D. 2009. The role of domoic acid in abortion and premature parturition of california sea lions (*Zalophus californianus*) on San Miguel island, California. J. Wildl. Dis. 45: 91–108.

Gosselin, S., Fortier, L. and Gagné, J.A. 1989. Vulnerability of marine fish larvae to the toxic dinoflagellate *Protogonyaulax tamarensis*. Mar. Ecol. Prog. Ser. 57: 1–10.

Gubbins, M.J., Guezennec, E.A., Eddy, F.B., Gallacher, S. and Stagg, R.M. 2001. Paralytic shellfish toxins and glutathione s-transferases in artificially intoxicated marine organisms. pp. 387–391. *In*: Hallegraeff, G., Blackburn, S., Bolch, C.J. and Lewis, R.J. (eds.). Harmful Algal Blooms 2000. Intergovernmental Oceanographic Commission of UNESCO.

Gunter, G., Smith, W. and Williams, R.H. 1947. Mass mortality of marine animals on the lower west coast of Florida, November 1946–January 1947. Science 105(2723): 256–257.

Gunter, G., Williams, R.H., Davis, C.C. and Smith, F.G.W. 1948. Catastrophic mass mortality of marine animals and coincident phytoplankton bloom on the west coast of Florida, November 1946 to August 1947. Ecol. Monogr. 18: 309–324. doi:10.2307/1948575.

Haberkorn, H., Lambert, C., Le Goic, N., Guéguen, M., Moal, J., Palacios, E. Lassus, P. and Soudant, P. 2010. Effects of *Alexandrium minutum* exposure upon physiological and hematological variables of diploid and triploid oysters, *Crassostrea gigas*. Aquatic Toxicology, Elsevier 97(2): 96–108.

Hallegraeff, G.M. 1993. A review of harmful algal blooms and their apparent global increase. Phycologia 32: 79–99.

Hallegraeff, G.M., Anderson, D.M. and Cembella, A.D. 2003. Manual on Harmful Marine Microalgae. Unesco Publishing, Paris, France.

Hallegraeff, G.M. 2014. Harmful algae and their toxins: Progress, paradoxes and paradigm shifts. pp. 3–20. *In:* Rossini, G.P. (ed.). Toxins and Biologically Active Compounds from Microalgae. CRC Press. doi:10.1201/b16569-3.

Hansen, P. 1989. The red tide dinoflagellate *Alexandrium tamarense*: effects on behaviour and growth of a tintinnid ciliate. Mar. Ecol. Prog. Ser. 53: 105–116. doi:10.3354/meps053105.

Hansen, P.J., Cembella, A.D. and Moestrup, Øjvind. 1992. The marine dinoflagellate *Alexandrium Ostenfeldii*: Paralytic shellfish toxin concentration, composition, and toxicity to a tintinnid ciliate. J. Phycol. doi:10.1111/j.0022-3646.1992.00597.x.

Hernández, M., Robinson, I., Aguilar, A., González, L.M., López-Jurado, L.F., Reyero, M.I., Cacho, E., Franco, J., López-Rodas, V. and Costas, E. 1998. Did algal toxins cause monk seal mortality? Nature 393: 28–29. doi:10.1038/29906.

Huntley, M., Sykes, P., Rohan, S. and Marin, V. 1986. Chemically-mediated rejection of dinoflagellate prey by the copepods *Calanus pacificus* and *Paracalanus parvus*: mechanism, occurrence and significance. Mar. Ecol. Prog. Ser. 28: 105–120. doi:10.3354/meps028105.

Huntley, M.E., Ciminiello, P. and Lopez, M.D.G. 1987. Importance of food quality in determining development and survival of *Calanus pacificus* (Copepoda: Calanoida). Mar. Biol. 113: 103–113.

Ives, J.D. 1987. Possible mechanisms underlying copepod grazing responses to levels of toxicity in red tide dinoflagellates. J. Exp. Mar. Bio. Ecol. 112: 131–144. doi:10.1016/0022-0981(87)90113-4.

Jones, T.O., Whyte, J.N.C., Ginther, N.G., Townsend, L.D. and Iwama, G.K. 1995. Haemocyte changes in the Pacific oyster, *Crassostrea gigas*, caused by exposure to domoic acid in the diatom *Pseudonitzschia pungens f. multiseries* 101.

Jones, T.O., Whyte, J.N.C., Townsendb, L.D., Gintherb, N.G. and Iwamaa, G.K. 1995. Effects of domoic acid on haemolymph pH, PCO, and PO, in the Pacific oyster, *Crassostrea gigas* and the California mussel, *Mytilus californianus* 31: 43–55.

Kao, C.Y. and Nishiyama, A. 1965. Actions of saxitoxin on peripheral neuromuscular systems. J. Physiol. 180: 50–66. doi:10.1016/0041-0101(66)90089-4.

Kreuder, C., Mazet, J.A., Bossart, G.D., Carpenter, T.E., Holyoak, M., Elie, M.S. and Wright, S.D. 2002. Clinicopathologic features of suspect brevetoxicosis in double-crested cormorants (*Phalacrocorax auritus*) along the Florida Gulf coast. J. Zoo. Wildl. Med. 33: 8.

Laabir, M. and Gentien, P. 1999. Survival of the toxic dinoflagellates after gut passage in the Pacific oyster *Crassostrea gigas* Thunberg. J. Shellfish Res. 18: 217–222.

Landsberg, J. 2002. The effects of harmful algal blooms on aquatic organisms. Rev. Fish. Sci. 10(2): 113–390.

Landsberg, J.H., Balazs, G.H., Steidinger, K.A., Baden, D.G., Work, T.M. and Russell, D.J. 1999. The potential role of natural tumor promoters in marine turtle fibropapillomatosis. J. Aquat. Anim. Health 11: 199–210.

Landsberg, J.H., Flewelling, L.J. and Naar, J. 2009. *Karenia brevis* red tides, brevetoxins in the food web, and impacts on natural resources: Decadal advancements. Harmful Algae 8: 598–607. doi:10.1016/j.hal.2008.11.010.

Lassus, P., Bardouil, M., Beliaeff, B., Masselin, P., Naviner, M. and Truquet, P. 1999. Effect of a continuous supply of the toxic dinoflagellate *Alexandrium minutum* Balim on the feeding behavior of the Pacific oyster (*Crassostrea gigas* Thunberg). J. Shellfish Res. 18: 211–216.

Lefebvre, K.A., Powell, C.L., Busman, M., Doucette, G.J., Moeller, P.D.R., Silver, J.B., Miller, P.E., Hughes, M.P., Singaram, S. and Silver, M.W. 1999. Detection of domoic acid in northern anchovies and California sea lions associated with an unusual mortality event. Nat. Toxins 7: 85–92.

Lefebvre, K.A., Dovel, S.L. and Silver, M.W. 2001. Tissue distribution and neurotoxic effects of domoic acid in a prominent vector species, the northern anchovy *Engraulis mordax*. Mar. Biol. 138: 693–700. doi:10.1007/s002270000509.

Lefebvre, K.A., Noren, D.P., Schultz, I.R., Bogard, S.M., Wilson, J. and Eberhart, B.T.L. 2007. Uptake, tissue distribution and excretion of domoic acid after oral exposure in coho salmon (*Oncorhynchus kisutch*). Aquat. Toxicol. 81: 266–274. doi:10.1016/j.aquatox.2006.12.009.

Lefebvre, K.A., Bill, B.D., Erickson, A., Baugh, K.A., O'Rourke, L., Costa, P.R., Nance, S. and Trainer, V.L. 2008. Characterization of intracellular and extracellular saxitoxin levels in both field and cultured *Alexandrium* spp. samples from Sequim Bay, Washington. Mar. Drugs 6: 103–116. doi:10.3390/md20080006.

Lefebvre, K.A., Robertson, A., Frame, E.R., Colegrove, K.M., Nance, S., Baugh, K.A., Wiedenhoft, H. and Gulland, F.M.D. 2010. Clinical signs and histopathology associated with domoic acid poisoning in northern fur seals (*Callorhinus ursinus*) and comparison of toxin detection methods. Harmful Algae 9: 374–383. doi:10.1016/j.hal.2010.01.007.

Lefebvre, K., Dovel, S. and Silver, M. 2001. Tissue distribution and neurotoxic effects of domoic acid in a prominent vector species, the northern anchovy *Engraulis mordax*. Mar. Biol. 138: 693–700.

Lefebvre, K., Elder, N., Hershberger, P., Trainer, V., Stehr, C. and Scholz, N. 2005. Dissolved saxitoxin causes transient inhibition of sensorimotor function in larval Pacific herring (*Clupea harengus pallasi*). Mar. Biol. 147: 1393–1402. doi:10.1007/s00227-005-0048-8.

Lesser, M.P. and Shumway, S.E. 1993. Effects of toxic dinoflagellates on clearance rates and survival in juvenile bivalve molluscs. J. Shellfish Res. 12: 377–381.

Levasseur, M., Michaud, S., Bonneau, E., Cantin, G., Auger, F., Gagne, A. and Claveau, R. 1996. Overview of the August 1996 red tide event in the St. Lawrence, effects of a storm surge. pp. 76. *In*: Penney, R.W. (ed.). Proceedings of the Fifth Canadian Workshop on Harmful Marine Algae No. 2138, Canadian Technical Report of Fisheries and Aquatic Science.

Leverone, J.R., Shumway, S.E. and Blake, N.J. 2007. Comparative effects of the toxic dinoflagellate *Karenia brevis* on clearance rates in juveniles of four bivalve molluscs from Florida, USA. Toxicon. 49: 634–645. doi:10.1016/j.toxicon.2006.11.003.

Li, S.-C., Wang, W.-X. and Hsieh, D.P.H. 2002. Effects of toxic dinoflagellate *Alexandrium tamarense* on the energy budgets and growth of two marine bivalves. Mar. Environ. Res. 53: 145–160. doi:10.1016/S0141-1136(01)00117-9.

Liu, H., Kelly, M.S., Campbell, D.A., Dong, S.L., Zhu, J.X. and Wang, S.F. 2007. Exposure to domoic acid affects larval development of king scallop *Pecten maximus* (Linnaeus, 1758). Aquat. Toxicol. 81: 152–158. doi:10.1016/j.aquatox.2006.11.012.

Liu, H., Kelly, M.S., Campbell, D.A., Fang, J. and Zhu, J. 2008. Accumulation of domoic acid and its effect on juvenile king scallop *Pecten maximus* (Linnaeus, 1758). Aquaculture 284: 224–230. doi:10.1016/j.aquaculture.2008.07.003.

Lopes, V.M., Lopes, A.R., Costa, P. and Rosa, R. 2013. Cephalopods as vectors of harmful algal bloom toxins in marine food webs. Mar. Drugs 11. doi:10.3390/md11093381.

MacQuarrie, S.P. and Bricelj, V.M. 2008. Behavioral and physiological responses to PSP toxins in *Mya arenaria* populations in relation to previous exposure to red tides. Mar. Ecol. Prog. Ser. 366: 59–74. doi:10.3354/meps07538.

Manfrin, C., Dreos, R., Battistella, S., Beran, A., Gerdol, M., Varotto, L., Lanfranchi, G., Venier, P. and Pallavicini, A. 2010. Mediterranean mussel gene expression profile induced by okadaic acid exposure. Environ. Sci. Technol. 44: 8276–8283. doi:10.1021/es102213f.

Marsden, I.D. and Shumway, S.E. 1992. Effects of the toxic dinoflagellate *Alexandrium tamarense* on the greenshell mussel *Perna canaliculus*. New Zeal. J. Mar. Freshw. Res. 26: 371–378.

Mase, B., Jones, W., Ewing, R., Bossart, G., Van Dolah, F., Leighfield, T., Busman, M., Litz, J., Roberts, B. and Rowles, T. 2000. Epizootic in bottlenose dolphins in the florida panhandle: 1999–2000. pp. 522–525. *In*: International Association for Aquatic Animal Medicine Conference.

Matsuyama, Y., Usuki, H., Uchida, T. and Kotani, Y. 2001. Effects of harmful algae on the early planktonic larvae of the oyster, *Crassostrea gigas*. pp. 411–415. *In*: Hallegraeff, G., Blackburn, S., Bolch, C.J. and Lewis, R.J. (eds.). Harmful Algal Blooms 2000 Intergovernmental Oceanographic Commission of UNESCO.

McCarthy, M., O'Halloran, J., O'Brien, N.M. and van Pelt, F.F.N.A.M. 2014. Does the marine biotoxin okadaic acid cause DNA fragmentation in the blue mussel and the pacific oyster? Mar. Environ. Res. 101: 153–160. doi:10.1016/j.marenvres.2014.09.009.

Mckernan, D.L. and Scheffer, V.B. 1942. Unusual numbers of dead birds on the Washington coast. Condor. 44: 264–266.

Mello, D.F., Silva, P.M. Da, Barracco, M.A., Soudant, P. and Hégaret, H. 2013. Effects of the dinoflagellate *Alexandrium minutum* and its toxin (saxitoxin) on the functional activity and gene expression of *Crassostrea gigas* hemocytes. Harmful Algae 26: 45–51. doi:10.1016/j.hal.2013.03.003.

Miles, C.O., Wilkins, A.L., Stirling, D.J. and MacKenzie, A.L. 2000. New analogue of gymnodimine from a Gymnodinium species. J. Agric. Food Chem. 48: 1373–6.

Mortensen, A.M. 1985. Massive fish mortalities in the Faroe Islands caused by a *Gonyaulax excavata* red tide. pp. 165–170. *In*: Anderson, D.M., White, A.W. and Baden, D.G. (eds.). Toxic Dinoflagellates. Elsevier, New York.

Navarro, J.M. and Contreras, A.M. 2010. An integrative response by *Mytilus chilensis* to the toxic dinoflagellate *Alexandrium catenella*. Mar. Biol. 157: 1967–1974. doi:10.1007/s00227-010-1465-x.

Negri, A., Stirling, D., Quilliam, M., Blackburn, S., Bolch, C., Burton, I., Eaglesham, G., Thomas, K., Walter, J. and Willis, R. 2003. Three novel hydroxybenzoate saxitoxin analogues isolated from the dinoflagellate *Gymnodinium catenatum*. Chem. Res. Toxicol. 16: 1029–1033. doi:10.1021/tx034037j.

Nisbet, I.C.T. 1983. Paralytic shellfish poisoning: Effects on breeding terns. Condor. 85: 338. doi:10.2307/1367071.

Nogueira, I., Lobo-da-Cunha, A., Afonso, A., Rivera, S., Azevedo, J., Monteiro, R., Cervantes, R., Gago-Martinez, A. and Vasconcelos, V. 2010. Toxic effects of domoic acid in the seabream *Sparus aurata*. Mar. Drugs 8: 2721–2732. doi:10.3390/md8102721.

O'Shea, T.J., Rathbun, G.B., Bonde, R.K., Buergelt, C.D. and Odell, D.K. 1991. An epizootic of Florida manatees associated with a dinoflagellate bloom. Mar. Mammal Sci. 7: 165–179.

Perez-mendes, P., Cinini, S.M., Medeiros, M.A., Tufik, S. and Mello, L.E. 2005. Behavioral and Histopathological Analysis of Domoic Acid Administration in Marmosets 46: 148–151.

Perreault, F., Matias, M.S., Oukarroum, A., Matias, W.G. and Popovic, R. 2012. Okadaic acid inhibits cell growth and photosynthetic electron transport in the alga *Dunaliella tertiolecta*. Sci. Total Environ. 414: 198–204. doi:10.1016/j.scitotenv.2011.10.045.

Prego-Faraldo, M.V., Valdiglesias, V., Laffon, B., Eirín-López, J.M. and Méndez, J. 2015. *In vitro* analysis of early genotoxic and cytotoxic effects of okadaic acid in different cell types of the mussel *Mytilus galloprovincialis*. J. Toxicol. Environ. Heal. Part A 78: 814–824. doi:10.1080/15287394.2015.1051173.

Prince, E.K., Lettieri, L., Mccurdy, K.J. and Kubanek, J. 2006. Fitness consequences for copepods feeding on a red tide dinoflagellate: deciphering the effects of nutritional value, toxicity, and feeding behavior. Oecologia. 479–488. doi:10.1007/s00442-005-0274-2.

Prince, E.K., Myers, T.L. and Kubanek, J. 2008. Effects of harmful algal blooms on competitors: Allelopathic mechanisms of the red tide dinoflagellate *Karenia brevis*. Limnol. Oceanogr. 53: 531–541. doi:10.4319/lo.2008.53.2.0531.

Quick, J.A. and Henderson, G.E. 1974. Effects of *Gymnodinium breve* red tide on fishes and birds: a preliminary report on behavior, anatomy, hematology and histopathology. pp. 85–113. *In*: Amborski, R.L., Hood, M.A. and Miller, R.R. (eds.). Proceedings of the Gulf Coast Regional Symposium on Diseases of Aquatic Animals. Louisiana State University, Louisiana Sea Grant.

Reyero, M., Cacho, E., Martõ, A. and Marina, A. 2000. Evidence of Saxitoxin Derivatives as Causative Agents in the 1997 Mass Mortality of Monk Seals in the Cape Blanc Peninsula 315: 311–315.

Riley, C.M., Holt, S.A., Holt, G.J., Buskey, E.J. and Arnold, C.R. 1989. Mortality of larval red drum (*Sciaenops ocellatus*) associated with a *Ptychodiscus brevis* red tide. Contrib. Mar. Sci. 31: 137–146.

Ritchie, J.M. and Rogart, R.B. 1977. The binding of saxitoxin and tetrodotoxin to excitable tissue. Rev. Physiol. Biochem. Pharmacol. 79: 1–50. doi:10.1007/BFb0037088.

Rounsefell, G.A. and Nelson, W.R. 1966. Red-tide research summarized to 1964 including an annotated bibliography. Special Scientific Report No. 535, United States Department of Interior.

Sadananda, M. and Bischof, H.-J. 2002. Enhanced fos expression in the zebra finch (*Taeniopygia guttata*) brain following first courtship. J. Comp. Neurol. 448: 150–164. doi:10.1002/cne.10232.

Salierno, J.D., Snyder, N.S., Murphy, A.Z., Poli, M., Hall, S., Baden, D. and Kane, A.S. 2006. Harmful algal bloom toxins alter c-Fos protein expression in the brain of killifish, *Fundulus heteroclitus*. Aquat. Toxicol. 78: 350–357. doi:10.1016/j.aquatox.2006.04.010.

Samson, J.C., Shumway, S.E. and Weis, J.S. 2008. Effects of the toxic dinoflagellate, *Alexandrium fundyense* on three species of larval fish: a food-chain approach. J. Fish. Biol. 72: 168–188. doi:10.1111/j.1095-8649.2007.01698.x.

Satake, M., Ofuji, K., Naoki, H., James, K.J., Furey, A., McMahon, T., Silke, J. and Yasumoto, T. 1998. Azaspiracid, a new marine toxin having unique spiro ring assemblies, isolated from irish mussels, Mytilus edulis. J. Am. Chem. Soc. 120: 9967–9968. doi:10.1021/JA981413R.

Scholin, C.A., Gulland, F., Doucette, G.J., Benson, S., Busman, M., Chavez, F.P., Cordaro, J., DeLong, R., De Vogelaere, A. and Harvey, J. 2000. Mortality of sea lions along the central California coast linked to a toxic diatom bloom. Nature 403: 80–84.

Schreiber, R.W., Dunstan, F.M. and Dinsmore, J.J. 1975. Lesser scaup mortality in tampa Bay Florida 1974. Florida F. Nat. 3: 13–15.

Schröder, H.C., Breter, H.J., Fattorusso, E., Ushijima, H., Wiens, M., Steffen, R., Batel, R. and Müller, W.E.G. 2006. Okadaic acid, an apoptogenic toxin for symbiotic/parasitic annelids in the demosponge *Suberites domuncula*. Appl. Environ. Microbiol. 72: 4907–4916. doi:10.1128/AEM.00228-06.

Shaw, B.A., Andersen, R.J. and Harrison, P.J. 1997. Feeding deterrent and toxicity effects of apo-fucoxanthinoids and phycotoxins on a marine copepod (*Tigriopus californicus*) 273–280.

Shumway, S. and Cucci, T.L. 1987. The effects of the toxic dinoflagellate *Protogonyaulax tamarensis* on the feeding and behaviour of bivalve molluscs. Aquat. Toxicol. 10: 9–27. doi:10.1016/0166-445X(87)90024-5.

Shumway, S.E., Pierce, F.C. and Knowlton, K. 1987. The effect of *Protogonyaulax tamarensis* on byssus production in *Mytilus edulis* L., *Modiolus modiolus* Linneaus, 1758 and *Geukensia demissa* Dillwyn. Comp. Biochem. Physiol. 87: 1021–1023.

Shumway, S.E. 1990. A review of the effects of algal blooms on shellfish and aquaculture. J. World Aquac. Soc. 21: 65–104. doi:10.1111/j.1749-7345.1990.tb00529.x.

Shumway, S.E., Allen, S.M. and Dee Boersma, P. 2003. Marine birds and harmful algal blooms: sporadic victims or under-reported events? Harmful Algae 2: 1–17. doi:10.1016/S1568-9883(03)00002-7.

Sievers, A.M. 1969. Comparative toxicity of *Gonyaulax monilata* and *Gymnodinium breve* to annelids, crustaceans, molluscs and fish. J. Protozool. 16: 401–404.

Sierra-Beltrán, A., Palafox-Uribe, M., Grajales-Montiel, J., Cruz-Villacorta, A. and Ochoa, J.L. 1997. Sea bird mortality at Cabo San Lucas, Mexico: evidence that toxic diatom blooms are spreading. Toxicon. 35(3): 447–453.

Silvagni, P.A., Lowenstine, L.J., Spraker, T., Lipscomb, T.P. and Gulland, F.M.D. 2005. Pathology of domoic acid toxicity in California sea lions (*Zalophus californianus*). Vet. Pathol. 42: 184–91. doi:10.1354/vp.42-2-184.

Simon, J.L. and Dauer, D.M. 1972. A quantitative evaluation of red-tide induced mass mortalities of benthic invertebrates in Tampa Bay, Florida. Environ. Lett. 3: 229–234.

Smayda, T.J. 1997. What is a bloom? A commentary. Limnol. Oceanogr. 42 (5, par 2): 1132–1136.

Smayda, T.J. 1997. Harmful algal blooms: their ecophysiology and general relevance to phytoplankton blooms in the sea. Limnol. Oceanogr. 42: 1137–1153. doi:10.4319/lo.1997.42.5_part_2.1137.

Soliño, L., de la Iglesia, P., García Altares, M. and Diogène, J. 2014. The chemistry of ciguatoxins: From the first records to 176 current challenges of monitoring programs. pp. 176–207. *In*: Toxins and Biologically Active Compounds from Microalgae, Volume 1. CRC Press. doi:10.1201/b16569-10.

Steidinger, K.A., Burklew, M.A. and Ingle, M. 1973. The effects of *Gymnodinium breve* toxin on estuarine animals. pp. 179–202. *In*: Martin, D.F. and Padilla, G.M. (eds.). Marine Pharmacognosy: Action of Marine Toxins at the Cellular Level. Academic Press, New York.

Suganuma, M., Fujiki, H., Suguri, H., Yoshizawa, S., Hirota, M., Nakayasut, M., Ojikaf, M., Wakamatsu, K. and Yamadat, K. 1988. Okadaic acid: An additional non-phorbol-12-tetradecanoate-13-acetate-type tumor promoter. Biochemistry 85: 1768–1771.

Summerson, H.C. and Peterson, C.H. 1990. Recruitment failure of the bay scallop, *Argopecten irradians concentricus*, during the first red tide, *Ptychodiscus brevis*, outbreak recorded in North Carolina. Estuaries 13: 322–331. doi:10.2307/1351923.

Suzuki, T., Ota, H. and Yamasaki, M. 1999. Direct evidence of transformation of dinophysistoxin-1 to 7-O-acyl-dinophysistoxin-1 (dinophysistoxin-3) in the scallop *Patinopecten yessoensis*. Toxicon. 37: 187–198. doi:http://dx.doi.org/10.1016/S0041-0101(98)00182-2.

Svensson, S. and Förlin, L. 1998. Intracellular effects of okadaic acid in the blue mussel *Mytilus edulis*, and rainbow trout *Oncorhynchus mykiss*. Mar. Environ. Res. 46: 449–452. doi:10.1016/S0141-1136(97)00099-8.

Svensson, S., Särngren, A. and Förlin, L. 2003. Mussel blood cells, resistant to the cytotoxic effects of okadaic acid, do not express cell membrane p-glycoprotein activity (multixenobiotic resistance). Aquat. Toxicol. 65: 27–37. doi:10.1016/S0166-445X(03)00097-3.

Sykes, P.F. and Huntley, M.E. 1987. Acute physiological reactions of *Calanus pacificus* to selected dinoflagellates: direct observations. Mar. Biol. 94: 19–24. doi:10.1007/BF00392895.

Takada, N., Iwatsuki, M., Suenaga, K. and Uemura, D. 2000. Pinnamine, an alkaloidal marine toxin, isolated from *Pinna muricata*. Tetrahedron Lett. 41: 6425–6428. doi:10.1016/S0040-4039(00)00931-X.

Teegarden, G.J. 1999. Copepod grazing selection and particle discrimination on the basis of PSP toxin content. Mar. Ecol. Prog. Ser. 181: 163–176. doi:10.1097/00006534-199703000-00052.

Teegarden, G.J., Campbell, R.G. and Durbin, E.G. 2001. Zooplankton feeding behavior and particle selection in natural plankton assemblages containing toxic *Alexandrium* spp. Mar. Ecol. Prog. Ser. 218: 213–226. doi:10.3354/meps218213.

Thronson, A. and Quigg, A. 2008. Fifty-five years of fish kills in coastal Texas. Estuaries and Coasts 31: 802–813. doi:10.1007/s12237-008-9056-5.

Tran, D., Haberkorn, H., Soudant, P., Ciret, P. and Massabuau, J.C. 2010. Behavioral responses of *Crassostrea gigas* exposed to the harmful algae *Alexandrium minutum*. Aquaculture 298: 338–345. doi:10.1016/j.aquaculture.2009.10.030.

Tran, D., Ciutat, A., Mat, A., Massabuau, J.C., Hégaret, H., Lambert, C., Le Goic, N. and Soudant, P. 2015. The toxic dinoflagellate *Alexandrium minutum* disrupts daily rhythmic activities at gene transcription, physiological and behavioral levels in the oyster *Crassostrea gigas*. Aquat. Toxicol. 158: 41–49. doi:10.1016/j.aquatox.2014.10.023.

Turner, J. and Tester, P. 1997. Toxic marine phytoplankton, zooplankton grazers, and pelagic food webs. Limnol. Oceanogr. 42: 1203–1214. doi:10.4319/lo.1997.42.5_part_2.1203.

Turner, J.T. and Tester, P.A. 1989. Zooplankton feeding ecology: copepod grazing during an expatriate red tide. pp. 359–374. *In*: Cosper, E.M., Bricelj, V.M. and Carpenter, E.J. (eds.). Novel Phytoplankton Blooms. Causes and Impacts of Recurrent Brown Tides and Other Unusual Blooms, Springer.

Turner, J.T., Lincoln, J.A., Tester, P.A., Bates, S.S. and Leger, C. 1996. Do toxic phytoplankton reduce egg production and hatching success of the copepod *Acartia tonsa*? Eos (Washington, DC) 76: OS12G-.

Turriff, N., Runge, J.A. and Cembella, A.D. 1995. Toxin accumulation and feeding-behavior of the planktonic copepod *Calanus finmarchicus* exposed to the red-tide dinoflagellate *Alexandrium excavatum*. Mar. Biol. 123: 55–64. doi:10.1007/BF00350323.

Uye, S. 1986. Impact of copepod grazing on the red-tide flagellate *Chattonella antiqua*. Mar. Biol. Int. J. Life Ocean. Coast. Waters 92: 35–43. doi:10.1007/BF00392743.

Uye, S. and Takamatsu, K. 1990. Feeding interactions between planktonic copepods and red-tide flagellates from japanese coastal waters. Mar. Ecol. Prog. Ser. 59: 97–107. doi:10.3354/meps059097.

Vargo, G., Atwood, K., Deventer, M. Van and Harris, R. 2006. Beached bird surveys on shell key, pinellas county, Florida. Florida F. Nat. 34: 21–27.

Vasconcelos, V., Azevedo, J., Silva, M. and Ramos, V. 2010. Effects of marine toxins on the reproduction and early stages development of aquatic organisms. Mar. Drugs 8: 59–79. doi:10.3390/md8010059.

Villareal, T.A., Hanson, S., Qualia, S., Jester, E.L.E., Granade, H.R. and Dickey, R.W. 2007. Petroleum production platforms as sites for the expansion of ciguatera in the northwestern Gulf of Mexico. Harmful Algae 6: 253–259. doi:10.1016/J.HAL.2006.08.008.

Wang, L., Liang, X., Huang, Y., Li, S. and Ip, K. 2008. Transcriptional responses of xenobiotic metabolizing enzymes, HSP70 and Na1/K1-ATPase in the liver of rabbitfish (*Siganus oramin*) Intracoelomically Injected with Amnesic Shellfish Poisoning Toxin 363–371. doi:10.1002/tox.

Wells, M.L., Trainer, V.L., Smayda, T.J., Karlson, B.S.O., Trick, C.G., Kudela, R.M., Ishikawa, A., Bernard, S., Wulff, A., Anderson, D.M. and Cochlan, W.P. 2015. Harmful algal blooms and climate change: Learning from the past and present to forecast the future. Harmful Algae 49: 68–93. doi:10.1016/j.hal.2015.07.009.

White, A.W. 1981. Sensitivity of marine fishes to toxins from the red-tide dinoflagellate *Gonyaulax excavata* and implications for fish kills. Mar. Biol. 65: 255–260. doi:10.1007/BF00397119.

Wiens, M., Luckas, B., Brümmer, F. and Shokry, M. 2003. Okadaic acid: a potential defense molecule for the sponge *Suberites domuncula*. Mar. Biol. 142: 213–223. doi:10.1007/s00227-002-0886-6.

Wikfors, G.H. and Smolowitz, R.M. 1995. Experimental and histological studies of four life-history stages of the eastern oyster, *Crassostrea virginica*, exposed to a cultured strain of the dinoflagellate *Prorocentrum minimum*. Biol. Bull. 188: 313–328. doi:10.2307/1542308.

Windust, A.J., Wright, J.L.C. and McLachlan, J.L. 1996. The effects of the diarrhetic shellfish poisoning toxins, okadaic acid and dinophysistoxin-1, on the growth of microalgae. Mar. Biol. 126: 19–25. doi:10.1007/bf00571373.

Work, T., Beale, A., Fritz, L., Quilliam, M., Silver, M., Buck, K. and Wright, J. 1993. Domoic acid intoxication of brown pelicans and cormorants in santa cruz, California. pp. 643–649. *In*: Smayda, T.J. and Shimizu, Y. (eds.). Toxic Phytoplankton Blooms in the Sea.

Yan, T., Zhou, M., Fu, M., Wang, Y., Yu, R. and Li, J. 2001. Inhibition of egg hatching success and larvae survival of the scallop, *Chlamys farreri*, associated with exposure to cells and cell fragments of the dinoflagellate *Alexandrium tamarense*. Toxicon. 39: 1239–1244. doi:10.1016/S0041-0101(01)00080-0.

Yan, T., Zhou, M., Fu, M., Yu, R., Wang, Y. and Li, J. 2003. Effects of the dinoflagellate *Alexandrium tamarense* on early development of the scallop *Argopecten irradians concentricus*. Aquaculture 217: 167–178. doi:10.1016/S0044-8486(02)00117-5.

Zabka, T.S., Goldstein, T., Cross, C., Mueller, R.W., Kreuder-Johnson, C., Gill, S. and Gulland, F.M.D. 2009. Characterization of a Degenerative Cardiomyopathy Associated with Domoic Acid Toxicity in California Sea Lions (*Zalophus californianus*) 119: 105–119.

Zaman, L., Arakawa, O., Shimosu, A., Onoue, Y., Nishio, S., Shida, Y. and Noguchi, T. 1997. Two new isomers of domoic acid from a red alga, *Chondria armata*. Toxicon. 35: 205–12.

5

Overview of Phytoplankton Indicators and Biomarkers as Key-Tools for Trace Element Contamination Assessment in Estuaries

Maria Teresa Cabrita,[1,*] *Bernardo Duarte,*[2] *Carla Gameiro,*[2] *Ana Rita Matos,*[3] *Isabel Caçador*[2] and *Rita M. Godinho*[1]

INTRODUCTION

Estuaries are among the most productive ecosystems on Earth, providing a wide range of resources, benefits and natural services. Recycling of nutrients and other materials, such as pollutants, is the main natural service provided by estuaries, placing these ecosystems among the most valuable on a global scale (Constanza et al. 1997, Millennium Ecosystem Assessment 2005). As major centres of human population and pivotal points of trade, industry, energy production, fisheries and tourism, estuaries are severely impacted by human activities leading to a decline in the water quality, consequently threatening estuarine organisms as a whole. This is particularly relevant in estuaries where human populations rely on estuarine resources for food and recreation. Trace elements, such as chromium (Cr), cobalt (Co), nickel (Ni), copper (Cu), zinc (Zn), cadmium (Cd), mercury (Hg) and lead (Pb) are among the most persistent pollutants found in estuaries (Deforest et al. 2007).

[1] Portuguese Institute of Sea and Atmosphere (IPMA), Av. de Brasília, 1449-006 Lisboa, Portugal; Present affiliation: Centro de Estudos Geográficos (CEG), Instituto de Geografia e Ordenamento do Território (IGOT), University of Lisbon, Rua Branca Edmée Marques, 1600-276 Lisbon, Portugal.
[2] MARE – Marine and Environmental Sciences Centre, Faculty of Sciences of the University of Lisbon, Campo Grande 1749-016 Lisboa, Portugal.
[3] BioISI—Biosystems and Integrative Sciences Institute, Plant Functional Genomics Group, Departamento de Biologia Vegetal, Faculdade de Ciências da Universidade de Lisboa, Campo Grande, 1749-016 Lisboa, Portugal.
* Corresponding author: tcabrita@campus.ul.pt

The main sources of trace element pollution in estuaries, linked to anthropogenic activities, are industrial, wastewater and domestic effluents (Fu and Wang 2011), atmospheric inputs, boating and dredging activities (Förstner and Wittmann 1979, Eggleton and Thomas 2004). Impacts of these activities mainly include the transfer of trace elements to the water column (Nriagu 1990, Nayar et al. 2004). The impact of this mobilisation on estuaries is of particular concern due to the persistence, biogeochemical recycling and toxicity of several elements (Deforest et al. 2007, Pan and Wang 2012). Although dissolved trace element concentrations in estuarine waters are relatively low due to particle adsorption (Hoffmann et al. 2012), contamination has been detected in different trophic levels of estuarine food webs (Wang 2002). This may have dramatic consequences for primary productivity and commercially important fish stocks.

Phytoplankton are considered a promising indicator tool for trace elements, as they can accumulate considerable amounts of these elements (González Dávila 1995), and are at the base of estuarine food webs. Because of their ubiquitous presence in estuaries, phytoplankton play an important role in trace element bioaccumulation, transport and recycling in these systems. Recently, some estuarine phytoplankton species have been identified as effective indicators of trace element contamination under both natural conditions and extreme contamination events (Cabrita et al. 2013, 2014, 2016), due to their rapid response to changes in element availability in the water column (GonzálezDávila 1995, Sunda and Huntsman 1998, Rainbow 2006). In particular, the diatom *Phaeodactylum tricornutum* has been widely used as a model species in several ecotoxicology and trace element assessment works (e.g., Horvatić and Peršić 2007, Cabrita et al. 2013, 2014, 2016) because this species reacts rapidly to trace element fluctuations and is representative of estuarine and coastal phytoplankton communities which are frequently dominated by diatoms.

Once incorporated into phytoplankton cells, trace element have been found to cause impairment of fundamental physiological processes (Anderson and Morel 1978, Shaw 1990, Sunda and Huntsman 1998, Küpper et al. 2002, Cabrita et al. 2016), affecting cell growth (Thomas et al. 1980, Brand et al. 1986, Cabrita et al. 2016) and photosynthetic performance (Cid et al. 1995, Küpper et al. 1996, Cabrita et al. 2016), thus providing the opportunity to identify specific indicators and biomarkers of trace element stress, in representative estuarine phytoplankton species. Besides these physiologically-based indicators, the phytoplankton community as a whole and individual taxa, known to be altered by trace elements (Hollibaugh et al. 1980, Sunda 1989), can also be used as sensitive and efficient indicators of trace element contamination (Cabrita et al. 2014), although these have been generally disregarded within the scope of element pollution assessments. The variety of phytoplankton responses to trace element loading may thus be used to extract adequate indicators and biomarkers to be applied for the early detection of trace element contamination in estuaries. Phytoplankton indicators and biomarkers effectively integrate both physical-chemical status and biological quality elements, effectively enabling a holistic perspective of the adverse impact of pollution on the health status of the estuarine organisms, populations and ecosystem. In estuaries and coastal waters where trace element contamination is an enduring problem, a comprehensive identification of phytoplankton indicators and biomarkers as key-tools for the early detection of element stress is an emerging requirement. Moreover, the value of phytoplankton biomarkers has been increasingly emphasised within the scope of the Water Framework Directive (WFD, European Commission 2000) and the Marine Strategy Framework Directive (MSFD, European Commission 2008). This offers the opportunity to promote an ecosystem-based approach including both phytoplankton indicators and biomarkers of element stress as valuable integrative key-tools to monitoring programmes in the future.

This chapter provides an integrated overview of the overall alterations in the water column induced by trace element contamination in estuarine systems, using the Tagus Estuary as a case-study, in order to identify reliable and efficient phytoplankton indicators and biomarkers of element stress to be employed for the early detection and monitoring of trace element contamination in estuarine systems. Impacts of trace element contamination on phytoplankton are addressed using dredging events such as drivers of trace element contamination in estuaries, as well as laboratory experiments simulating *in situ* conditions of trace element loading. Suitable sentinel and biomonitoring phytoplankton species are also suggested along with techniques and methodologies able to address field experimental constraints.

We present a straightforward and logical step-by-step approach for setting up a successful plan to obtain suitable phytoplankton indicators and biomarkers of element stress, following the pathway of trace elements from their sources in the water column to their allocation within phytoplankton cells, and subsequent effects (Fig. 1). The first step is the chemistry of trace elements in the contaminated estuarine environment which controls their bioavailability. The second step is the trace element accumulation into phytoplankton to provide proof that the bioavailable trace elements are actually taken up and in what relative proportions, and the evaluation of phytoplankton as a biomonitoring tool. Next, intracellular trace element partitioning in phytoplankton is tackled to understand consequences to cellular toxicity, tolerance mechanisms, and element fate. The following step is the overall effect of trace element overload on phytoplankton fundamental processes, such as growth and photosynthesis, to determine potential physiologically-based indicators. As the trace element combined effects on individual cells define the overall effect on phytoplankton populations and communities, community and individual taxa indicators can also be extracted. Finally, phytoplankton biomarkers are investigated and identified, providing further evidence

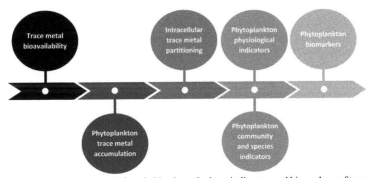

Fig. 1. Step-by-step approach to obtain suitable phytoplankton indicators and biomarkers of trace element stress.

supporting the use of phytoplankton as key-tool for the early detection of element stress. Steps are illustrated with relevant results obtained from our research work, providing other detailed sources supporting our findings. The phytoplankton indicators and biomarkers of element stress herein identified can potentially be applied to element contaminated estuarine and coastal systems worldwide.

Trace Element Bioavailability in Estuaries

Estuaries receive trace elements, mainly through atmospheric inputs, industrial, wastewater and domestic effluents, boating, harbour and dredging activities (Eggleton and Thomas 2004, Fu and Wang 2011). Despite rigorous environmental regulations requiring improved effluent quality, and recent developments of innovative, cheaper and more effective technologies for reducing the amount of trace elements in effluents (Barakat 2011), historical contamination is still a serious threat to estuarine ecosystems (Johnston and Roberts 2009). Estuarine sediments have accumulated trace elements mainly from historical industrial pollution sources for long periods of time and are still being contaminated by those elements that are currently used in industrial processes and remain in the effluents produced (Förstner and Wittmann 1979, Cundy et al. 2003). Disturbance resulting from storms, strong currents, tides, bioturbation, trawling and dredging can cause periodic remobilisation of contaminated sediments into the estuarine water column (Calmano et al. 1993, Eggleton and Thomas 2004) where sediment particles rapidly bind to particulate matter and ultimately sink again into the estuarine bottom sediments (Cundy et al. 2003). However, during resuspension, trace elements are released by sediments into the water column (Eggleton and Thomas 2004). For instance, dredging activities, which are recurrent operations to maintain the water depth of navigation and access channels (Kenny and Rees 1994), are frequently a major source of trace elements in estuaries. Large variations in trace element concentrations can be observed during dredging events in the dissolved and suspended particulate fractions (Cabrita et al. 2014). The exchanges of trace elements between dissolved and particulate fractions may occur in short-time scales, namely within the first few minutes following sediment resuspension (Vale et al. 1998, Caetano et al. 2003). The exposure of the dredged anoxic sediments to a different chemical environment, such as the oxygenated estuarine waters, can induce several reactions on particle surfaces, resulting in desorption and transformation of trace elements into more bioavailable forms, depending on their relative solubility in such conditions (Zhuang et al. 1994, Caetano et al. 2003). In particular, sulphide-bound complexes tend to be rapidly oxidised when particles are exposed to oxygen (Calmano et al. 1993). The result is the release of cations to solution, followed by the formation of fresh iron-oxides concomitantly with the incorporation of other elements that have been mobilised to the solution. The desorption rates of elements adsorbed to sulphides may vary. For instance, Cu, Hg and Pb are released faster than Zn to the water column (Caille et al. 2003). The enhanced dissolved trace elements transferred to the water column may then accumulate in the estuarine biota, including phytoplankton.

A prerequisite to adequately assessing impacts of trace element loading on phytoplankton is the measurement of the variation of trace element bioavailability in the water column over space and time, during disturbance events. Only environmentally bioavailable trace elements in the estuarine water column, rather than total element concentrations, are relevant to accumulation and have the potential for toxicity in phytoplankton (Rainbow 2006). Diffusive gradients in thin films (DGTs) have been recently used as a possible approach to directly assess *in situ* environmental bioavailability of trace elements in estuarine waters (INAP 2002), minimising problems associated with extremely low concentrations typical of some trace elements and contamination risks during collection and analysis. The DGT passively accumulates labile element species from solution while deployed *in situ*, avoiding contamination problems related to conventional water collection and filtration procedures. Labile element species are trapped in binding agents (Chelex100 resins) after diffusion through a diffusive gel (type APA, 0.8 mm thickness, open pore) (Zhang and Davison 1999). The Chelex resin used in DGT is selective for free or weakly complexed species, thus providing a proxy for the element labile fraction in solution and consequently, bioavailability (further details in INAP 2002). Trace element concentrations are then directly quantified in resin eluates, typically by Inductively Coupled Plasma Mass Spectrometry (ICP-MS). The DGT exposure period typically varies between hours to days, providing a time-integrated measure of labile-element species over the chosen deployment time interval (Davison and Zhang 1994). This is particularly relevant as bioavailability of trace elements commonly fluctuates over time in estuaries due to differential inputs of river and sea water. Extremely low trace element bioavailability (close to detectability limits) may occur, particularly for some elements, shifting to much higher levels during the sediment disturbance period (Cabrita et al. 2014), depending on the source of contamination (e.g., amount of suspended sediments, and the degree, frequency and duration of contamination) and the physical, chemical and biological characteristics of the estuarine area (Vale et al. 1998). Particularly concerning dredging events, operational conditions and rhythm of dredging, sediment composition and trace element fractionation (Newell et al. 1998, Vale et al. 1998), coupled with fluctuations of the currents with the tide (Vale and Sundby 1987), all contribute to large variations of total trace element concentrations and bioavailability found during the dredging period. Difficulties in recording trace element transformations during resuspension events have been identified by Vale et al. (1998), highlighting the complexity and quickness of fractionation of trace elements in solids, via dredging simulated laboratory experiments. Therefore, appropriate time-scales must be chosen in order to assess the trace element contamination event as accurately as possible, and in agreement with the particular objectives of each study. It is important to bear in mind that each DGT measurement represents a sole time-integrated measure of labile-element species over a chosen deployment time interval, and that extended measurements in time are a fundamental requirement in order to have a realistic picture of the impact of sediment disturbance. In our studies investigating trace element bioavailability changes resulting from dredging conducted in the Tagus Estuary (Cabrita et al. 2014), DGTs were used with a 48-h exposure period, three times before dredging, weekly during the dredging period and twice a few months after dredging had stopped, for an accurate comparison of water column changes. A significant increase of trace element bioavailability was found during dredging in comparison with non-dredging periods, indicating the escape of trace elements as contaminated sediments were dredged and transferred to the dredger vessel (Fig. 2). Comparison between non-dredging and dredging conditions in terms of concentration medians, highlighted increases of 12 times for Cr (0.046 to 0.56 µM), 10 times for Hg (0.0050 to 0.051 µM), 6 times for Pb (0.021 to 0.13 µM), 5 times for Cu (0.33 to 1.6 µM), and less than 3 times for Zn (2.8 to 8.9 µM) and Cd (0.016 to 0.027 µM) (Cabrita et al. 2014). Despite the differences between these contrasting conditions, levels of dissolved elements like Cr, Cu, Cd and Pb varied within broader intervals during dredging (10 times for Cu, 8 times for Hg and Pb and 3 times for Zn) than under no dredging conditions (less than 5 times for the same elements). This was most likely due to rapid chemical interactions between the particle surface and dissolved fraction, in addition to rapid sedimentation of dredged sediments and dilution due to water tidal renewal occurring between consecutive dredging procedures, as mentioned above. These results are consistent with previous research that identified that the dissolved fraction only indicate punctual alterations of the water quality, failing integration in time (Zhou et al. 2008). Even tough trace element bioavailability only provides a rough indication of the potential for ecotoxicological impact, it should not be disregarded as a valuable complementary measure providing chemical environmental context, as long as natural drivers (typical of impacted estuaries) are taken into account.

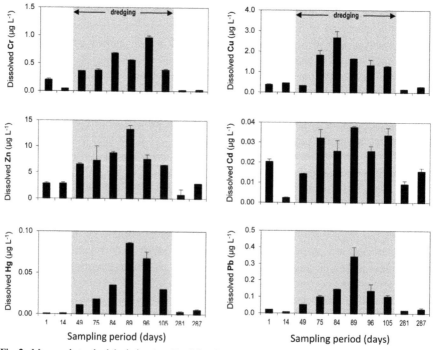

Fig. 2. Mean and standard deviation (n = 3) of dissolved Cr, Cu, Zn, Cd, Hg and Pb (µg L^{-1}) in the water column, at an experimental site on the Tagus Estuary, during non-dredging and dredging conditions. Grey area indicates the dredging period (modified from Cabrita et al. 2014).

Accumulation of Trace Elements by Phytoplankton

Trace element environmental bioavailability does not directly translate into accumulated element levels in the receiving phytoplankton (Sunda 1989, Rainbow 2006), because the way trace elements exert a biological effect is largely dependent on the environmental chemical conditions and on the organisms themselves (Campbell et al. 2006). Therefore, the next logical step to evaluate impacts of trace element contamination on phytoplankton, is to measure the amount of trace elements accumulated within the phytoplankton cells. This will provide evidence that bioavailable trace elements are taken up and serves as the foundation for understanding potential toxicological impacts of trace elements. Because accumulation of trace elements in immobilised cells is a time-integrated process, phytoplankton species are considered preferable and valuable indicators of anthropogenic disturbances in estuarine systems (Reynolds 1998). Interaction between trace elements and phytoplankton is a two-way process where trace elements control phytoplankton growth, biomass and species composition, and concurrently phytoplankton regulate the distribution, chemical speciation, and cycling of these elements in the water through cell uptake and recycling processes, downward flux of biogenic particles, release of organic chelators and mediation of redox reactions (Sunda and Huntsman 1998, Vijver et al. 2004, Sunda 2012) (Fig. 3). These complex relationships have a profound influence on the biogeochemistry, distribution and fate of trace elements in the estuaries (Beardall and Stojkovic 2006). Trace element content of phytoplankton thus allows for the anticipation of the consequences of anthropogenically driven

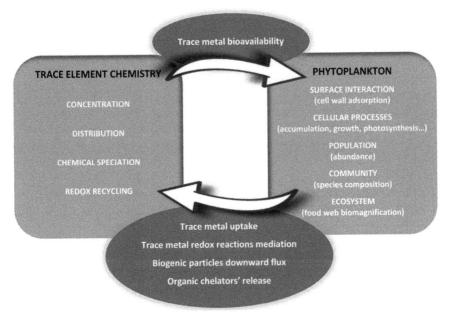

Fig. 3. Conceptual diagram of the interactions between trace element chemistry and phytoplankton in estuarine and coastal areas.

element enhancement to phytoplankton organisms and community, and to trace element biogeochemistry in disturbed estuaries.

Phytoplankton can accumulate considerable amounts of trace elements, whether essential or not, due to their high binding affinity to elements (GonzálezDávila 1995). Trace elements are required by phytoplankton to maintain crucial cell biochemical and physiological processes. A thorough review of trace element functions in phytoplankton was recently produced by Quigg (2016). For instance, Mn is involved in the O_2-evolving complex of photosystem II (PS II), and is part of the superoxide dismutase usually found in diatoms (Wolfe-Simon et al. 2005). Cobalt is component of vitamin B_{12} with functions on C and H transfer reactions, while Ni is involved in the hydrolysis of urea, a cofactor in enzymes and in the Ni form of superoxide dismutase (Wolfe-Simon et al. 2005). Cu is associated with both photosynthetic and respiratory electron transport chains, and is also a component of the Cu-Zn form of superoxide dismutase. Zinc has many important metabolic functions, as constituent of more than 150 enzymes, due to its ability to function as a Lewis acid. The only known biological function of Cd in phytoplankton is being an element cofactor in carbonic anhydrase in diatoms (Lane et al. 2000, 2005). Despite their metabolic role on phytoplankton, essential trace elements, particularly Zn, Cu and Ni, are known to become toxic to phytoplankton cells at elevated concentrations (Sunda 1989). Other trace elements are nonessential, such as Hg and Pb with unknown metabolic roles, and have been found to be exceedingly bioaccumulative and extremely toxic even in very low concentrations (Sunda 1989, Jaishankar et al. 2014). These toxic elements enter phytoplankton cells using the same transport systems as essential elements (Sunda and Huntsman 1998), displacing essential elements from their metabolic sites, causing cell malfunctioning and toxicity. The mechanisms of trace element uptake and accumulation in phytoplankton have been thoroughly presented and explained (e.g., Sunda 1989, GonzálezDávila 1995, Sunda and Huntsman 1998, Wang and Chen 2006, Sunda 2012). Briefly, trace elements occurring as element cations in the estuarine water column are passively adsorbed and actively assimilated by phytoplankton. Adsorption to the phytoplankton cell surface takes place through interactions between element ions and element-functional groups (for instance carboxyl, phosphate, hydroxyl, amino and sulphur) located in the cell wall. Subsequently, trace elements are transferred across the cell membrane, and typically enter the cells by binding to receptor sites of membrane transport proteins responsible for the transport of essential elements (e.g., Mg, Fe, Mn, Zn, Co and Cu) (Wang and Chen 2006). Once inside the cell cytoplasm, trace element ions are compartmentalized into different subcellular organelles (Twining et al. 2003, Godinho et al. 2014).

To date, most studies are carried out under laboratory controlled conditions and mostly focus on trace element accumulation in individual phytoplankton species (e.g., Fisher et al. 1984, Jin et al. 2012) and more rarely on natural phytoplankton assemblages (e.g., Connell and Sunders 1999) exposed to one element in concentrations that frequently lack environmental and ecological relevance. Furthermore, in disturbed estuaries, contamination rarely occurs for single trace elements, being commonly due to mixtures of elements. Effects of combined elements under natural environmental conditions in phytoplankton are also few (e.g., Braek et al. 1976,

Thomas et al. 1980, Cabrita et al. 2016). Surprisingly, field studies monitoring *in situ* accumulation of trace elements in phytoplankton in element contaminated areas (e.g., Dwivedi et al. 2010) are, to the best of our knowledge, insufficient. This is probably because measuring trace element accumulation in phytoplankton during sediment disturbance events (e.g., dredging) may itself be far more complex than originally expected. The sizes of resuspended sediment particles are frequently similar to those of phytoplankton cells which makes the collection of phytoplankton samples, with sediment content at the lowest possible level, extremely difficult to obtain during those events.

Immobilisation of phytoplankton cells may partially overcome this problem providing an indication of trace element accumulation in phytoplankton species representative of the phytoplankton community (Cabrita et al. 2013, 2014). This technique is extremely useful because it allows for the easy manipulation and maintenance of the microalgae under *in situ* conditions (Bozeman et al. 1989, Twist et al. 1997) and avoids sediment particle accumulation in the cell samples. Immobilised cells retain their respiratory and photosynthetic activities, maintain contact with the surrounding water environment, are able to accumulate available trace elements, and simultaneously, are prevented from being washed away or grazed on by herbivores during *in situ* exposure. Calcium alginate is an ideal gel for entrapment of phytoplankton cells. Immobilisation in calcium alginate gel is a simple, inexpensive, rapid and nontoxic method for immobilisation of phytoplankton cells. However, the use of alginate-immobilised species has proven to be a challenging task in estuaries due to the gel instability occurring with exposure to saline waters (Cabrita et al. 2013). The main reason appears to be the presence of non-gelling cations (Na^+, Mg^+ and K^+), which are more abundant in estuarine and seawater than in freshwater, causing mechanical instability of the alginate matrix (Fraser and Bickerstaff 1997). Additionally, water turbulence has also been found to be an influencing factor (Cabrita et al. 2013). Consequently, successful immobilisation of phytoplankton cells in Ca-alginates for application in estuaries requires a tailormade approach tackling restrictions associated mainly with salinity and turbulence and concomitantly ensuring sustained cell growth. Chemical characteristics of alginate beads must be modified to ensure gel stability in estuarine water, namely the guluronic (G):mannuronic (M) acid ratio of the alginate polymer chain, and the type and concentration of the hardening agent (Martinsen et al. 1989). Moreira et al. (2006) have shown that a guluronic acid rich alginate was crucial to ensure bead stability of *Phaeodactylum tricornutum* in a saltwater milieu. The hardening agent for microalgae immobilisation purposes is usually $CaCl_2$, used as a cation source (Ca^{2+}) (Cabrita et al. 2013) because it provides no toxicity and produces more stable alginate gels (i.e., Cu, Sr, Ba and Pb). Increasing Ca^{2+} concentration strengthens the linkage of Ca^{2+} to the alginate chains, and increases gel thickness (Chai et al. 2004). However, compactness of the gel structure should be balanced with suitability for cell growth. Too much compactness leads to the decrease of cell growth rates (Cabrita et al. 2013) and to the formation of higher pore sizes within the gel structure (Gåserød et al. 1998) increasing the risk of cell leakage. Our studies have demonstrated that $CaCl_2$ concentration of 5 percent (w/v) provided the best conditions for alginate beads hardening in order to maintain the gel stability in estuarine water for time periods long enough to detect environmental impacts (e.g., eight days) and concomitantly to avoid cell growth inhibition (Cabrita et al. 2013). Beads produced with this $CaCl_2$ concentration were more resistant to salinity and better withstood natural turbulence conditions than beads gelled with lower $CaCl_2$ concentrations. The replacement of Ca^{2+} in the gel matrix by Na^+, Mg^+ and K^+ present in the estuarine water was slow enough to avoid loss of the alginate matrix integrity and thus decreased the risk of gel dissolution. This is particularly relevant for *in situ* assays designed for the detection of trace element impacts in mesotidal and macrotidal estuaries where tidal currents are a continuing source of water turbulence. Furthermore, alginate beads hardened with $CaCl_2$ concentration of 5 percent (w/v) maintained a high diffusional resistance (Donati and Paoletti 2009) which is fundamental for contact of immobilised cells with the surrounding water environment during exposure throughout disturbance events. In fact, free ions and elements bonded to organic ligands in the water column, easily and quickly diffuse through the alginate gel matrix and actually reach the immobilised cells (Chen et al. 1993, Cabrita et al. 2013). A possible experimental design for this type of *in situ* assessment is schematically presented in Fig. 4, and has been thoroughly explained in Cabrita et al. (2013, 2014). Briefly, beads are placed in modified 24-well plates (PBEs, plates for bead exposure), with the bottom of each well and the top of the plate lids replaced by a 50 μm size nylon mesh net to ensure

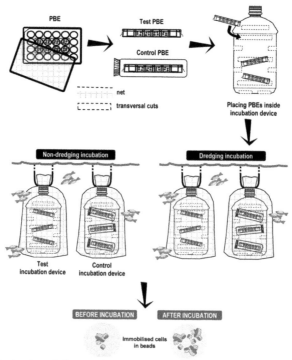

Fig. 4. Schematic drawing of the experimental design for the *in situ* assessment of trace element accumulation in phytoplankton exposed to dredging in estuarine and coastal areas, using immobilised *Phaeodactylum tricornutum* in alginate beads, showing the construction of plates for bead exposure (PBE), of test and control PBEs and incubation devices and *in situ* incubation device setup (modified from Cabrita et al. 2013).

full water circulation within the PBEs. Control and test devices must be produced. The PBEs are inserted in transversely cut 5 L polyethylene bottles which are placed into 200 μm nylon mesh bags closed with nylon thread in order to simultaneously allow entrance of estuarine water and prevented damage. The bags with the 5 L bottles inside are fastened to a buoy attached to a fixed platform to keep them at subsurface depth regardless of water level tidal fluctuations during the entire exposure period. The beads are then recovered and the cells demobilised (for further details, see Cabrita et al. 2013). The obtained cell biomass is free of sediments which enables the accurate determination of the accumulation of trace elements in the cells.

Results obtained from the application of this technique are relevant if the chosen phytoplankton species are considered as good proxies of the estuarine phytoplankton communities concerning element accumulation efficiency. For instance, the model diatom *P. tricornutum* was used in our studies on the accumulation of trace elements in phytoplankton during dredging events, because diatoms are a major component of the phytoplankton communities in estuaries and other coastal systems, and also this ubiquitous marine species has been shown to rapidly respond to trace element changes in the environment (Cabrita et al. 2014). This species accumulated considerable amounts of trace elements, such as Cr, Cu, Zn, Cd, Hg and Pb, during the dredging period in comparison with periods where dredging was not taking place (Cabrita et al. 2014), providing the much-needed indication that elements were being consumed and, furthermore, accumulated in relatively high concentrations (Fig. 5).

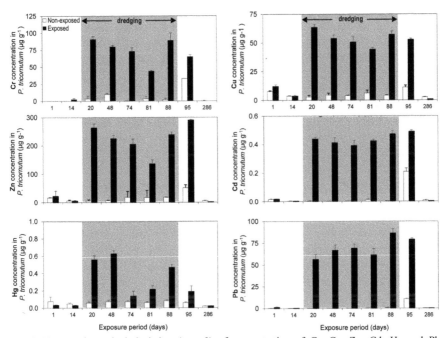

Fig. 5. Mean and standard deviation (n = 3) of concentration of Cr, Cu, Zn, Cd, Hg and Pb (μg g^{-1} d.w.) in non-exposed (open box) and exposed (shaded box) immobilised *Phaeodactylum tricornutum*, incubated *in situ* in the water column, at an experimental site on the Tagus Estuary, during non-dredging and dredging conditions. Grey area indicates the dredging period (modified from Cabrita et al. 2014).

These results also allowed for the identification of the differences in element relative accumulation, and connection with their bioavailability in the water column. For instance, some trace elements, such as Cr, Hg and Pb, were highly accumulated in the cells although their bioavailability in the water column was relatively low in comparison with that of other elements. Zinc was, by far, the highest accumulated element, as well as the most abundant element in the water. This shows that accumulation of trace elements in immobilised cells is a time-integrated process, and further highlights phytoplankton species as valuable indicators of trace element changes, as previously pointed out by Reynolds (1998). Additionally, internal concentrations of trace elements in phytoplankton were also found to be good indicators of element exposure, as highlighted by Luoma and Rainbow (2005).

Intracellular Trace Element Partitioning in Phytoplankton

Once evidence has been obtained that trace elements are accumulated in the phytoplankton cells, the question remains whether elements are simply being adsorbed to the cell wall or whether are they actively internalized into the cells. This is an important issue to further evaluate the impacts of trace elements on phytoplankton. The intracellular elemental distribution provides information about the physiological state and biogeochemical role of diatoms, and reflects biochemical demands and environmental availability (Twining et al. 2003, Godinho et al. 2014, 2015). Understanding trace element uptake, accumulation and compartmentalization in whole phytoplankton cells exposed to element overload will allow for the examination of cellular toxicity and tolerance mechanisms (Godinho et al. 2014, 2015). Furthermore, the trace element content of phytoplankton constitutes a direct measurement of environmental changes, and influences the distribution and geochemical fate of various trace elements (Beardall and Stojkovic 2006), which has implications for broader estuarine biogeochemical cycles.

The traditional bulk size-fractionation techniques used to measure phytoplankton elemental composition may present ambiguous results due to the presence of considerable detrital matter (Twining et al. 2003). The determination of intracellular elemental distributions in isolated cells is challenging, involving extraction and purification of different organelles and the use of specific labelling probes, with associated risks of modifying the cell physiological state and production of artefacts (Fahrni 2007). Alternatively, nuclear microprobe techniques are a powerful tool that allows for qualitative imaging of the distribution and quantitative determination of intracellular concentrations (Ortega et al. 2009). The high trace element sensitivity and spatial resolution of the microprobe has been shown to be well-suited for studying the interactions of trace elements and single cells (Ortega et al. 2009, Godinho et al. 2014). The simultaneous assessment of cell morphology and elemental partitioning enables to identify boundaries that allow distinguishing between adsorption to the cell walls from absorption into compartments, and to infer element function and imbalances in exposure condition. The analysis of morphology, elemental concentrations and their distribution in individualized phytoplankton cells with nuclear microscopy techniques requires cellular integrity to be maintained and no interference from culture media. Major obstacles to the analysis of estuarine or marine diatoms are the high medium salinity and the presence of extracellular exudates, mainly polysaccharides, produced by microalgae in both natural media and *in vitro*, which severely hamper cell identification and element analysis. Cells must be completely isolated from the surrounding media components. Furthermore, the use of fixative agents, such as glutaraldehyde, to avoid cell disruption, have been proved inappropriate for elemental determination and for intracellular morphological analysis, as trace elements are washed out from cells (Godinho et al. 2015). Methodological protocols for suitable sample preparation of diatom cells to elemental analysis have been thoroughly established to overcome these problems (Godinho et al. 2014, 2015).

Response to trace element exposure, namely Ni, Cu and Zn, has been evaluated in two diatom species *Coscinodiscus eccentricus* and *Coscinodiscus wailesii*, representative of estuarine phytoplankton, simultaneously using Particle induced X-ray emission (PIXE), Rutherford backscattering spectrometry (RBS) and scanning transmission ion microscopy (STIM) to obtain morphological and quantitative elemental distribution data (Godinho et al. 2014, 2015). Results showed that elements were actually internalised into the cells and not only adsorbed to the cell walls. Concentrations of Cu and Zn in exposed cells were higher than in non-exposed cells, and Ni was virtually absent in controls. The compartmentalisation of these trace elements also contrasted between exposed and control cells (Fig. 6), highlighting their role in several metabolic pathways. Ni was mainly stored in the vacuole of exposed cells which may indicate the ability of these diatom species to remove excess Ni efficiently. Although marine phytoplankton has been shown to require Ni for specific enzymatic processes, such as nitrogen recycling from urea carried out by ureases (Follmer 2008), absence of this element in controls merely indicates that levels of Ni in the cells may have been much lower relative to the concentrations of the other measured elements and therefore difficult to detect.

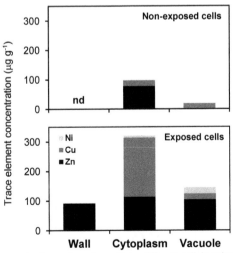

Fig. 6. Trace element concentration values ($\mu g\ g^{-1}$) in different cell locations (wall, cytoplasm and vacuole) of *Coscinodiscus eccentricus*, incubated in the lab, under control and exposed conditions. The notation "nd" means not detected (modified from Godinho et al. 2014).

Copper was significantly accumulated in the cytoplasm of exposed cells but not mobilized to the vacuole suggesting cell susceptibility to Cu toxicity, which is in line with other studies reporting Cu toxicity to phytoplankton (e.g., Thomas et al. 1980). In particular, Cu has been shown to inhibit photosystem II performance and cell growth (Reiriz et al. 1994), and cause chloroplast damage and oxidative stress (Wang and Zheng 2008). Unlike Ni and Cu, Zn was equally distributed in the cell wall, cytoplasm and vacuole in exposed cells, whereas in controls, this element was essentially associated with the cytoplasm. Furthermore, the concentration of Zn in cytoplasm was not significantly different in exposed cells. This feature denotes a tight control of the cell on maintaining Zn levels in the cytoplasm and an ability of the cell to mobilize this element to the vacuole. This could mean that cells had a mechanism of tolerance to Zn. It may have been a way for the cell to reduce toxicity associated with this element as high concentrations have been shown to inhibit cell growth and have a toxic effect (Sunda and Huntsman 2000). Alternatively, Zn has been shown to be a high demand element for the cells (Sunda and Huntsman 2005) and vacuole compartmentalisation may just constitute a way to store this element for later use (i.e., luxury uptake). The significant enrichment of the cell wall with Zn in exposed diatoms, indicates that Zn may also be adsorbed or retained in frustules.

The differentiated intracellular accumulation patterns, namely, Ni and Cu in the cell organic fraction, and Zn in the cell inorganic fraction (diatom frustule), imply distinct environmental fates. Cu and Ni may be recycled in the water column and remain available with potential for biomagnification through the estuarine food web, whereas Zn may be deposited in the bottom sediments through frustule sedimentation (Safi and Hayden 2010). The latter mechanism may constitute a considerable biological sink of Zn in estuaries, due to the short life and large abundance of diatoms and the elevated amounts of internalised Zn.

Although these results are promising, further work is needed to obtain quantitative profiles of the distribution of other trace elements in exposed phytoplankton cells. This will clearly define trace element signatures of cellular contents and exudates that will contribute to the understanding of the complex relationships between the phytoplankton and the trace element impacted estuarine environments.

Phytoplankton Indicators of Trace Element Pollution

With evidence that trace elements are internalised in different phytoplankton cell compartments, the following step is to evaluate the effects of element overload in phytoplankton cells. Several phytoplankton processes are affected by internalised trace elements which enable phytoplankton to be used as sensitive, informative, efficient and reliable bioindicators. The use of phytoplankton as bioindicators offers many advantages, mostly related to the fact that they are efficient scavengers of trace elements (Gonzalez-Dávila 1995), rapidly respond to environmental trace element changes (Cabrita et al. 2014), are present throughout the water column and are relatively tolerant to a wide range of temperature, salinity and eutrophic conditions. Biological processes within individual phytoplankton cells, as well as species or communities may all potentially act as bioindicators and be successfully used to assess changes in environmental conditions over time. These responses at different levels of phytoplankton organization are time-integrated, reflecting present and past environmental conditions, and, moreover, multi-effect integrated because in estuaries, the source of contamination is commonly a mixture of trace elements. In fact, the sum of the effects of trace elements on individual microalgal cells determines their overall effect on phytoplankton populations and communities. The phytoplankton responses to enhanced trace elements can be both quantitative (e.g., reduction in phytoplankton growth) or qualitative (e.g., absence of a trace element sensitive microalgae species from the phytoplankton standing stock) and should be concurrently used to provide a more integrated insight on the effects of trace element overload.

Growth and Photosynthesis as Indicators of Trace Element Pollution

Among biological processes within phytoplankton individual cells, cell growth and photosynthesis have been shown to be highly effective indicators of trace element effects (Cid et al. 1995, Cabrita et al. 2014, 2016). Among the effects causing impairment of cell growth (Thomas et al. 1980, Brand et al. 1986, Cabrita et al. 2016) and photosynthetic performance (Cid et al. 1995, Küpper et al. 1996, Cabrita et al. 2016), are the disturbance of cell membrane permeability (De Filippis 1979), disruption of nutrient uptake processes (Shaw 1990, Sunda and Huntsman 1998), inhibition of oxygen consumption in respiration, reduction of photosynthetic electron transport and carbon fixation (Anderson and Morel 1978, Küpper et al. 1996), and inhibition of enzyme reactions or protein synthesis (Sunda and Huntsman 1998) are among the effects causing impairment of cell growth (Thomas et al. 1980, Brand et al. 1986, Cabrita et al. 2016) and photosynthetic performance (Cid et al. 1995, Küpper et al. 1996, Cabrita et al. 2016).

Growth responses to trace elements vary between different phytoplankton species (Whitton 1968, Stauber and Florence 1985, Kumar et al. 2014) due to different uptake and absorption efficiencies, cell compartmentalization strategies and tolerance and detoxification mechanisms, which are dependent on the trace element type and concentration, as well as the cell physiological status (Sunda and Huntsman 1998). In general, trace elements have been found to suppress phytoplankton growth to various degrees (Torres et al. 1997, Markina and Aizdaicher 2006, Horvatić and Peršić 2007). Our research regarding trace element effects was focused on *P. tricornutum* used as a sentinel and model phytoplankton species for trace element contamination (Cabrita et al. 2013 and 2014, Cabrita et al. 2016). Results obtained under both *in situ* (Cabrita et al. 2014) and laboratory simulated element exposure conditions (Cabrita et al. 2016), generally corroborated those previous findings, showing an accentuated decrease in *P. tricornutum* growth rates unequivocally attributed to a combination of trace elements (Fig. 7). Regarding the effect of each trace element acting individually on growth, Hg was the most toxic element causing growth inhibition, followed by Pb, Co, Cr and all elements combined (Mix) (Cabrita et al. 2016) (Fig. 7). Although Pb and Hg do not take part in the phytoplankton metabolic pathways, *P. tricornutum* cells incorporated these elements (Cabrita et al. 2014, 2016). Internalized Hg in cells can have harmful repercussions to cell growth and division (Hannan and Patouillet 1972, Deng et al. 2013), by binding to cytosolic ligands, blocking functional groups of essential biomolecules (e.g., enzymes), and allocating into organelles (Le Faucheur et al. 2014). Lead is also a severe cell toxicant, even in low concentrations (Moreira et

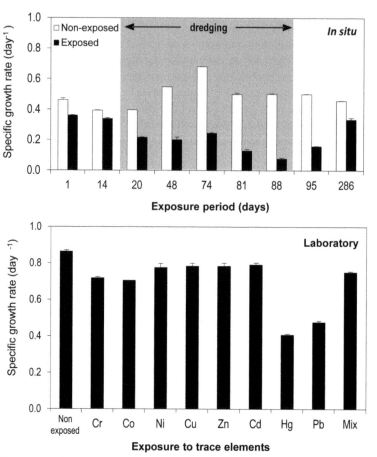

Fig. 7. Mean and standard deviation (n = 3) of specific growth rate (day⁻¹) of immobilised *Phaeodactylum tricornutum*, incubated *in situ* in the water column, at the experimental site on the Tagus estuary, during no dredging and dredging conditions, and of *P. tricornutum* non-exposed and exposed cells to Cr, Co, Ni, Cu, Zn, Cd, Hg, Pb and Mix (mixture of all trace elements combined). Grey area indicates the dredging period (modified from Cabrita et al. 2014, 2016).

al. 2001) it is able to reduce cell growth (Irmer 1985). As for other elements, such as Ni, Zn and Cu, their effect on growth was minor or even undetectable in *P. tricornutum*, as has been shown for several other phytoplankton species (Fisher 1981), possibly because they are essential components of microalgae metabolic pathways (Sunda 1989), and levels of these elements found during dredging were not apparently within toxic levels able to induce severe cell damage. Tolerance to Zn, Cu and Ni, as reported for several other phytoplankton species (Fisher 1981), comply with these elements being essential components in many diatom metabolic pathways (Sunda 1989). In the case of Ni, tolerance may be explained by the low cell binding capacity of Ni found in several microalgae species (Horvatić and Peršić 2007), preventing Ni to reach toxic concentrations within the cells. Nevertheless, high levels of these elements have been found to be toxic to microalgae, as previously reported for *P. tricornutum* and other microalgae under high levels and chronic conditions (Sunda 1989, Horvatić and

Peršić 2007). The values of "no observable effect concentration" (NOEC) and "lowest observable effect concentration" (LOEC) obtained from previous chronic toxicity tests with *P. tricornutum*, were < 1.5 μg L^{-1} and 1.5 μg L^{-1}, for Cu, respectively (Levy et al. 2007, 2008), and a NOEC value of 2700 μg L^{-1} was reported for Zn (Bodar 2007). For Ni and the other elements, no NOEC or LOEC values were found for *P. tricornutum* in the available literature. Copper and Zn levels in the water during exposure experiments were lower than these reported chronic threshold levels which further suggests that the observed effects of trace elements on *P. tricornutum* cell growth were probably not triggered by Zn and Cu, and also possibly not by Ni due to its low cell binding capacity, but rather by the other elements present in the water (e.g., Cr, Hg and Pb). Regarding Cd, the effects on growth ranged from a significant stimulation at the beginning of exposure (Cabrita et al. 2016), to no detectable or insignificant decrease after prolonged exposure (Cabrita et al., unpublished data). Growth enhancement with Cd addition has also been previously observed for *P. tricornutum* (Lee and Morel 1995). Stimulation triggered by Cd may be associated with a requirement of *P. tricornutum* cells for Cd, reportedly necessary for the formation cadmium carbonic anhydrase in diatoms (Lane et al. 2000, 2005) which plays a key role in the acquisition of inorganic carbon for photosynthesis in phytoplankton. Gene sequences for Cd carbonic anhydrase were found in *P. tricornutum* (Park et al. 2007). Carbonic anhydrase can either use Zn or Cd as its element centre (Xu et al. 2008), and the expression of cadmium carbonic anhydrase was controlled by pH changes, and induced by the addition of Cd (Park et al. 2008), which were conjoint features of trace element contamination in the estuarine waters and may have been the reason for growth enhancement found in *P. tricornutum*. Although Cd levels may have increased within the cells with continued exposure, tolerance to Cd toxicity appeared to be a feature of *P. tricornutum* (Torres et al. 1997), mainly due to the ability of this species to incorporate Cd induced sulfide ions in Cd-phytochelatin complexes and produce nanometer sized phytochelatin-coated CdS nanocrystalites (Scarano and Morelli 2003).

Regarding photosynthesis, degradation and biosynthesis inhibition of photosynthetic pigments (De Filippis and Pallaghy 1994), structural changes in the chlorophyll a (Chl *a*) molecule (Küpper et al. 1996, Aggarwal et al. 2012), changes in pigment profiles (Duarte et al. 2012), and inhibition of photosystem II (PS II) (Küpper et al. 2002) are triggered by some elements. These effects have been found in several plant species, including phytoplankton. These alterations dramatically affect the overall photosynthetic performance of phytoplankton cells (Cid et al. 1995, Küpper et al. 1996, Cabrita et al. 2016), which will inevitably have consequences for the primary productivity of estuarine systems impacted by trace element contamination. Chlorophyll fluorescence intensity measurements have been extensively used to understand plant photosynthetic responses to trace element induced stress (Duarte et al. 2012, Santos et al. 2014, Cabrita et al. 2016, Anjum et al. 2016 and references herein). Furthermore, Chl *a* fluorescence is a sensitive indicator of the status of photochemical reactions (Buschmann 2007). Among several fluorescence techniques, Laser-Induced Fluorescence (LIF) and Pulse Modulated Amplitude (PAM) Fluorometry have been shown to be highly sensitive, time-saving and powerful non-destructive techniques, enabling direct optical probing of phytoplankton cells. *In vivo* assessment of trace element stress in phytoplankton is thus possible without requiring the addition of reagents or cell fixatives. Adding to that, these techniques yield nearly instant results so that many replicate measurements can be performed on the same sample. The possibility for measurements to be performed under natural conditions may be an advantage to get rapid large-scale assessment of an entire region under surveillance and to identify stressed zones, which can be studied in detail afterwards, utilizing more expensive and time-consuming methods.

In our studies, the application of LIF technique to the assessment of the effect of trace elements in *P. tricornutum* provided reliable and suitable Chl *a* fluorescence-based indicators for trace element stress (Cabrita et al. 2016). *In vivo* Chl *a* fluorescence spectra of this diatom typically included two maxima, one in the red (685 to 690 nm) and one in the far-red (710 to 740 nm) regions of the spectrum, comparable to fluorescence emission bands already observed in other phytoplankton species (Barbini et al. 1998), intertidal microphytobenthos (Vieira et al. 2011), and higher plants (Lichtenthaler and Rinderle 1988). Most of this emission is linked to the pigment protein PS II-associated Chl *a* (Govindjee 1995), with the red band arising from the main electronic transitions, and the far-red band resulting from vibrational sublevels whose relative intensities are increased *in vivo* through self-absorption (Franck et al. 2002). Additionally, some fluorescence of the photosystem I (PS I), which is highest around 730 nm, also contributes to the fluorescence emitted at the far-red band (Govindjee 1995, Pfündel 1998). Exposure to trace elements induced alterations in the fluorescence emission intensity of Chl *a* as well as deviations in the maximum-emission wavelength, in comparison to non-exposed *P. tricornutum* cells. These alterations indicating Chl *a* concentration variations have been pointed as one of the earliest indicators of physiological status in microalgae (Baker 2008), as shown for Hg as an example (Fig. 8). The significant decrease in fluorescence intensity (mostly in the far-red compared to the red region) combined with large deviations in wavelength emission maxima allocated in the red as compared to the far-red region, observed in exposed cells, suggest Chl *a* damage, structural changes in Chl *a* with the possibility of formation of low fluorescent or non-fluorescent trace element-substituted Chl *a* (Küpper et al. 1996), a decrease in re-absorption due to the decline in Chl *a*, and also imply that PS II chlorophylls were more affected by trace elements than the PS I chlorophylls. These alterations were observed for most of the trace elements tested (Co, Ni, Cu, Zn, Cd, Hg, Pb and all elements combined), in particular for Co, Hg and all elements combined, supported by similar findings observed in higher plants and microalgae (Küpper et al. 1996, Pandey and Gopal 2011). For instance, sublethal Cd and Hg levels strongly inhibit chlorophyll biosynthesis (De Filippis and Pallaghy 1976) and induce a Chl *a* decline in several microalgae species (Maurya et al. 2008). Cobalt, Cd, Hg and Pb have also been shown to cause structural and functional damage in photosynthesis (Aggarwal et al. 2012), due to substitution of magnesium (Mg) in the Chl *a* molecule by Ni, Cu, Zn, Cd, Hg and Pb that caused a decline in the net photosynthesis (Küpper et al. 1996, Küpper et al. 2002). The red/far red emission fluorescence ratio (F685/ F735 ratio) obtained from the fluorescence spectra significantly increased for all elements, in particular for Co, Hg and all elements combined (Mix), except for Zn,

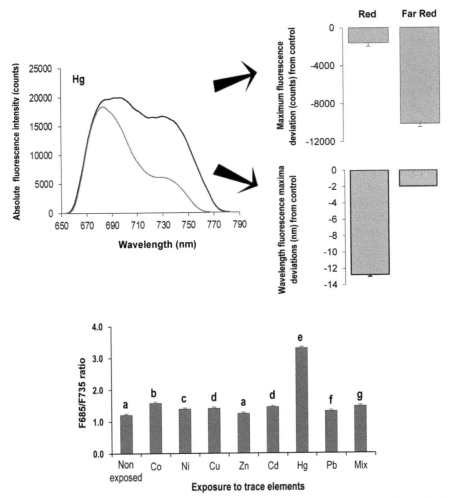

Fig. 8. Absolute fluorescence spectra (counts) of non-exposed (black line) and Hg-exposed (gray line) *Phaeodactylum tricornutum* cells, between 650 and 790 nm, maximum fluorescence deviation (counts) and wavelength fluorescence maxima deviations (nm) from control (non-exposed cells), and red/far red fluorescence ratio (F685/F735 ratio) in non-exposed and exposed cells to Co, Ni, Cu, Zn, Cd, Hg, Pb and Mix (mixture of all trace elements combined) (mean ± standard deviation, n = 6). Different letters above standard deviation bars indicate values significantly different (p < 0.05) from non-exposed values (modified from Cabrita et al. 2016).

mostly due to a decline in fluorescence intensity at 735 nm as illustrated here for Hg (Fig. 8). This suggests that trace elements incorporated in the cells not only induced alterations in the PSII efficiency, but also affect the distribution of excitation energy between the two photosystems (state transitions) (Wollman 2001). The increase in red/far red fluorescence ratios in response to trace element toxicity is comparable to different types of stress inflicted on many plant species showing breakdown or reduced synthesis of Chl *a* and damage in the photosynthetic apparatus, caused by nutrient stress (Subhash and Mohanan 1997), Cd exposure (Maurya and Gopal 2008), herbicides and senescence processes (D'Ambrosio et al. 1992). Overall, these results clearly showed deviation in wavelength emission maxima and F685/F735 ratio as reliable and suitable Chl *a* fluorescence-based indicators that can be used for trace element stress detection and monitoring in estuarine systems, using *P. tricornutum*. Furthermore, these indicators discriminated the toxic action of the different trace elements, highlighting Co and Hg as the most damaging elements to this species.

Pulse Modulated Amplitude (PAM) Fluorometry was used in parallel in experiments to provide further insight on trace element induced changes throughout the photosynthetic pathway, namely in the PS II activity and overall efficiency during light harvesting, the behaviour of the Electron Transport Chain (ETC) and also the entire energetic transduction pathway from the PS II-captured photons to the PS I acceptor side. This complementary approach allowed for the extraction of other efficient and instantaneous indicators that can potentially be employed for the early detection of trace element stress in phytoplankton. The stepwise flow of energy through PS II reflects the alterations of fluorescence response over time, in the shape of a curve (known as the Kautsky curve, the chlorophyll fluorescence transient or the OJIP-transient curve). The Kautsky curve provides an overall imaging of the energetic processes occurring inside the cells. The chlorophyll fluorescence intensity rises from a minimum level (the O level, O is for origin), in less than 1s, to a maximum level (the P-level, P is for Peak) and J and I are two intermediate levels. From this curve, specific OJIP-test parameters were extracted and investigated as potential indicators of element stress that can be employed for the early detection of trace element contamination in estuarine systems (Strasser and Strasser 1995). Table 1 summarizes the parameters computed from the fluorometric data. Corroborating LIF results; PAM Fluorometry data also highlighted the damaging effects of Hg, Co, Cr, Pb, Hg and all trace elements combined (Mix) and pointing to Hg as the most deleterious element to *P. tricornutum*, affecting all fluorometric parameters. The other elements had more restricted effects. Here we explain the damaging effects of these elements, using Hg as an example (Fig. 9). Mercury induced a distinct decrease on the PS II activity of cells and a striking decline in the reaction centres (RC) closure net rate and, consequently, a decrease in the light absorption energy flux, affecting negatively the partial performance due to the light reactions for primary photochemistry and shifting the equilibrium constant for the redox reactions between PS II and PS I. Considering a lower light absorption by the RC, a rise in the dissipation energy was still apparent. Additionally, a concurrent shift in the balance towards the PS I activity was found, triggered by an enhancement in the electron transport from formation of plastoquinol (PQH_2) to the reduction of PS I end electron acceptors, reducing them efficiently, although an increase in the number of oxidized RC and in the electronic transport energy flux was noticed. The lack of efficiency verified in the PS II in Hg-exposed cells was mostly due to the low connectivity between PS II antennae, leaving a large part of the quinone pool (Q) in the oxidized state. This excessive energy dissipation was visible in the appearance of the K-step in the Kautsky curve. The K step in the OJIP curves, at about 300 µs, is always associated with the damage to the PS II donor side (Strasser et al. 2000, Chen and Cheng 2009). Strasser et al. (2000) found that changing the oxygen evolving complex (OEC) enables alternative internal electron donors to donate electrons to PS II, creating a short-lived increase

Table 1: Fluorometric analysis parameters and their description.

Photosystem II Efficiency	Description
F'_0 and F_0	Basal Fluorescence under weak actinic light in light and dark-adapted samples.
F'_M and F_M	Maximum Fluorescence measured after a saturating pulse in light and dark-adapted samples.
F'_v and F_v	Variable fluorescence light ($F'_M - F'_0$) and dark ($F_M - F_0$) adapted samples.
PSII Operational and Maximum Quantum Yield	Light and dark-adapted Quantum yield of primary photochemistry, equal to the efficiency by which an absorbed photon trapped by the PSII reaction centre will result in reduction of Q_A to Q^-_A.
Rapid Light Curves (RLCs)	
rETR	Relative electron transport rate at each light intensity (rETR = QY × PAR × 0.5).
ETR_{max}	Maximum ETR obtained from the RLC after which photo-inhibition can be observed.
E_k	The onset of light saturation.
a	Photosynthetic efficiency, obtained from the initial slope of the RLC.
Energy Fluxes (Kautsky curves)	
Area	Corresponds to the oxidized quinone pool size available for reduction and is a function of the area above the Kautsky plot.
W	Variable fluorescence intensity normalized to the J-step ($W = F_t - F_0)/(F_J - F_0$).
W_K	Amplitude of the K-step ($W_K = V_K - V_J$).
φ_{P0}	Maximum Yield of Primary Photochemistry.
φ_{E0}	Probability that an absorbed photon will move an electron into the ETC.
φ_{D0}	Quantum yield of the non-photochemical reactions.
ψ_0	Probability of a PS II trapped electron to be transported from Q_A to Q_B.
M_0	Net rate of PS II RC closure.
P_G	The Grouping Probability is a direct measure of the connectivity between the two PSII units (Strasser and Stribet 2001).
ABS/CS	Absorbed energy flux.
TR/CS	Trapped energy flux.
ET/CS	Electron transport energy flux.
DI/CS	Dissipated energy flux.
RC/CS	Number of available reaction centres per leaf cross section.
RE_0/RC	Electron transport from PQH_2 to the reduction of PSI end electron acceptors.
δR_0	Efficiency with which an electron can move from the reduced intersystem electron acceptors to PSI to reduce the end electron acceptors.
TR_0/DI_0	The contribution or partial performance due to the light reactions for primary photochemistry.
$(\psi_0/1-\psi_0)$	Contribution of the dark reactions from QA^- to primary photochemistry.
$(\delta R_0/1-\delta R_0)$	Contribution of PSI, reducing its end acceptors.
$(\psi_{E0}/1-\psi_{E0})$	Equilibrium constant for the redox reactions between PSII and PSI.

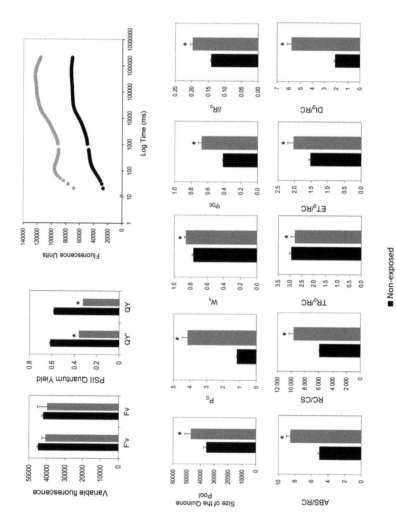

Fig. 9. PS II variable fluorescence and quantum yields, Kautsky curves and associated parameters, in light and dark-adapted *Phaeodactylum tricornutum* non-exposed and exposed cells to Hg (mean ± standard deviation, n = 3), * above standard deviation bars indicates values significantly different (p < 0.05) from control (modified from Cabrita et al. submitted).

in the fraction of Pheo⁻ and Q_{A^-}, which causes an increase in the K fluorescence step. Krüger et al. (2014) suggest that this K-band indicates accessibility of internal non-water electron donors to PS II (e.g., ascorbate red) competing with the OEC. The difference in fluorescence at the J-step between exposed and control cells lower than zero was probably due to decelerated electron donation from H_2O to an increased activity of PS I or stimulated reactive oxygen compounds (ROS) formation, as suggested by Krüger et al. (2014). Similarly, the 30 ms (Iband) was significantly reduced in exposed cells compared to control ones, suggesting an enhancement of the final reduction of end electron acceptors, such as Fd_{red} and NADP⁺, due to a higher oxidized state of the pool mixture of plastoquinone, cytochrome *b/f* and plastocyanin (Yusuf et al. 2010) and higher activity of PS I (Krüger et al. 2014). The significant alterations found in these fluorometric parameters with relevant physiological and ecological meaning indicate that they can be used as a useful suite of indicators for future trace element contamination assessment programmes.

Phytoplankton Community and Individual Taxa as Indicators of Trace Element Pollution

The structure of the phytoplankton community as well as several phytoplankton species have been widely used as indicators of ecological condition for the evaluation of a range of impacts, for instance, eutrophication (Tas et al. 2009) and climatic change (Paerl et al. 2010), but has seldom been applied in trace element contamination assessments. Previous studies showed that phytoplankton indicators provided valid information on the level of eutrophication in a dredged lake (Wang et al. 2012, Xu and Pan 2013). Our studies have identified much needed phytoplankton indicators, at the community and individual taxa levels, for the monitoring of estuarine systems contaminated with elements, taking into account the local hydrologic conditions and the natural variability of the estuarine phytoplankton assemblages in a given area (Cabrita 2014).

Our results indicated diatom:other groups ratio and benthic:pelagic diatom ratio as efficient and reliable indicators of trace element contamination (Cabrita 2014) (Fig. 10). The consistent raise in the diatom:other groups ratio found during increased trace element in the water column, triggered by dredging, was mostly due to benthic diatoms, as highlighted by the benthic:pelagic diatom ratio, and seemed to have been generated by inputs of suspended particulates that occurred during dredging. Similarly, De Jonge (2000) reported enhanced contribution of resuspended benthic cells to phytoplankton related to increased background turbidity in the Ems Estuary, during dredging events. High contribution of benthic diatoms to the phytoplankton community was also observed during periods of high perturbation in shallow areas (Lucas 2003). This implies that benthic microalgae were not settling on the bottom sediments, but otherwise remained in the water column, growing and even blooming, in spite of the elevated levels of trace elements triggered by dredging. Trace element resistance has been reported for benthic diatoms growing in contaminated sediments (Tuovinen et al. 2012), with diatom density decreasing only under considerably severe contamination (Cattaneo et al. 2008). Adaptation strategies to high trace element

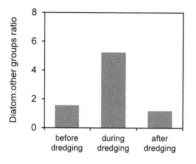
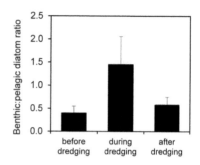

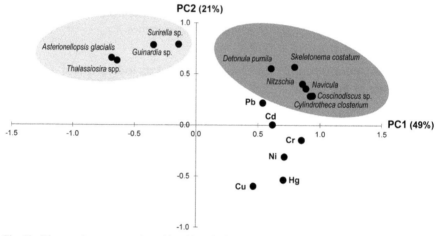

Fig. 10. Diatom:other groups ratio and benthic:pelagic diatom ratio in surface water, at an experimental site on the Tagus Estuary, before, during and after dredging periods, and projection of Cr, Ni, Cu, Cd, Hg, Pb dissolved concentrations and abundance of *Asterionellopsis glacialis*, *Coscinodiscus* sp., *Cylindrotheca closterium*, *Detonula pumila*, *Guinardia* sp., *Navicula* sp., *Nitzschia* sp., *Skeletonema costatum*, *Surirella* sp. and *Thalassiosira* sp., obtained from the Principal Component Analysis (PCA). Two diatom assemblages are highlighted in light grey and dark grey. Percentage of total variance is indicated in brackets close to principal component axes (modified from Cabrita 2014).

concentrations occurring in benthic diatoms that inhabit contaminated sediments were possibly a competitive advantage of the suspended diatoms during dredging in relation to the pelagic phytoplankton adapted to a relatively less contaminated environment.

Species richness (Margalef's index) and species diversity indices (Simpson's diversity, Shannon-Wiever's, and Warwick and Clarke's taxonomic diversity and distinctness indices) failed to discriminate variations in the phytoplankton community induced by dredging, possibly because the species richness was maintained regardless of occurrence of dredging (Cabrita 2014). Similar results were obtained by Chadwick and Canton (1984) for a stream invertebrate community subject to element mine drainage, suggesting that those indices can be useful for detecting environmental perturbations that favour tolerant species over more sensitive ones with resulting reduction in species richness. Although sensitive species reduced or even disappeared due to increasing availability of trace elements driven by dredging, the displacement of the benthic species from the bottom sediments into the water column surely contributed to sustain phytoplankton species diversity. Even the distinctness index, found to be a more sensitive indicator of community perturbation in relation to anthropogenic disturbances (e.g., Brown et al. 2002), was unable to highlight community changes. This points to the importance of a comprehensive analysis of the phytoplankton community when assessing pollution impacts, particularly those driven by dredging, that seem to not compromise species richness.

Identifying phytoplankton taxa as potential indicators of trace element changes induced by dredging was also carried out, requiring care in order to correctly ensure that changes in species individual contributions associated with dredging are discriminated from seasonal variations. In our studies, dredging coincided with the spring and summer period when pelagic species such as *Asterionellopsis glacialis* and *Thalassiosira* spp. have been found to occur in relatively high concentration, or even to bloom. Nevertheless, trace elements were found to promote the shaping of the phytoplankton community composition, as already found in other studies (Hollibaugh et al. 1980, Horvatić and Peršić 2007). Phytoplankton responded essentially through a shift in the microalgae composition towards species less susceptible to trace elements. This response was in line with previous findings in other systems subject to trace element enhancement (Moore et al. 1979, Monteiro et al. 1995, Jing et al. 2011). In fact, most of the species developing highest abundance during dredging have been found resistant to elevated trace element levels. The pelagic *Skeletonema costatum* has been shown to be tolerant to Cu, Zn and Pb (Fisher 1981, Rukminasari and Sahabuddin 2012), *Coscinodiscus* sp. to a wide range of trace elements (Rick and Dürselen 1995), and the benthic diatoms, *Navicula* sp. was shown resistant to Cu (Thomas and Seibert 1977, Ivorra et al. 1999) and *Nitzschia* sp. to Cu, Zn, Cd and Pb (Thomas and Seibert 1977, Moore 1981, Santos et al. 2013). The increased abundance of those benthic diatoms in the water column during dredging (Fig. 10) suggests that these species may be suitable indicators of dredging induced changes. An additional outcome of the trace element enhancement during dredging was the decrease and even the disappearance of the most sensitive microalgae species. The dissociation of the pelagic diatoms *Asterionellopsis glacialis*, *Thalassiosira* spp., *Guinardia* sp. and *Surirella* sp. from the trace element enhancement generated by dredging (Fig. 10) implied low tolerance of these species to elements, as reported for some of these taxa (*A. glacialis*, *Thalassiosira* spp.) in preceding studies (Tortell and Price 1996). Specifically, the decline of *A. glacialis* and *Thalassiosira* spp. and the fact that they are part of the phytoplankton standing stock and may form blooms in the Tagus Estuary (Cabrita 1997, Gameiro et al. 2004, Brogueira et al. 2007) suggest that the absence of these species would be considered a good indicator of trace element changes associated with the occurrence of dredging in this estuarine area in particular. However, the wide range of trace element sensitivity among phytoplankton species and the varying sediment contamination levels from different areas requires site-specific evaluations for the identification of the best taxa indicators, bearing in mind the species composition of the local phytoplankton assemblages and their temporal variation patterns.

Overall, these results point to diatom:other groups ratio, benthic:pelagic diatom ratio and individual taxa as reliable and sensitive indicators of trace element stress that can be used, in combination with the other formerly mentioned indicators, for the early detection of element impacts during disturbance events in estuaries.

Phytoplankton Biomarkers of Trace Element Pollution

During the past 25 years, several biomarkers for trace element pollution have been developed with the aim of supporting and complementing more conventional environmental biomonitoring. In Europe, conventions and legislation have been developed, aiming to prevent deterioration or restore adversely affected coastal and marine areas, although the potential of biomarkers is not always stressed. The Water Framework Directive (WFD, European Commission 2000) provides a major driver for achieving sustainable management and use of water bodies in the EU Member States, to ensure the progressive reduction of aquatic pollution. The directive, focusing on the two main elements "good ecological status" and "good chemical status", offers a key legislative opportunity to promote and implement an integrated approach including biomarkers, for the protection of estuarine and coastal waters. The Marine Strategy Framework Directive (MSFD, European Commission 2008) already promotes the application of an integrated ecosystem-based approach, with the incorporation of biomarkers. Furthermore, the use of biomarkers as a tool for the assessment of ecosystem health is being increasingly endorsed by the OSPAR Commission of Western Europe and also by Baltic Marine Environment Protection Commission (HELCOM). Biomarkers are thus envisioned as valuable integrative tools that will become fully integrated in the monitoring programmes in future.

The value of biomarkers relies on their capacity to effectively integrate both physical-chemical status and biological quality elements, which may provide an early warning of trace element induced stress in contaminated estuaries and a holistic perspective of pollution adverse consequences on the health of the estuarine organisms, populations and ecosystem (Depledge et al. 1993, Moore et al. 2004). A wide array of biomarkers have been proposed to assess trace element exposure and effects in estuarine waters, focused on possible sentinel organisms, mostly invertebrate and fish species (e.g., Vieira et al. 2009, Caçador et al. 2012, Vasanthi et al. 2012). Considered as effective biomarkers of trace element contamination (Cabrita et al. 2013, 2014, 2016), and placed at the base of the estuarine food web, phytoplankton can provide sensitive and effective biomarkers that can be employed for the early detection of trace element contamination in estuarine systems (Volterra and Conti 2000, Torres et al. 2008). Trace elements internalised into phytoplankton cells are known to cause impairment of fundamental physiological processes (Overnell 1975, Sunda and Huntsman 1998, Cabrita et al. 2016) which may be seen as an opportunity to identify specific biomarkers of element stress, in representative estuarine phytoplankton species. Phytochelatins (Perales-Vela et al. 2006), heat-shock proteins (Spijkerman et al. 2007) and a wide range of oxidative stress response and defence molecules (Pinto et al. 2003), have been highlighted as powerful phytoplankton biomarkers for trace element contamination (see review by Torres et al. 2008). Our work focused on *P. tricornutum* as a model sentinel species has identified trace element substituted chlorophyll *a* (Me-Chl *a*) molecules as unequivocal biomarkers of element stress (Cabrita et al. submitted). Exposure to Cu, Zn and Cd induced the formation of Cu-, Zn- and Cd-Chl *a* in the cells exposed to those elements (Fig. 11). Results showed that Me-Chl *a* accounted for approximately 40 percent of the total Chl *a* cell content, promoting energy dissipation which caused impairment of the photosynthesis energetic pathway in stressed cells compared to non-exposed ones. Trace elements, such as Ni, Cu, Zn, Cd, Hg or Pb, have been found to replace the central atom of the chlorophyll molecule, magnesium, under *in vivo* conditions, in element exposed plant cells (Küpper et al. 1996). Most of the Me-Chls are unsuitable for photosynthesis because the first excitation state becomes unstable, preventing the resonance energy transfer from the antenna pigment complexes to the reaction centres in the thylakoids. As a result, *in vitro* fluorescence quantum yields are much lower in Me-Chls than in Mg-Chls (Watanabe and Kobayashi 1988). Furthermore, the Me-Chls' capacity to release electrons from the singlet excited state is considerable decreased compared to all Mg-Chls (Watanabe et al. 1985). A significant fraction of Me-Chls in total photosynthetic pigments, occurring during trace element exposure, will thus lead to a complete breakdown of photosynthesis (Küpper et al. 1996).

The concentrations of Me-Chl *a* also allowed researchers to identify differences in the element relative toxicity when cells were exposed to a mixture of trace elements, and highlighted the complexity of element combined effects on phytoplankton.

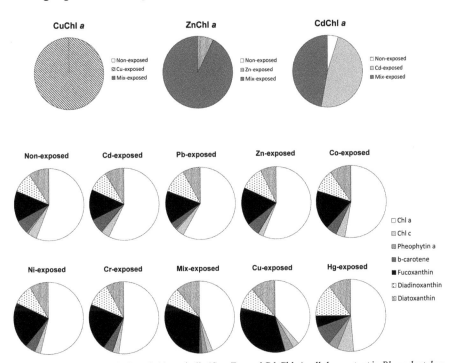

Fig. 11. Trace element substituted chlorophylls (Cu-, Zn- and Cd-Chl *a*) cellular content in *Phaeodactylum tricornutum* non-exposed and exposed cells to Cu, Zn, Cd and Mix (mixture of all trace elements combined), and pigment concentration patterns in cells exposed to trace elements (Cr, Co, Ni, Cu, Zn, Cd, Hg, Pb and Mix) (modified from Cabrita et al. submitted).

In estuaries, elements are commonly discharged as mixtures, and synergistic and antagonistic interactions between multiple trace elements are expected, with each element showing a different relative toxicity (Thomas et al. 1980, González-Dávila 1995). In cells exposed to element mixture containing Cu, Zn and Cd, only Zn- and Cd-Chl *a* were found (Fig. 11), illustrating that Cu was only toxic when tested singly and toxicity was mainly due to Zn and Cd. Zinc induced a lower concentration of Zn-Chl *a* when acting alone than in cells exposed to element mixture and Cd-Chl *a* concentrations were similar regardless of single or combined element exposure (Fig. 11). These effects suggests competition for the chlorophyll Mg binding site and a prevalent effect of Zn and Cd over Cu, in *P. tricornutum*. These results are in line with previous studies. For instance, Kawakami et al. (2006) showed similar Cu, Zn and Cd combined effects on the production of phytochelatins and glutathione in *Phaeodactylum tricornutum*. Similarly, competitive antagonism has been found for Cu and Zn in another diatom species (Rueter and Morel 1981). The application of trace element substituted chlorophylls as sensitive biomarkers of element stress allows to identify which elements are most likely to produce synergistic or antagonistic effects and points out to the most critical elements in the mixture of elements which may be extremely useful in the assessment of element contamination in natural conditions.

Alongside trace element substituted chlorophylls, naturally occurring phytoplankton pigments can also be used as suitable biomarkers of trace element contamination, contributing to the better understanding of the effects of trace elements on the photosynthetic features of phytoplankton. Trace elements have been found to affect pigment content and profile in phytoplankton because they inhibit photosynthetic pigments biosynthesis and cause pigment degradation (De Filippis and Pallaghy 1994) resulting in changes in pigment profiles (Duarte et al. 2012). These changes may occur very fast and the evaluation of the pigment content after trace element exposure generally reflects integration in time, so that element-regulated pigment interconversion intermediate steps may be overlooked. Of course, this could be minimized with a continuous sampling plan which may not always be applicable. Otherwise, interpretation of pigment profiles should be made with extreme caution which makes naturally occurring phytoplankton pigments less unequivocal biomarkers than Me-Chls. Furthermore, photosynthetic pigments in phytoplankton also change rapidly during growth (exponential, stationary and declining phases). In *P. tricornutum*, a rise in the carotenoid:chlorophyll *a* ratio, associated with variations in the percentage of individual carotenoids occurs with the shift from logarithmic to the stationary phase (Carreto and Catoggio 1976). The fucoxanthin content decreases whereas diadinoxanthin content is enhanced and β-carotene remains nearly constant with cell aging. Therefore, experiments regarding phytoplankton responses to trace element stress in terms of pigment variations should only be performed during the growth exponential phase to ensure that observed changes in pigments are essentially associated with responses to trace element stress rather than to other factors such as cell aging or culture nutrient depletion. In our previous studies, cell aging has been detected in *P. tricornutum* cultures after six days, obscuring the physiological responses to environmental stressors (data not published). Our research conducted with *P. tricornutum* as a sentinel species, to investigate pigments as biomarkers of element stress, showed significant changes in the levels of several pigments, including Chl *a* and *c*, Pheophytin *a*, *β*-carotene, Fucoxanthin, Diadinoxanthin and Diatoxanthin (Fig. 11), which were induced by trace elements internalised in the phytoplankton cells. Under trace element stress conditions, changes in the pigment profiles of *P. tricornutum* occurred. A common feature found for most trace elements, particularly detected for Cu, Cr and Mix, was the decrease in Chl *a* and other pigments involved in photosynthesis, in particular *β*-carotene, concomitant with elevated levels of Fucoxanthin, suggesting that the Xanthophyll cycle was triggered with the conversion of *β*-carotene into Fucoxanthin to allow for thermal dissipation of the excess energy at the antenna side of the photosynthetic apparatus (Xia et al. 2013). In addition, an enhancement of Diadinoxanthin and Diatoxanthin levels was also found, more visible for cells exposed to Cr, Co, Cu, Hg, Pb and all elements combined (Mix), possibly related to decrease in the effective absorption cross section of PS II as a result of energy dissipation in the reaction centre. The apparent activation of the diadinoxanthin cycle involving the interconversion of two carotenoids, Diadinoxanthin and Diatoxanthin, has been observed in several species of diatoms (Xia et al. 2013, Masmoudi et al. 2013), including *P. tricornutum* (Olaizola et al. 1994, Grouneva et al. 2009). For instance, in *P. tricornutum* subject to dithiothreitol stress, changes in the non-photochemical quenching appeared to result from energy dissipation in the reaction centre and were found to be associated with decreased photochemical efficiency (Olaizola et al. 1994). This defense feature underlies *P. tricornutum* strategy to cope with the damaging effect of Cr, Co, Cu, Hg, Pb and all trace elements combined (Mix) on the photochemical apparatus. These results point to pigments as sensitive biomarkers of trace element stress that can be extrapolated to other phytoplankton species, with application to efficient biomonitoring programs in estuaries.

Our search for efficient phytoplankton biomarkers led us to investigate fatty acids (FA) as potential biomarkers of trace element stress. Phytoplankton are major producers of fatty acids in diverse lipid classes (Guschina and Harwood 2009). They are able to synthetize linoleic—(C18:2) and linolenic—(C18:3) acids (C18:3), of the classes omega 6 (ω-6) and omega-3 (ω-3), respectively, which are essential fatty acids (EFA) for vertebrates. Since most organisms at higher trophic levels of estuarine food webs, such as fish and humans, have limited ability to produce long chain polyunsaturated fatty acids (LC-PUFA) from EFA, they rather obtain them from the diet, relying on their *de novo* production by estuarine phytoplankton (Parrish 2009). Also, the FA composition of cell membranes is a key factor allowing cells to deal with changes in environmental conditions (Upchurch 2008). Therefore, we hypothesised that trace elements could potentially modify phytoplankton fatty acid composition, which could then be used as potential biomarkers, and thus acquire reinforced relevance in future biomonitoring studies. As a first approach, the effect of Ni on *P. tricornutum* fatty acid composition was investigated, and a reduction of polyunsaturated fatty acid eicosapentaenoic acid (C20:5, EPA) percentage induced by this element was found (Fig. 12) (Matos et al. 2016). In spite of this promising result pointing to EPA as a sensitive biomarker of trace element overload, further research is needed to obtain quantitative profiles of fatty acids in phytoplankton cells exposed to other trace elements, individually and combined, in order to undoubtedly

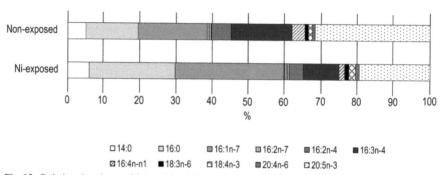

Fig. 12. Relative abundance of fatty acids of *Phaeodactylum tricornutum* non-exposed and exposed cells to Ni, namely saturated and mono-unsaturated fatty acids (white, and lighter shades of grey), plastidial fatty acids (vertical and oblique stripe patterns and darker shades of grey), omega-6 fatty acids (black and dense dotted pattern), omega-3 fatty acids (checkerboard pattern) (modified from Matos et al. 2016).

assert fatty acids as trace element biomarkers to be incorporated in estuarine and coastal biomonitoring programs in the future.

Conclusions

Capitalising our ongoing effort to provide suitable phytoplankton indicators and biomarkers of element stress with potential application to element contaminated estuarine and coastal systems worldwide, a step-by-step approach was set, following the pathway of trace elements from their sources in the water column to their allocation within phytoplankton cells, and subsequent effects.

A trace element-specific suite of phytoplankton indicators and biomarkers was produced. Trace element accumulation and intracellular partitioning, growth, photosynthesis (Chl *a* fluorescence, F685/F735 ratio and OJIP-test parameters), diatom:other groups ratio, benthic:pelagic diatom ratio and individual taxa, were found effective and sensitive indicators of trace element stress alongside trace element substituted chlorophylls and pigments as reliable biomarkers that can be used in combination for the early detection of trace element impacts during disturbance events in estuarine and coastal waters. The potential of fatty acids as efficient phytoplankton biomarkers of trace element stress was disclosed, highlighting the need for further research before fatty acids can be proclaimed as efficient phytoplankton biomarkers of trace element stress. This overview pave the way for further, improved assessments of trace element contamination events in estuarine and coastal systems, which, hopefully, will necessarily lead to more robust decision-making regarding effective environmental monitoring and protection programmes.

Acknowledgments

The authors would like to thank to the "Fundação para a Ciência e Tecnologia (FCT)" for funding the research in the Marine and Environmental Sciences Centre (MARE) throughout the project UID/MAR/04292/2013 and the Biosystems and Integrative

Sciences Institute (BioISI) throughout the project UID/MULTI/04046/2013. B. Duarte and M.T. Cabrita investigation were supported by FCT through Posdoctoral grants SFRH/BPD/115162/2016 and SFRH/BPD/50348/2009, respectively.

References

Aggarwal, A., Sharma, I., Tripathi, B.N., Munjal, A.K., Baunthiyal, M. and Sharma, V. 2012. Metal toxicity and photosynthesis. pp. 229–236. *In*: Itoh, S., Mohanty, P. and Guruprasad, K.N. (eds.). Photosynthesis: Overviews on Recent Progress and Future Perspectives. IK International Publishing House.

Anderson, D.M. and Morel, F.M.M. 1978. Copper sensitivity of *Gonyaulax tamarensis*. Limnol. Oceanogr. 23: 283295.

Anjum, N.A., Duarte, B., Caçador, I., Sleimi, N., Duarte, A.C. and Pereira, E. 2016. Biophysical and biochemical markers of metal/metalloid-impacts in salt marsh halophytes and their implications. Front. Environ. Sci. 4: 24.

Baker, N.R. 2008. Chlorophyll fluorescence: a probe of photosynthesis *in vivo*. Annu. Rev. Plant Biol. 59: 89–113.

Barakat, M.A. 2011. New trends in removing heavy metals from industrial wastewater. Arab. J. Chem. 4: 361–377.

Barbini, R., Colao, F., Fantoni, R., Micheli, C., Palucci, A. and Ribezzo, S. 1998. Design and application of a lidar fluorosensor system for remote monitoring of phytoplankton. ICES J. Mar. Sci. 55: 793–802.

Beardall, J. and Stojkovic, S. 2006. Microalgae under global environmental change: implications for growth and productivity, populations and trophic flow. Science Asia 32: 1–10.

Bodar, C.W.M. 2007. Environmental risk limits for zinc. RIVM letter report 11235/2007, 32 pp.

Bozeman, J., Koopman, B. and Bitton, G. 1989. Toxicity testing using immobilized algae. Aquat. Toxicol. 14: 345–352.

Braek, G.S., Jensen, A. and Mohus, A. 1976. Heavy metal tolerance of marine phytoplankton. III. Combined effects of copper and zinc ions on cultures of four common species. J. Exp. Mar. Biol. Ecol. 25: 37–50.

Brand, L.E., Sunda, W.C. and Guillard, R.R.L. 1986. Reduction of marine phytoplankton reproduction rates by copper and cadmium. J. Exp. Mar. Biol. Ecol. 96: 225–250.

Brogueira, M.J., Oliveira, M.R. and Cabeçadas, G. 2007. Phytoplankton community structure defined by key environmental variables in Tagus estuary, Portugal. Mar. Environ. Res. 64: 616–628.

Brown, B.E., Clarke, K.R. and Warwick, R.M. 2002. Serial patterns of biodiversity change in corals across shallow reef flats in Ko Phuket, Thailand, due to the effects of local (sedimentation) and regional (climatic) perturbations. Mar. Bio. 141: 21–29.

Buschmann, C. 2007. Variability and application of the chlorophyll fluorescence emission ratio red/far-red of leaves. Photosynth. Res. 92: 261–271.

Cabrita, M.T. 1997. Inorganic nitrogen dynamics in the Tagus estuary (Portugal): spatial and temporal variation in input and uptake of nitrate and ammonium. Ph.D. Thesis, University of Lisbon, Portugal.

Cabrita, M.T., Raimundo, J., Pereira, P. and Vale, C. 2013. Optimizing alginate beads for the immobilisation of *Phaeodactylum tricornutum* in estuarine waters. Mar. Environ. Res. 87-88: 37–43.

Cabrita, M.T. 2014. Phytoplankton community indicators of changes associated with dredging in the Tagus estuary (Portugal). Environ. Pollut. 191: 17–24.

Cabrita, M.T., Raimundo, J., Pereira, P. and Vale, C. 2014. Immobilised *Phaeodactylum tricornutum* as biomonitor of trace element availability in the water column during dredging. Environ. Sci. Pollut. Res. 21: 3572–2581.

Cabrita, M.T., Gameiro, C., Utkin, A.B., Duarte, B., Caçador, I. and Cartaxana, P. 2016. Photosynthetic pigment laser-induced fluorescence indicators for the detection of changes associated with trace element stress in the diatom model species *Phaeodactylum tricornutum*. Environ. Monit. Assess. 188(5): 285–297.

Cabrita, M.T., Duarte, B., Gameiro, C., Godinho, R.R., Caçador. Submitted. Bio-optical features and trace element substituted chlorophylls as early detection biomarkers of metal exposure in the model diatom *Phaeodactylum tricornutum*. Special Issue (SI) "Biomarkers: Tools/ endpoints to detect environmental and toxicological stressors", Ecological Indicators Journal.

Caçador, I., Costa, J.L., Duarte, B., Silva, G., Medeiros, J.P., Azeda, C., Castro N, Freitas, J., Pedro, S., Almeida, P.R., Cabral, H. and Costa, M.J. 2012. Macroinvertebrates and fishes as biomonitors of heavy metal concentration in the Seixal Bay (Tagus estuary): which species perform better? Ecol. Indic. 19: 184–190.

Caetano, M., Madureira, M.J. and Vale, C. 2003. Metal remobilisation during resuspension of anoxic contaminated sediment: short-term laboratory study. Water Air Soil Poll. 143: 23–40.

Caille, N., Tiffreau, C., Leyval, C. and Morel, J.L. 2003. Solubility of metals in an anoxic sediment during prolonged aeration. Sci. Total Environ. 301: 239–250.

Calmano, W., Hong, J. and Forstner, U. 1993. Binding and mobilisation of heavy metals in contaminated sediments affected by pH and redox potential. Water Sci. Technol. 28: 22335.

Campbell, P.G.C., Chapman, P. and Hale, B.A. 2006. Risk assessment of metals in the environment. pp. 102–130. *In*: Harrison, R.M. and Hester, R.E. (eds.). Chemicals in the Environment: Assessing and Managing Risk. The Royal Society of Chemistry.

Carreto, J.I. and Catoggio, J.A. 1976. Variations in pigment contents of the diatom *Phaeodactylum tricornutum* during growth. Mar. Biol. 36: 105–112.

Cattaneo, Y., Couillard, Y. and Wumsam, S. 2008. Sedimentary diatoms along a temporal and spatial gradient of metal contamination. J. Paleolimnol. 40: 115127.

Chadwick, J.W. and Canton, S.P. 1984. Inadequacy of diversity indices in discerning metal mine drainage effects on a stream invertebrate community. Water Air Soil Poll. 22: 217223.

Chai, Y., Mei, L.H., Wu, G.L., Lin, D.Q. and Yao, S.J. 2004. Gelation conditions and transport properties of hollow calcium alginate capsules. Biotechnol. Bioeng. 87: 228233.

Chen, D., Lewandowski, Z., Roe, F. and Surapaneni, P. 1993. Diffusivity of Cu^{2+} in calcium alginate gel beads. Biotechnol. Bioeng. 41: 755760.

Chen, L.S. and Cheng, L. 2009. Photosystem II is more tolerant to high temperature in apple (*Malus domestica* Borkh.) leaves than in fruit peel. Photosynthetica. 47: 112120.

Cid, A., Herrero, C., Torres, E. and Abalde, J. 1995. Copper toxicity on the marine microalga *Phaeodactylum tricornutum*: effects on photosynthesis and related parameters. Aquat. Toxicol. 31: 165174.

Connell, D. and Sunders, J. 1999. Variation in cadmium uptake by estuarine phytoplankton and transfer to the copepod *Eurytemora affinis*. J. Mar. Biol. 133: 259265.

Constanza, R., D'Arge, R., de Groot, R., Farber, S., Grasso, M., Hannon, B., Limburg, K., Naeem, S., O'Neill, R.V., Paruelo, J., Raskin, R.G., Sutton, P. and van den Belt, M. 1997. The value of the world's ecosystem services and natural capital. Nature 387: 253260.

Cundy, A.B., Croudace, I.W., Cearreta, A. and Irabien, M.J. 2003. Reconstructing historical trends in metal input in heavily-disturbed, contaminated estuaries: studies from Bilbao, Southampton Water and Sicily. Appl Geochem. 18: 311325.

D'Ambrosio, N., Szábo, K. and Lichtenthaler, H.K. 1992. Increase of the chlorophyll fluorescence ratio F690/F735 during the autumnal chlorophyll breakdown. Radiat. Environ. Biophys. 31: 5162.

Davison, W. and Zhang, H. 1994. *In situ* speciation measurements of trace components in natural waters using thin-film gels. Nature 367: 546548.

De Filippis, L.F. and Pallaghy, C.K. 1976. The effect a sublethal concentration of mercury and zinc on *Chlorella*. I. Growth characteristic and uptake of metals. Z. Pflanzenphysiol. 78: 197207.

De Filippis, L.F. 1979. The effect of heavy metal compounds on the permeability of *Chlorella* cells. Z. Pflanzenphysiol. 92: 3949.

De Filippis, L.F. and Pallaghy, C.K. 1994. Heavy metals: Sources and biological effects. pp. 3177. *In*: Rai, L.C., Gaur, J.P. and Soeder, C.J. (eds.). Advances in Limnology Series: Algae and Water Pollution. E. Scheizerbartsche Press, Verlagsbuchhandlung, Stuttgart.

De Jonge, V.N. 2000. Importance of temporal and spatial scales in applying biological and physical process knowledge in coastal management, an example for the Ems estuary. Cont. Shelf Res. 20: 16551686.

Deforest, D., Brix, K. and Adams, W. 2007. Assessing metal bioaccumulation in aquatic environments: The inverse relationship between bioaccumulation factors, trophic transfer factors and exposure concentration. Aquat. Toxicol. 84: 236246.

Deng, C.N., Zhang, D.Y., Pan, X.L., Chang, F.Q. and Wang, S.Z. 2013. Toxic effects of mercury on PSI and PSII activities, membrane potential and transthylakoid proton gradient in *Microsorium pteropus*. J. Photochem. Photobiol. B 127: 1–7.

Depledge, M.H., Amaral-Mendes, J.J., Daniel, B., Halbrook, R.S., Kloepper-Sams, P., Moore, M.N. and Peakall, D.B. 1993. The conceptual basis of biomarker approach. pp. 15–29. *In*: Peakall, D.B. and Shugart, L.R. (eds.). Biomarkers-research and Application in the Assessment of Environmental Health. Advances in Marine Pollution, Springer, Berlin.

Donati, I. and Paoletti, S. 2009. Material properties of alginates. pp. 154. *In*: Rehm, B.H.A. (ed.). Alginates: Biology and Applications. Microbiology Monographs, Springer-Verlag Berlin, Heidelberg.

Duarte, B., Silva, V. and Caçador, I. 2012. Hexavalent chromium reduction, uptake and oxidative biomarkers in *Halimione portulacoides*. Ecotoxicol. Environ. Saf. 83: 17.

Dwivedi, S., Srivastava, S., Mishra, S., Kumar, A., Tripathi, R.D., Rai, U.N., Dave, R., Tripathi, P., Charkrabarty, D. and Trivedi, P.K. 2010. Characterization of native microalgal strains for their chromium bioaccumulation potential: Phytoplankton response in polluted habitats. J. Hazard Mater. 173: 95101.

Eggleton, J. and Thomas, K.V. 2004. A review of factors affecting the release and bioavailability of contaminants during sediment disturbance events. Environ. Int. 30: 973980.

European Commission. 2000. Directive 2000/60/EC of the European Parliament and of the Council of 23 October 2000 establishing a framework for Community action in the field of water policy as amended by Decision 2455/2001/EC and Directives 2008/32/EC, 2008/105/EC and 2009/31/EC (Water Framework Directive). Official Journal of the European Communities L327 22.12.2000.

European Commission. 2008. Directive 2008/56/EC of the European Parliament and of the Council of 17 June 2008 establishing a framework for Community actions in the field of marine environmental policy (Marine Strategy Framework Directive). Official Journal of the European Communities L164/19 25.06.2008.

Fahrni, C.J. 2007. Biological applications of X-ray fluorescence microscopy: exploring the subcellular topography and speciation of transition metals. Curr. Opin. Chem. Biol. 11: 121127.

Fisher, N.S. 1981. On the selection for heavy metal tolerance in diatoms from the Derwent Estuary, Tasmania. Aust. J. Mar. Fresh. Res. 32: 555561.

Fisher, N.S., Bohé, M. and Teyssié, J.-L. 1984. Accumulation and toxicity of Cd, Zn, Ag, and Hg in four marine phytoplankters. Mar. Ecol. Prog. Ser. 18: 201–213.

Follmer, C. 2008. Insights into the role and structure of plant ureases. Phytochem. 69: 18–28.

Förstner, U. and Wittmann, G.T.W. 1979. Metal Pollution in the Aquatic Environment. Springer-Verlag, Berlin, Heidelberg, New York.

Franck, F., Juneau, P. and Popovic, R. 2002. Resolution of the photosystem I and photosystem II contributions to chlorophyll fluorescence of intact leaves at room temperature. Biochim. Biophys. Acta 1556: 239246.

Fraser, J.E. and Bickerstaff, G.F. 1997. Entrapment in calcium alginate. pp. 61–66. *In*: Methods in Biotechnology, Immobilization of Enzymes and Cells. Methods in Biotechnology. Humana Press Inc, Totowa N.J.

Fu, F. and Wang, Q. 2011. Removal of heavy metal ions from wastewaters: a review. J. Environ. Manage. 92: 407–418.

Gameiro, C., Cartaxana, P., Cabrita, M.T. and Brotas, V. 2004. Variability in chlorophyll and phytoplankton composition in an estuarine system. Hydrobiologia. 525: 113–124.

Gåserød, O., Sannes, A. and Skjåk-Bræk, G. 1998. Microcapsules of alginate-chitosan-II. A study of capsule stability and permeability. Biomaterials 20: 773783.

Godinho, R., Cabrita, M.T., Alves, C. and Pinheiro, T. 2014. Imaging of intracellular metal partitioning in marine diatoms exposed to metal pollution: consequences to cellular toxicity and metal fate in the environment. Metallomics 6: 1626–1631.

Godinho, R., Cabrita, M.T., Alves, C. and Pinheiro, T. 2015. Changes of the elemental distributions in marine diatoms as a reporter of sample preparation artefacts. A nuclear microscopy application. Nuclear Instruments and Methods in Physics Research Section B: Beam Interactions with Materials and Atoms 348: 265–268.

GonzálezDávila, M. 1995. The role of phytoplankton cells on the control of heavy metal concentration in seawater. Mar. Chem. 48: 215–236.

Govindjee. 1995. Sixty-three years since Kautsky: Chlorophyll a fluorescence. Aust. J. Plant Physiol. 22: 131–160.

Grouneva, I., Jakob, T., Wilhelm, C. and Goss, R. 2009. The regulation of xanthophyll cycle activity and of non-photochemical fluorescence quenching by two alternative electron flows in the diatoms *Phaeodactylum tricornutum* and *Cyclotella meneghiniana*. Biochim. Biophys. Acta Bioenerg. 1787: 929–938.

Guschina, I.A. and Harwood, J.L. 2009. Algal lipids and effect of environment on their biochemistry. pp. 1–24. *In*: Arts, M.T., Brett, M.T. and Kainz, M.J. (eds.). Lipids in Aquatic Ecosystems. New York: Springer.

Hannan, P.J. and Patouillet, C. 1972. Effect of mercury on algal growth rates. Biotechnol. Bioeng. 14: 93–101.

Hoffmann, L.J., Breitbarth, E., Boyd, P.W. and Hunter, K.A. 2012. Influence of ocean warming and acidification on trace metal biogeochemistry. Mar. Ecol. Prog. Ser. 470: 191–205.

Hollibaugh, J.T., Siebert, D.L.R. and Thomas, W.H. 1980. A comparison of acute toxicity of ten heavy metals to phytoplankton from Saanich Inlet, BC, Canada. Estuarine, Coastal Shelf Sci. 10: 93105.

Horvatić, J. and Peršić, V. 2007. The Effect of Ni^{2+}, Co^{2+}, Zn^{2+}, Cd^{2+} and Hg^{2+} on the growth rate of marine diatom *Phaeodactylum tricornutum* Bohlin: microplate growth inhibition test. Bull. Environ. Contam. Tox. 79: 494498.

INAP. 2002. Diffusive gradients in thin-films (DGT). A technique for determining bioavailable metal concentrations. International Network for Acid Prevent.

Irmer, G. 1985. Zum einfluß der apparatefunktion auf die bestimmung von streuquerschnitten und lebensdauern aus optischen phononenspektren. Exp. Tech. Phys. 33: 501–506.

Ivorra, N., Hettelaar, J., Tubbing, G.M.J., Kraak, M.H.S., Sabater, S. and Admiraal, W. 1999. Translocation of microbenthic algal assemblages used for *in situ* analysis of metal pollution in rivers. Arch. Environ. Contam. Toxicol. 37: 19–28.

Jaishankar, M., Tseten, T., Anbalagan, N., Blessy, B., Mathew, B.B. and Beeregowda, K.N. 2014. Toxicity, mechanism and health effects of some heavy metals. Interdiscip Toxicol. 7: 60–72.

Jin, Z., DaYong, Z., YongBan, J.I. and QingLong, W.U. 2012. Comparison of heavy metal accumulation by a bloom-forming cyanobacterium, *Microcystis aeruginosa*. Chin Sci. Bull. 57: 37903797.

Jing, L., Peishia, Q., Yun, M., Hao, Z. and Bing, J. 2011. Metal stress on the phytoplankton community of the Zhalong wetland (China). Procedia Environ. Sci. 10: 19671973.

Johnston, E.L. and Roberts, D.A. 2009. Contaminants reduce the richness and evenness of marine communities: a review and meta-analysis. Environ. Pollut. 157: 1745–1752.

Kawakami, S.K., Gledhill, M. and Achterberg, E.P. 2006. Effects of metal combinations on the production of phytochelatins and glutathione by the marine diatom *Phaeodactylum tricornutum*. Biometals 19: 51–60.

Kenny, A.J. and Rees, H.L. 1994. The effects of marine gravel extraction on the macrobenthos early post-dredging recolonization. Mar. Pollut. Bull. 28: 442447.

Krüger, G.H.J., De Villiers, M.F., Strauss, A.J., de Beer, M., van Heerden, P.D.R., Maldonado, R. and Strasser, R.J. 2014. Inhibition of photosystem II activities in soybean (*Glycine max*) genotypes differing in chilling sensitivity. S. Afr. J. Bot. 95: 8596.

Kumar, S.D., Santhanam, P., Ananth, S., Devi, A.S., Nandakumar, R., Prasath, B.B. Jeyanthi, S., Jayalakshmi, T. and Ananthi, P. 2014. Effect of different dosages of zinc on the growth and biomass in five marine microalgae. Int. J. Fish. Aquac. 6: 1–8.

Küpper, H., Küpper, F. and Spiller, M. 1996. Environmental relevance of heavy metal substituted chlorophylls using the example of water plants. J. Exp. Bot. 47: 259–266.

Küpper, H., Setlik, I., Spiller, M., Küpper, F.C. and Prasil, O. 2002. Heavy-metal-induced inhibition of photosynthesis: targets of *in vivo* heavy metal chlorophyll formation. J. Phycol. 38: 429–441.

Lane, T.W. and Morel, F.M.M. 2000. A biological function for cadmium in marine diatoms. PNAS 97: 4627–4631.

Lane, T.W., Saito, M.A., George, G.N., Pickering, I.J., Prince, R.C. and Morel, F.M.M. 2005. Biochemistry: A cadmium enzyme from a marine diatom. Nature 435: 42.

Le Faucheur, S., Campbell, P.G.C., Fortin, C. and Slaveykova, V. 2014. Interactions between mercury and phytoplankton: Speciation, bioavailability, and internal handling. Environ. Toxicol. Chem. 33: 1211–1224.

Lee, J.G. and Morel, F.M.M. 1995. Replacement of zinc by cadmium in marine phytoplankton. Mar. Ecol. Prog. Ser. 127: 305–309.

Levy, J.L., Stauber, J.L. and Jolley, D.F. 2007. Sensitivity of marine microalgae to copper: The effect of biotic factors on copper adsorption and toxicity. Sci. Total Environ. 387: 141–154.

Levy, J.L., Angel, B.M., Stauber, J.L., Poon, W.L., Simpson, S.L., Cheng, S.H. and Jolley, D.F. 2008. Uptake and internalisation of copper by three marine microalgae: comparison of copper-sensitive and copper-tolerant species. Aquat. Toxicol. 89: 82–93.

Lichtenthaler, H.K. and Rinderle, U. 1988. The role of chlorophyll fluorescence in the detection of stress conditions in plants. Crit. Rev. Anal. Chem. 19: S29–S85.

Lucas, C. 2003. Observations of resuspended diatoms in the turbid tidal edge. J. Sea Res. 50: 301308.

Luoma, S.N. and Rainbow, P.S. 2005. Why is metal bioaccumulation so variable? Biodynamics as a unifying concept. Environ Sci. Technol. 39: 1921–1931.

Markina, Zh.V. and Aizdaicher, N.A. 2006. Content of photosynthetic pigments, growth, and cell size of microalgae *Phaeodactylum tricornutum* in the copper-polluted environment. Russ. J. Plant Physiol. 53: 305–309.

Martinsen, A., Skjåk Bræk, G. and Midsrød, O. 1989. Alginate as immobilization material. I. Correlation between chemical and physical properties of alginate gel beads. Biotechnol. Bioeng. 33: 7989.

Masmoudi, S., Nguyen-Deroche, N., Caruso, A., Ayadi, H., Morant-Manceau, A. and Tremblin, G. 2013. Cadmium, copper, sodium and zinc effects on diatoms: from heaven to hell—A review. Cryptogam. Algol. 34: 185–225.

Matos, A.R., Gameiro, C., Duarte, B., Caçador, I. and Cabrita, M.T. 2016. Effects of nickel on the fatty acid composition of the diatom *Phaeodactylum tricornutum*. *In*: Frontiers in Marine Science Conference Abstract Book. IMMR—International Meeting on Marine Research 2016 (DOI: 10.3389/conf.FMARS.2016.04.00033).

Maurya, R. and Gopal, R. 2008. Laser-induced fluorescence ratios of *Cajanus cajan* L. under the stress of cadmium and its correlation with pigment content and pigment ratios. Appl. Spectrosc. 62: 433–438.

Maurya, R., Prasad, S.M. and Gopal, R. 2008. LIF technique offers the potential for the detection of cadmium-induced alteration in photosynthetic activities of *Zea mays* L. J. Photochem. Photobiol. C 9: 29–35.

Millennium Ecosystem Assessment (MEA). 2005. Ecosystems and Human Wellbeing: synthesis. Island Press, Washington, DC.

Monteiro, M.T., Oliveira, R. and Vale, C. 1995. Metal stress on the plankton communities of Sado river (Portugal). Water Res. 29: 695–701.

Moore, J.W., Sutherland, D.J. and Beaubien, V.A. 1979. Algal and invertebrate communities in three sub-Arctic lakes receiving mine wastes. Water Res. 13: 1193–1202.

Moore, J.W. 1981. Epipelic algal communities in a eutrophic northern lake contaminated with mine wastes. Water Res. 15: 97–105.

Moore, M.N., Depledge, M.H., Readman, J.W. and Leonard, P. 2004. An integrated biomarker-based strategy for ecotoxicological evaluation of risk in environmental management. Mutat. Res. 552: 247–268.

Moreira, E.G., Vassilieff, I. and Vassilieff, V.S. 2001. Developmental lead exposure. Behavioral alterations in the short and long term. Neurotoxicol. Teratol. 23: 489–495.

Moreira, S.M., Moreira-Santos, M., Guilhermino, L. and Ribeiro, R. 2006. Immobilization of the marine microalga *Phaeodactylum tricornutum* in alginate for *in situ* experiments: Bead stability and suitability. Enzyme Microb. Tech. 38: 135–141.

Nayar, S., Goh, B.P.L. and Chou, L.M. 2004. Environmental impact of heavy metals from dredged and resuspended sediments on phytoplankton and bacteria assessed in *in situ* mesocosms. Ecotoxicol. Environ. Saf. 59: 349–369.

Newell, R.C., Seiderer, L.J. and Hitchcock, D.R. 1998. The impact of dredging works in coastal waters: A review of the sensitivity to disturbance and subsequent recovery of biological resources on the seabed. Oceanogr. Mar. Biol. An Annual Review 36: 127–178.

Nriagu, J.O. 1990. Global metal pollution–Poisoning the biosphere? Environment 32: 7–33.

Olaizola, M., La Roche, J., Kolber, Z. and Falkowski, P.G. 1994. Non-photochemical fluorescence quenching and the diadinoxanthin cycle in a marine diatom. Photosynth. Res. 41: 357–370.

Ortega, R., Devés, G. and Carmona, A. 2009. Bio-metals imaging and speciation in cells using proton and synchrotron radiation X-ray microspectroscopy. J. R. Soc. Interface 6: S649–S658.

Overnell, J. 1975. The effect of heavy metals on photosynthesis and loss of cell potassium in two species of marine alga *Dunaliella tertiolecta* and *Phaeodactylum tricornutum*. Mar. Biol. 29: 99103.

Paerl, H.W., Rossignol, K.L., Hall, N., Peierls, B.L. and Wetz, M.S. 2010. Phytoplankton community indicators of short- and long-term ecological change in the anthropogenically and climatically impacted Neuse River estuary, North Carolina, USA. Estuaries Coasts 33: 485–497.

Pan, K. and Wang, W.-X. 2012. Trace metal contamination in estuarine and coastal environments in China. Sci. Total Environ. 421-422: 3–16.

Pandey, J.K. and Gopal, R. 2011. Laser-induced chlorophyll fluorescence and reflectance spectroscopy of cadmium treated *Triticum aestivum* L. Plants. Spectroscopy: An International Journal 26: 129–139.

Park, H., Bongkeun, S. and Morel, F.M.M. 2007. Diversity of the cadmium containing carbonic anhydrase (CDCA) in marine diatoms and natural waters. Environ. Microbiol. 9: 403–413.

Park, H., McGinn, P.J. and Morel, F.M.M. 2008. Expression of cadmium carbonic anhydrase of diatoms in seawater. Aquat. Microb. Ecol. 51: 183–193.

Parrish, C.C. 2009. Essential fatty acids in aquatic food webs. pp. 309–326. *In*: Arts, M.T., Brett, M.T. and Kainz, M.J. (eds.). Lipids in Aquatic Ecosystems. New York: Springer.

Perales-Vela, H.V., Peña-Castro, J.N. and Cañizares-Villanueva, R.O. 2006. Heavy metaldetoxification in eukaryotic microalgae. Chemosphere 64: 1–10.

Pfündel, E. 1998. Estimating the contribution of photosystem I to total leaf chlorophyll fluorescence. Photosynth. Res. 56: 185–195.

Pinto, E., Sigaud-Kutner, T.C.S., Leitão, M.A.S., Okamoto, O.K., Morse, D. and Colepicolo, P. 2003. Heavy metal-induced oxidative stress in algae. J. Phycol. 39: 1008–1018.

Quigg, A. 2016. Micronutrients. pp. 211–231. *In*: Borowitzka, M.A., Beardall, J. and Raven, J.D. (eds.). The Physiology of Microalgae. Springer.

Rainbow, P.S. 2006. Biomonitoring of trace metals in estuarine and marine environments. Australas. J. Ecotoxicol. 12: 107–122.

Reiriz, S., Cid, A., Torres, E., Abalde, J. and Herrero, C. 1994. Different responses of the marine diatom *Phaeodactylum tricornutum* to copper toxicity. Microbiologia 10: 263–272.

Reynolds, C.S. 1998. Plants in motion: Physical-biological interaction in the plankton. pp. 535560. *In:* Imberger, J. (ed.). Physical Processes in Lakes and Oceans. American Geophysical Union.

Rick, H.-J. and Dürselen, C.-D. 1995. Importance and abundance of the recently established species *Coscinodiscus wailesii* Gran & Angst in the German Bight. Helgoländer Meeresuntersuchungen 49: 355–374.

Rueter Jr., J.G. and Morel, F.M.M. 1981. The interaction between zinc deficiency and copper toxicity as it affects the silicic acid uptake mechanisms in *Thalassiosira pseudonana*. Limnol. Oceanogr. 26: 67–73.

Rukminasari, N. and Sahabuddin, S. 2012. Distribution and concentration of several types of heavy metal correlated with diversity and abundance of microalgae at Tallo River, Makassar, South Sulawesi, Indonesia. Int. J. Plant Anim. Environ. Sci. 2: 162–168.

Safi, K.A. and Hayden, B. 2010. Differential grazing on natural planktonic populations by the mussel *Perna canaliculus*. Aquat. Biol. 11: 113–125.

Santos, D., Duarte, B. and Caçador, I. 2014. Unveiling Zn hyperacumulation in *Juncus acutus*: implications on the electronic energy fluxes and on oxidative stress with emphasis on non-functional Zn-chlorophylls. J. Photochem. Photobiol. B 140: 228–239.

Santos, J., Almeida, S.F.P. and Figueira, E. 2013. Cadmium chelation by frustulins: a novel metal tolerance mechanism in *Nitzschia palea* (Kützing) W. Smith. Ecotoxicology 22: 166–173.

Scarano, G. and Morelli, E. 2003. Properties of phytochelatin-coated CdS nanocrystallites formed in a marine phytoplanktonic alga (*Phaeodactylum tricornutum*, Bohlin) in response to Cd. Plant Sci. 165: 803–810.

Shaw, A.J. 1990. Heavy Metal Tolerance in Plants: Evolutionary Aspects. CRC, Boca Raton, FL, USA.

Spijkerman, E., Barua, D., Gerloff-Elias, A., Kern, J., Gaedke, U. and Heckathorn, S.A. 2007. Stress responses and metal tolerance of (*Chlamydomonas acidophila*) in metal-enriched lake water and artificial medium. Extremophiles 11: 551–562.

Stauber, J.L. and Florence, T.M. 1985. Interactions of copper and manganese: a mechanism by which manganese alleviates copper toxicity to the marine diatom *Nitzschia closterium* (Ehrenberg) W. Smith. Aquat. Toxicol. 7: 241–254.

Strasser, B.J. and Strasser, R.J. 1995. Measuring fast fluorescence transients to address environmental questions: The JIP-test. pp. 977–980. *In*: Mathis, P. (ed.). Photosynthesis: From Light to Biosphere. Kluwer Academic Publishers, Dordrecht, Netherlands.

Strasser, R.J., Srivastava, A. and Tsimilli-Michael, M. 2000. The fluorescence transient as a tool to characterize and screen photosynthetic samples. pp. 445–483. *In*: Yunus, M., Pathre, U. and Mohanty, P. (eds.). Probing Photosynthesis: Mechanisms, Regulation and Adaptation. Taylor and Francis Press, London.

Subhash, N. and Mohanan, C.N. 1997. Curve fit analysis of chlorophyll fluorescence spectra: Application to nutrient stress detection in sunflower. Remote Sens. Environ. 60: 347–356.

Sunda, W.G. 1989. Trace metal interactions with marine phytoplankton. Biol. Oceanogr. 6: 411442.

Sunda, W.G. and Huntsman, S.A. 1998. Processes regulating cellular metal accumulation and physiological effects: phytoplankton as model systems. Sci. Total Environ. 219: 165181.

Sunda, W.G. and Huntsman, S.A. 2000. Effect of Zn, Mn, and Fe on Cd accumulation in phytoplankton: implications for oceanic Cd cycling. Limnol. Oceanogr. 45: 1501–1516.

Sunda, W.G. and Huntsman, S.A. 2005. Effect of CO_2 supply and demand on zinc uptake and growth limitation in a coastal diatom. Limnol. Oceanogr. 50: 1181–1192.

Sunda, W. 2012. Feedback interactions between trace metal nutrients and phytoplankton in the ocean. Front. Microbiol. 3: 204.

Tas, S., Yilmaz, I.N. and Okus, E. 2009. Phytoplankton as an indicator of improving water quality in the Golden Horn estuary. Estuaries Coasts 32: 1205–1224.

Thomas, W.H. and Seibert, D.L.R. 1977. Effects of copper on the dominance and the diversity of algae: controlled ecosystem pollution experiment. Bull. Mar. Sci. 27: 23–33.

Thomas, W.H., Hollibaugh, J.T., Seibert, D.L.R. and Wallace Jr., G.T. 1980. Toxicity of a mixture of ten metals to phytoplankton. Mar. Ecol. Prog. Ser. 2: 213–220.

Torres, M.A., Barros, M.P., Campos, S.C., Pinto, E., Rajamani, S., Sayre, R.T. and Colepicolo, P. 2008. Biochemical biomarkers in algae and marine pollution: a review. Ecotoxicol. Environ. Saf. 71: 115.

Torres, E., Cid, A., Fidalgo, P., Herrero, C. and Abalde, J. 1997. Long-chain class III metallothioneins as a mechanism of cadmium tolerance in the marine diatom *Phaeodactylum tricornutum* Bohlin. Aquat. Toxicol. 39: 231–246.

Tortell, P.D. and Price, N.M. 1996. Cadmium toxicity and zinc limitation in centric diatoms of the genus *Thalassiosira*. Mar. Ecol. Prog. Ser. 138: 245–254.

Tuovinen, N., Weckström, K. and Salonen, V.-P. 2012. Impact of mine drainage on diatom communities of Orijärvi and Määrjärvi, lakes in SW Finland. Boreal Environ. Res. 17: 437–446.

Twining, B.S., Baines, S.B., Fisher, N.S., Maser, J., Vogt, S., Jacobsen, C., Tovar-Sanchez, A. and Sañudo-Wilhelmy, S.A. 2003. Quantifying trace elements in individual aquatic protist cells with a Synchrotron X-ray fluorescence microprobe. Anal. Chem. 75: 3806–3816.

Twinning, B.S. and Baines, S.B. 2013. The trace metal composition of marine phytoplankton. Ann. Rev. Mar. Sci. 5: 191–215.

Twist, H., Edwards, A.C. and Codd, G.A. 1997. A novel *in situ* biomonitor using alginate immobilised algae (*Scenedesmus subspicatus*) for the assessment of eutrophication in flowing surface waters. Water Res. 31: 2066–2072.

Upchurch, R.G. 2008. Fatty acid unsaturation, mobilization, and regulation in the response of plants to stress. Biotechnol. Lett. 30: 967–977.

Vale, C. and Sundby, B. 1987. Suspended sediment fluctuations in the Tagus estuary on semi-diurnal and fortnightly time scales. Estuarine, Coastal Shelf Sci. 25: 495–508.

Vale, C., Ferreira, A., Micaelo, C., Caetano, M., Pereira, E., Madureira, M.J. et al. 1998. Mobility of contaminants in relation to dredging operations in a mesotidal estuary (Tagus estuary, Portugal). Water Sci. Technol. 37: 25–31.

Vasanthi, A.L., Revathi, P., Arulvasu, C. and Munuswamy, N. 2012. Biomarkers of metal toxicity and histology of *Perna viridis* from Ennore estuary, Chennai, South East coast of India. Ecotoxicol. Environ. Saf. 84: 92–98.

Vieira, L.R., Gravato, C., Soares, A.M., Morgado, F. and Guilhermino, L. 2009. Acute effects of copper and mercury on the estuarine fish *Pomatoschistus microps*: linking biomarkers to behaviour. Chemosphere 76: 1416–1427.

Vieira, S., Utkin, A.B., Lavrov, A., Santos, N.M., Vilar, R., Marques da Silva, J. and Cartaxana, P. 2011. Effects of intertidal microphytobenthos migration on biomass determination via laser-induced fluorescence. Mar. Ecol. Prog. Ser. 432: 45–52.

Vijver, M.G., vanGestel, C.A.M., Lanno, R.P., vanStraalen, N.M. and Peijnenburg, W.J.G.M. 2004. Internal metal sequestration and its ecotoxicological relevance: a review. Environ. Sci. Technol. 38: 4705–4712.

Volterra, L. and Conti, M.E. 2000. Algae as biomarkers, bioaccumulators and toxin producers. Int. J. Environ. Pollut. 13: 92–125.

Wang, L. and Zheng, B. 2008. Toxic effects of fluoranthene and copper on marine diatom *Phaeodactylum tricornutum*. J. Environ. Sci. (China) 20: 1363–1372.

Wang, J. and Chen, C. 2006. Biosorption of heavy metals by *Saccharomyces cerevisiae*. Biotech. Adv. 24: 427–451.

Wang, W.-X. 2002. Interactions of trace metals and different marine food chains. Mar. Ecol. Prog. Ser. 243: 295–309.

Wang, Y., Zhang, W., Luo, Y., Wang, D., Pan, X. and Xu, T. 2012. Phytoplankton investigation and environmental assessment for dredged area in Caohai section of Dianchi lake, China. Proceedings of the 2nd International Conference on Remote sensing, Environment and transportation engineering, Jiangsu, China, 948–950.

Watanabe, T., Machida, K., Suzuki, H., Kobayashi, M. and Honda, K. 1985. Photoelectrochemistry of metallochlorophylls. Coord. Chem. Rev. 64: 207–224.

Watanabe, T. and Kobayashi, M. 1988. Chlorophylls as functional molecules in photosynthesis. Molecular composition *in vivo* and physical chemistry *in vitro*. Special Articles on Coordination Chemistry of Biologically Important Substances 4: 383–395.

Whitton, B.A. 1968. Effect of light on toxicity of various substances to *Anacystis nidzdans*. Plant Cell Physiol. 9: 23–26.

Wolfe-Simon, F., Grzebyk, D., Schofield, O. and Falkowski, P.G. 2005. The role and evolution of superoxide dismutases in algae. J. Phycol. 41: 453–65.

Wollman, F.A. 2001. State transitions reveal the dynamics and flexibility of the photosynthetic apparatus. EMBO J. 20: 3623–3630.

Xia, S., Wang, K., Wan, L., Li, A., Hu, Q. and Zhang, C. 2013. Production, characterization, and antioxidant activity of fucoxanthin from the marine diatom *Odontella aurita*. Mar. Drugs 11: 2667–2681.

Xu, Y., Feng, L., Jeffrey, P.D., Shi, Y. and Morel, FM.M. 2008. Structure and metal exchange in the cadmium carbonic anhydrase of marine diatoms. Nature 452(7183): 56–61.

Xu, T. and Pan, X. 2013. Phytoplankton investigation and environmental assessment for dredged area in Daqing estuary section of Dianchi Lake, China. Appl. Mech. Mater. 295-298: 726–729.

Yusuf, M.A., Kumar, D., Rajwanshi, R., Strasser, R.J., Tsimilli-Michael, M., Govindjee and Sarin, N.B. 2010. Overexpression of γ-tocopherol methyltransferase gene in transgenic *Brassica juncea* plants alleviates abiotic stress: Physiological and chlorophyll *a* fluorescence measurements. Biochim. Biophys. Acta 1797: 1428–1438.

Zhang, H. and Davison, W. 1999. Diffusional characteristics of hydrogels used in DGT and DET techniques. Anal. Chim. Acta 398: 329–340.

Zhou, Q., Zhang, J., Fu, J., Shi, J. and Jiang, G. 2008. Biomonitoring: An appealing tool for assessment of metal pollution in the aquatic ecosystem. Anal. Chim. Acta 606: 135–150.

Zhuang, Y., Allen, H.E. and Fu, G. 1994. Effect of aeration of sediment on cadmium binding. Environ. Toxicol. Chem. 13: 717–724.

6

Phytoremediation and Removal of Contaminants by Algae and Seagrasses

Bouchama Khaled,[1,4,*] *Rouabhi Rachid*[2] and *Bouchiha Hanen*[3]

INTRODUCTION

The marine environment is a complex system which is mainly influenced by various physical, chemical and biological processes (Bandekar and Haragi 2017). The open ocean is rather stable compared to the near shore waters, where the interaction with the terrestrial zone results in variations of different physic-chemical parameters near the shore waters (Bhadja and Kundu 2012). The high human population density in coastal regions has produced many economic benefits, including improved transportation links, industrial and urban development, revenue from tourism and food production, but the combined effects of booming population growth and economic and technological development are threatening the ecosystems that provide these economic benefits (Creel 2003). Now, one of the major concerns of the industrialized world is the high level of marine pollution with hazardous compounds (Naik and Dubey 2017). Pollutants are responsible for significant lethal and sub-lethal effects on marine life; pollutants impacts all trophic levels, from primary producers to apex predators, and thus interfere with the structure of marine communities and consequently, ecosystem functioning (Todd et al. 2010). Many pollutants are susceptible to interact with the physiological processes such as growth

[1] Ecology and environment department, Khenchela University, Algeria.
[2] Applied biology department, Tebessa University, Algeria.
[3] Biology department, El Tarf University, Algeria.
[4] Cellular Toxicology Laboratory, Annaba University, Algeria.
* Corresponding author: khaled.bouchama@yahoo.fr

and reproduction of marine biota; in fact, pollutants can alter their life and lead to serious disruptions such as reduction of populations and changes of the reproductive functions (Hamza-Chaffai 2014). In contaminated marine environments, several plant species have shown the ability to tolerate and accumulate organic and non-organic pollutants. Many of them can accumulate various contaminants and their concentrations are much higher than present in their environments (Fernandez and Gardinali 2016, Navarro et al. 2016, Qiu et al. 2017).

Pollution and Marine Biota

Marine and aquatic environment are reservoirs of biological diversity with an extreme importance for several animal species and plants; for a considerable number of them, water and wetlands are absolutely necessary for life cycle achievement (Gayet et al. 2016). However, they are severely affected by human activity and pollution, which greatly affect the aquatic environments that are very sensitive to different contamination typologies (Mille et al. 1998). Various forms of physical, chemical and biological contaminants are reported in polluted waters (Dhir 2013a). The major part of all pollution comes from terrestrial and coastal areas, either through runoff and discharges via water ways including the river influx (44%) or through the atmosphere, mainly due to flaring from offshore oil and gas activity and the entering to sea through atmospheric transport (33%) (IMO Center 2012). Only 12 percent of all pollution are due to maritime activity and shipping accidents, garbage and sewage dumping (10%), as well as the consequences of offshore drilling and mining (1%) (Clark et al. 1989, Roose et al. 2011, IMO Center 2012).

Many harmful contaminants find their way into the marine environment, being ecotoxicologically the most significant substances harmful to marine ecosystems (Tornero and Hanke 2016). It should be kept in mind that very few substances are added to the sea in a chemically pure state, but most are part of complex liquid or gaseous solutions (Kachel 2008). There are thousands of chemicals present in the marine environment due to anthropogenic activities (Gioia et al. 2011). Contaminants generally change marine ecosystem functioning by reducing productivity and increasing respiration; however, these effects vary according to the type of contaminant and the components of the system studied (e.g., particular trophic levels, functional groups or taxonomic groups) (Johnston et al. 2015). These xenobiotics can be divided into those organic and inorganic contaminants, according to their composition (Newman 2014).

Organic contaminants

Organic contaminants discharged to the aquatic environment are characterized by a high diversity with respect to their molecular structures and the resulting physic-chemical properties (Schwarzbauer 2006). Organic compounds are crudely divided into aromatic compounds include benzene or similar compounds in their structure and aliphatic compounds composed of carbon chains with various bonds and other non-aromatic structures (Newman 2014). Aquatic plants irrespective free-floating, submerged and emergent including seagrasses and algae possess immense potential for remediation of various organic contaminants (Dhir 2013b).

Polycyclic aromatic hydrocarbons (PAHs)

The PAHs are classified depending on their origin (Dahle et al. 2003, Hylland 2006). There are three major types: petrogenic, biogenic and pyrogenic (Dahle et al. 2003). Pyrogenic PAHs are formed by incomplete combustion of organic compounds, such as fossil fuels (Abdel-Shafy and Mansour 2016), while petrogenic PAHs are present in oil and by-products, closely related to shipping activities and sewage input (Dahle et al. 2003, Li et al. 2015b). Although petrogenic PAHs appear to be bioavailable to a large extent, pyrogenic PAHs are often associated with soot particles and less available for uptake by organisms (Hylland 2006). The third type, biogenic PAHs can be produced biologically by certain plants, fungi and bacteria or formed during the organic matter decomposition, and include perylene and retene (Blackman 1986, Abdel-Shafy and Mansour 2016). Polycyclic aromatic hydrocarbons have a relatively low solubility in water, but are highly lipophilic, and therefore interfere with cell membrane fluidity (Van Brummelen TC 1998, Boehm 2004). Consequently, the concentration of these compounds in seawater and sediments undoubtedly carries toxicological significance for both benthic and pelagic marine organisms (Wetzel and Van Vleet 2004). In ecotoxicologically terms, PAHs are carcinogenic and mutagenic to aquatic animals and humans (Pampanin and Sydnes 2013, Rubio-Clemente et al. 2014). Petrogenic, biogenic and pyrogenic entering the water system can first accumulate in sediments (Meador et al. 1995, Abdel-Shafy and Mansour 2016). They remobilize later in seawater, becoming bioavailable and finally accumulated in some marine biota such as gelatinous zooplankton (Almeda et al. 2013) and corals *Acropora* sp. and *Montipora* sp. (Ko et al. 2014). Inevitably polycyclic aromatic hydrocarbons will also be bioaccumulated through the food chain (Li et al. 2015a). Despite the toxic effect of those compounds, some species of algae and seagrasses have shown the ability to accumulate and biodegrade those contaminants; this capacity could play an important role in the fate of PAHs in coastal and estuarine systems. Algae are widely distributed in aquatic environments which may be a major sink for degradation and/or transformation of PAHs. These organisms oxidize PAHs under photoautotrophic conditions to form hydroxylated intermediates (Mueller et al. 1996). According to Hong et al. (2008), the brown diatom algae *Skeletonema costatum* and *Nitzschia* sp. are capable of accumulating and degrading phenanthrene and fluoranthene (two typical PAHs). Also, in a recent study, for the first time, the simultaneous removal, bioaccumulation and degradation of a mixture of three PAHs (phenanthrene, fluoranthene and pyrene) by the marine algae *Rhodomonas baltica* were observed (Arias et al. 2017). The transformation of PAHs in marine and freshwater algae is species specific and depends on the presence and activity of enzymes localized in the plant cells. The most important enzyme systems for detoxification are *o*-diphenol oxidase, cytochrome P450 and peroxidase (Kirso and Irha 1998). Previous studies on bioconcentration and transformation of PAHs by algae, have demonstrated that the removal of those organic contaminants results from various physicochemical processes, transformation, accumulation, metabolization, abiotic transformation through photodegradation, volatilization, sorption and adsorption (Kirso and Irha 1998, Matamoros et al. 2015, Ghosal et al. 2016, Gupta et al. 2017). Concerning seagrasses in general, when exposed to petrochemicals and PAHs at sub-lethal concentrations, the contaminants are incorporated into the tissue, resulting in reduced tolerance capacities to other stress factors, suffocation, reduce growth rates and lower flowering rates (Zieman et al. 1984, Kenworthy et al. 1993, Dean et al. 1998). Several studies were performed to investigate the PAHs concentration in tissues of different seagrass species and evaluated the use of these organisms as bioindicators, as well as for biodegradation and bioaccumulation purposes: *Posidonia oceanica* (Pergent et al. 2011, Apostolopoulou et al. 2014), *Thalassia testudinum* and *Halodule wrighti* (Lewis et al. 2007), *Syringodium filiforme* and *Thalassia testudinum* (Bouchon et al. 2016). Despite the toxic effects of PAHs, these studies revealed that the accumulation of those contaminants does not seem to be an environmental threat for seagrass beds, with some species showing a good potential as bioindicator. Effectively, there are numerous examples of PAHs contamination having no significant impact on the various seagrass after several years (see Kenworthy et al. (1993) and references herein).

Persistent organic pollutants (POPs)

Persistent organic pollutants (POPs) constitute a group of organic chemicals that contain bound chlorine or bromide atoms (Haynes and Johnson 2000). Among the most important classes of persistent organic pollutants are many families of chlorinated and brominated aromatics, including polychlorinated biphenyls (PCBs), dioxins as polychlorinated dibenzo-p-dioxins and-furans (PCDD/Fs) and different organochlorine pesticides (DDT and it's metabolites, toxaphene) (Jones and de Voogt 1999). Their synthesis is mostly directed towards industrial or agrochemicals uses (Jones and de Voogt 1999), while few may have natural origin, for example from volcanic eruptions (WHO 2009). Persistent organic pollutants enter the marine realm by two main ways: either by atmospheric deposition by way of dry or wet sedimentation or via surface effluent release, for the offshore areas, this being the primary route by which POPs enter the marine environment through riverine input (Song 2011). Under the Stockholm Convention, 90 signatory countries have agreed to reduce and/or eliminate the production, use, and release of the 12 POPs (e.g., aldrin, chlordane, DDT, dieldrin, heptachlor, mirex, etc.) of greatest concern to the global community (USEP Agency 2002). This refers to a group of substances that to varying extents resist photolytic, biological and chemical degradation, characterized by low water solubility and high lipid solubility, leading to their bioaccumulation in fatty tissues (IPCS 1995). For example, the accumulation of PCBs and DDT in *Selenastrum capricornutum* increases with the increase in total algal lipid content (Halling-Sørensen et al. 2000). The potential collateral effect of POPs to the nontarget organisms is not well understood as such contaminants behave differently individually, whereas the toxicity increases/decreases several folds due to synergistic and antagonistic effects of co-occurring POPs (Newman 2014). Also, they possess a very high bioaccumulation potential in aquatic and terrestrial organisms including humans (Beek 1999). Persistent organic pollutants (POPs) could hardly be degraded by microorganism under natural circumstances, since are biodegraded at very slow rates (Song 2011). In marine environment, different living species possess the ability to accumulate POPs, such as the two species of seagrasses *Thalassia testudinum* and *Halodule wrighti* (Lewis et al. 2007). As well *Chlorella* sp. (Geyer et al. 2000). Marine diatoms (*Phaeodactylum tricornutum, Thalassiosira nordenskio, Skeletonema costatum*) and algae (*Emiliana huxleyi*) as a consequence of their large surface area and significant organic carbon content, they sorb PCBs (Gerofke et al. 2005).

Several factors influence the amount of POPs accumulated in seagrasses and algae. Climate change will have an effect on the environmental fate and behaviour of contaminants by altering physical, chemical and biological drivers of partitioning between atmosphere, water, soil and biota, including reaction rates. For example, increasing temperatures and subsequent melting of snow and ice can remobilize POPs into water and the atmosphere (Noyes et al. 2009).

Emerging organic contaminants (EOCs)

Emerging and newly identified contaminants are a major concern for marine ecosystem (Jiang et al. 2014, Geissen et al. 2015). This term is used to cover not only newly developed compounds, but also newly compounds discovered in the environment often due to analytical developments (Richardson and Ternes 2011, Sauvé and Desrosiers 2014). The contaminants of emerging concern (CEC) can occur naturally, be manufactured or manmade or can be new materials which have now been discovered or are suspected to be present in various environmental compartments and whose toxicity or persistence are likely to significantly alter the metabolism of a living being (Sauvé and Desrosiers 2014). A diverse array of synthetic organic compounds are used by society in high amounts for a range of purposes including the production and preservation of food, industrial manufacturing processes and for human and animal healthcare (Lapworth et al. 2012). With effects ranging from acute to chronic in aquatic organisms, its accumulation in the ecosystem leads inevitably to losses of habitats and biodiversity, as well as threats to the human health (Sánchez-Avila et al. 2012). An increasing number of emerging contaminants studies are arising in the ecology and ecotoxicology literatures (Geissen et al. 2015, Thomaidi et al. 2015, Hurtado et al. 2017, Pintado-Herrera et al. 2017).

Endocrine-disrupting chemicals (EDCs)

According to the World Health Organization, endocrine disrupting chemicals EDCs and potential EDCs are mostly man-made, found in various materials such as pesticides, metals, additives or contaminants in food and personal care products (WHO 2013). They include insecticides, fungicides, herbicides, pharmaceuticals and industrial contaminants (Hotchkiss et al. 2008). Endocrine-disrupting chemicals are defined as contaminants with the ability to interfere with the endocrine system of different organisms causing important alterations in their normal development (Álvarez-Muñoz et al. 2016). Numerous studies on aquatic organisms such as algae and different types of seagrass have been recently explored for the removal of organic pollutants possessing endocrine disrupting capacity. Eliana Gattullo et al. (2012) observed that after 2–4 days of bisphenol exposure at low concentrations, there were no toxic effects detectable for the green alga *Monoraphidium braunii*, whilst a good removal efficiency was detected. In addition, the study of Solé and Matamoros (2016) indicated that the use of co-immobilized microalgae-based wastewater treatment systems increased the removal efficiency for nutrients and some EDCs from wastewater effluents. In recent years, several species of marine mammals and birds have been affected by EDCs (Tanabe 2002). Mangroves and seagrasses are known to support very high densities juvenile reef fish species and likely to offer enhanced survival for these individuals compared to those living in other habitats (Nagelkerken 2009). Seagrassbeds help to trap sediments as they settle out in the calm waters among shoots and roots, and can also play a role in the active removal of pollutants dispersed into adjacent waters (Waycott et al. 2009). For example, the accumulation of Diuron (herbicide used to control a wide variety of annual and perennial broadleaf and grassy weeds) by *Zostera capricorni* and *Halodule uninervis*, oftenly occurs in higher concentrations in plant tissues than sediment concentrations (Haynes et al. 2000). This filtration potential is thus recognised as an essential ecosystem service for protecting juvenile fish, benthic organisms and nursery areas (Burke et al. 2011).

Pharmaceuticals and personal care products (PPCPs)

Pharmaceuticals and personal care products (PPCPs) is currently one of the most concerned new forms of pollution (Xie 2017). Many (PPCPs) were found in aquatic environment, including personal care products (e.g., detergents and musks), pharmaceuticals (e.g., drugs, veterinary and human antibiotics and birth control substances) (Claessens et al. 2013, Dhir 2013b, Newman 2014). Personal care products, pharmaceuticals compounds and their bioactive metabolites are continually introduced to the aquatic environment as complex mixtures via a number of routes, but mostly by both untreated and treated sewage (Daughton and Ternes 1999). An increasing number of studies has confirmed the presence of various PPCPs in different environmental compartments, which raises concerns about the potential adverse effects to humans and wildlife (Ebele et al. 2017). Simvastatin, clofibric acid, diclofenac, carbamazepine, fluoxetine and triclosan represent some of the most commonly used and/or detected PPCPs in aquatic environments (DeLorenzo and Fleming 2008). Biodegradation, photodegradation and sorption are the main processes involved in PPCP removal during treatment but also depending on the properties of the compounds (Wang et al. 2017). There is a growing body of literature reporting the effects of PPCPs on freshwater organisms, but studies on the effects of PPCPs to marine and estuarine organisms are more limited (Prichard and Granek 2016). The study of Bai and Acharya (2017) evaluated the removal of five common PPCPs (trimethoprim, sulfamethoxazole, carbamazepine, ciprofloxacin and triclosan) from Lake Mead water mediated by the green alga *Nannochloris* sp., showed that the algae-mediated sorption contributed to 11 percent of the removal of trimethoprim and sulfamethoxazole, 13 percent of carbamazepine and 27 percent of the removal of triclosan from the lake water. Similar research studies were performed to investigate the capability and mechanisms of PPCPs removal by algae in laboratory conditions which showed a high ability to remove many PPCPs; for example, the biodegradation of 17b-estradiol by *Selenastrum capricornutum* and

Chlamydomonas reinhardtii (Hom-Diaz et al. 2015), biodegradation of Norgestrel by *Scenedesmus obliquus* (Peng et al. 2014), the biotransformation of Estriol by *Scenedesmus dimorphus* (Zhang et al. 2014) and the hydrolysis, photolysis and adsorption of 100 percent of 7-amino cephalosporanic acid by *Mychonastes* sp. in five days (Guo et al. 2016). However, several PPCPs exhibited low removal rates (carbamazepine, lorazepam, trimethoprim and verapamil) which can be attributed to their low biodegradability (Wang et al. 2017).

Surfactants

Surfactants are one of the most ubiquitous families of organic compounds (Cirelli et al. 2008). Although their main application is the formulation of household and industrial detergents, they are also used in the manufacturing of other products such as cosmetics and personal care products, paints, textile, dyes, polymers, agrochemicals and petroleum (Jackson et al. 2015, Álvarez-Muñoz et al. 2016). The global surfactant market volume size is more than 18 million tons per year; large quantities are continuously released into the environment, where they can or cannot be degraded depending on their structure (Cirelli et al. 2008). Surfactants contain both hydrophobic groups (their "tails") and hydrophilic groups (their "heads") making them soluble in both organic solvents and water; they are classified into nonionic, anionic, cationic or zwitterionic by the presence or absence of formally charged hydrophilic head groups (Cowan-Ellsberry et al. 2014). These compounds enter the aquatic environment after use, often going via sewage treatment works, where they may undergo biological degradation before entering rivers and other waters (Sumpter et al. 1996). The most-used surfactants and their acronyms are shown in Table 1.

Table 1. Acronyms of the most widely used surfactants (Ying 2006, Cirelli et al. 2008).

Common name	Acronym
Anionic surfactants	
Linear alkyl benzene sulfonates	LAS
Secondary alkane sulfonates	SAS
Alcohol ether sulfates (Alkyl ethoxy sulfates)	AES
Alcohol sulfates (Alkyl sulfates)	AS
Nonionic surfactants	
Alkylphenol ethoxylates	APE (or APEO)
Nonyl phenol ethoxylates	NPE (or NPEO)
Octyl phenol ethoxyales	OPE (or OPEO)
Alcohol ethoxyaltes (Alkyl ethoxyaltes)	AE (or AEO)
Cationic surfactants	
Quaternary ammonium-based compounds	QAC
Alkyl dimethyl (and trimethyl) ammonium halides	DMAC (and TMAC)
Alkyl benzyl dimethyl ammonium halides	BDMAC
Dialkyl dimethyl ammonium halides	DADMAC
Dihydrogenated tallow dimethyl ammonium chloride	DHTDMAC or DTDMAC
Ditallow trimethyl ammonium chloride	DTTMAC
Diethyl ester dimethyl ammonium chloride	DEEDMAC
Zwitterionic surfactants	
Cocamidopropyl betaine	CAPB
Cocamidopropyl hydroxysultaine	CAHS

It is also likely that surfactants, and their degradation products, bioaccumulate and/or bioconcentrate in aquatic organisms (Sumpter et al. 1996). Various species of algae (*Scendesmus quadriauda, Chlorella vulgaris, Ankistrodesmus acicularis, Chroococcus minutes* and the diatom *Navicula incerta*) have shown a ability for the removal and biodegradation of nonylphenol (NPs) (Liu et al. 2010, He et al. 2016). Also, the seagrass *Posidonia oceanica* biomass showed a high potential for adsorptive removal of anionic (NaDBS) and non-ionic (TX-100) surfactants from aqueous solutions. Ncibi et al. (2008) found that as a biological adsorbent to remove anionic and non-ionic surfactants, *Posidonia oceanica* biomass seems to be a promising technique.

Inorganic contaminants

The presence of inorganic contaminants is very common in the marine environment and polluted waters; these mainly include metals, ions/nutrients and radionuclides (Dhir 2013b). Metals are present in the marine ecosystems within its different compartments: water column, sediments and marine biota, where they are accumulated directly from water or food (Barka 2012). Nevertheless, many metals are released naturally into the environment through weathering and leaching processes, with human activities inducing severe shifts in these elements estuarine and coastal budgets (Karouna-Renier et al. 2007). Two metals types can be distinguished according to their physiological and toxic effects: the essential elements as macroelements (N, P, K, Mg, S, Ca) or microelements (Zn, Fe, Mn, Cu, B, Mo, Co, Ni, Cl, V), which are essential for plant growth in low concentrations (Prasad 2001, Mitra 2017); however, beyond certain threshold concentrations they become toxic (Mitra 2017). Some common micronutrients required by both plants and animals include aluminum (Al), boron (B), bromine (Br), chromium (Cr), cobalt (Co), copper (Cu), fluorine (F), gallium (Ga), iodine (I), iron (Fe), manganese (Mn), molybdenum (Mo), selenium (Se), silicon (Si), strontium (Sr), tin (Sn), titanium (Ti), vanadium (V), and Zinc (Zn) (Pidwirny 2017). Other elements are non-essential, namely arsenic (As), cadmium (Cd), mercury (Hg) and lead (Pb) and are not required but they have been studied extensively due to their potentially hazardous effects to plants, animals and microorganisms (Stevenson 1999, Adriano 2001). In fact, beyond a given threshold, all metals, whether essential or not, may have toxicological and ecotoxicological effects (Barka 2012). In different marine organisms, the behaviour of heavy metals is described in terms of their absorption, storage, excretion and regulation when different concentrations are available in the environment (Bryan 1971) as well as the route and duration of exposure, that is, acute or chronic exposures (Jaishankar et al. 2014). In polluted marine ecosystems, some types of seagrass and algae show a high capacity to absorb heavy metals as results of tolerance mechanisms such as the uptake of (Cd, Cu, Mn, Ni, Pb, Zi and Hg) by marine algae (*Centrocerous clavutum, Sargassum wightii, Colpemenia sinosa, Spyridia hypnoides, Valoniopsis pachynema, Ulva reticulata, Gelidialla acerosa* and *Turbinaria ornate*) and seagrasses (*Syringodium isoetifolium & Cymodocea serrulata*) (Sudharsan et al. 2012). Also *Halodule pinifolia* showed the capacity to absorb nickel (1.1 ppm) higher than the threshold value limit (0.02 ppm) for water medium (Espiritu and Paz-Alberto 2018). The uptake of metals in seaweeds and seagrass depends on the surface reaction in which metals absorbed through electrostatic attraction to negatives sites, this being independent of factors influencing metabolism such as temperature, light, pH or age of the plant, but it is dependent on the virtual abundance of elements in the surrounding water (Sánchez-Rodríguez et al. 2001). Concerning ions/nutrients, although in a strict sense not as toxic as the pollutants discussed above, nutrients can have severely damaging effects on the marine environment (Kachel 2008). The major ions present in contaminated water are ammonia, nitrite, nitrate, chloride, sulphate, phosphorus and cyanide (Dhir 2013b). These result mostly from nutrient application as fertilizers in agriculture and correspondent runoff to the sea due by riverine inputs (Kachel 2008). Contributions of high levels of nitrogen and phosphorus compounds, in particular, often result in eutrophication (Kachel 2008, Chislock et al. 2013). In India, the seagrass (*Cymodocea rotundata*) was used as biosorbent to remove excessive N and P from aqueous solution and dye from the textile wastewater. This study revealed that this species has a high potential in removing N and P from aqueous solution (Vasanthi et al. 2015).

Recently, there has been an enrichment of specific radionuclides in the environment due to manufacture and testing of nuclear weapons, extensive construction of nuclear power plants, commercial fuel reprocessing, nuclear waste disposal, uranium mining and enrichment and nuclear accidents (Mitra 2017). The term radiation refers to the propagation of energy through space, this being in the form of photons, ejected atomic nuclei or their fragments or subatomic particles. Additionally, radiation can also be generated within the nuclei of unstable elements: such nuclei are called radionuclides (Newman 2014). Exposure of biota to radiation and transfer of radionuclides in the environment are intimately linked, and occurs when radionuclides, present naturally in the environment or released through man's activities, decay releasing radiation of various types and energies (Brown et al. 2009). There are three natural and one artificial series of radionuclides in which one (parent) radionuclide decays to another (progeny) in a series of steps until the most stable element is formed, the series include the uranium–radium, thorium, actinium and neptunium series (Newman 2014). After the Fukushima disaster in Japan in 2011, many studies discussed the effects of radionuclides releases from the accident site in marine biota (Buesseler et al. 2012). Some of these studies detected Cs isotopes in zooplankton and mesopelagic fish, and also ^{131}I and [134,137]Cs in seaweed, mollusks and fish (i Batlle and Vandenhove 2014). According to Moss (1973), the presence of radionuclides in the marine environment is possibly due to radioactive discharges resulting from the practice of nuclear medicine.

Biology and Ecology of Seagrasses and Algae

Seagrass

Seagrass are a functional group of about 60 species of underwater marine flowering plants monocotyledons. They often grow in dense beds and extensive meadows creating a productive and diverse habitat used as shelter, nursery, spawning or food area by a large variety of animal species (Green and Short 2003). These

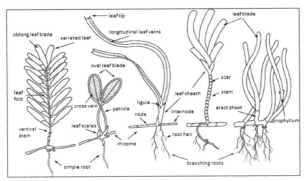

Fig. 1. Composite illustration demonstrating morphological features used to distinguish main seagrass taxonomic groups (McKenzie et al. 2009).

plants occupy sandy, muddy and rocky substrates in every sea in the world (Denny and Gaines 2007). As reported by Duarte (2002), seagrasses cover about 0.1–0.2 percent of the global ocean. They range from the tiny, 2–3 cm, rounded leaves of sea vine (e.g., *Halophila decipiens*) to the strap-like blades of eelgrass (e.g., *Zostera caulescens*) at more than 4 meter long (Green and Short 2003), the world's longest seagrass reaching more than 7 metres (Aioi et al. 1998). Like terrestrial plants, seagrass have leaves, roots, conducting tissues, flowers and seeds. Unlike terrestrial plants, however, seagrasses do not possess the strong, supportive stems (Fig. 1). Rather, seagrass blades are supported by the natural buoyancy of water, remaining flexible when exposed to waves and currents (Denny and Gaines 2007). Seagrass accomplish their underwater reproduction by producing filamentous pollen grains that can be transported by water currents (Mitra and Zaman 2015). Due to their morphology, seagrasses are sometimes confused with marine macroalgae (seaweeds) (Levinton 1995, Mitra and Zaman 2015). In contrast to seagrass categorized as vascular, seaweeds have little or no vascular tissues and they lack a true root system as they have holdfasts (Lobban and Harrison 1994). Additionally, seaweeds have spores and do not flower or produce fruit, while seagrasses produce flowers, seeds and fruit (Denny and Gaines 2007, Larkum et al. 2007). Seagrass provide a habitat for fish and shellfish and nursery areas to the larger ocean, and perform important physical functions of filtering coastal waters, dissipating wave energy and anchoring sediments (Green and Short 2003). A reduction in seagrass vigorous growth and production, and consequent major reductions in their area, will result in serious changes to coastal ecosystems. Artificial seagrass restoration projects increased fish and shrimp in number and diversity and provided feeding areas, highlighting their essential ecosystem role (Talbot and Wilkinson 2001).

Algae

Seagrasses do not grow in isolation but form an integral and often defining part of highly complex ecosystems. Although they themselves are an important standing stock of organic matter, the productivity of these ecosystems is usually enhanced by other primary producers, including macroalgae epiphytic and free-living microalgae

(Green and Short 2003). There is considerable variability in the organization of algae, from unicellular organisms only a few microns in size to thalli of complex structure, such as the large phaeophycean kelp *Macrocystis*, which can reach 100 meters in length. Nevertheless they all have rudimentary conducting tissues (Gallardo 2014). Algae can be classified into two main groups; first one is the microalgae, which includes blue green algae, dinoflagellates, bacillariophyta (diatoms), etc., and second one composed by macroalgae (seaweeds) which includes green, brown and red algae (El Gamal 2010). In contrast to seagrass, marine algae supports a great diversity of species. They have been estimated to include anywhere from 30,000 to more than one million species, most of which are marine algae (Guiry 2012). However, they are more primitive than seagrasses (Mellors and McKenzie 2008). Algae are biochemically and physiologically very similar to what is known as higher plants: they essentially have the same metabolic pathways, possess chlorophyll and produce similar proteins and carbohydrates (Gallardo 2014).

Accumulation of Contaminants

Aquatic plants have been shown to play important roles in wetland biogeochemistry through their active and passive circulation of elements (Vymazal and Kröpfelová 2008). In contaminated marine environment, several flora species have shown to have the ability to tolerate and accumulate organic and non-organic pollutants, and in some cases present concentrations much higher than those present in water and sediments (Singh et al. 2002, Ayangbenro and Babalola 2017). Contaminants are deposited in the various components of the aquatic environment (water, sediment and biota) and can be accumulated through the food chain (Binelli and Provini 2003). In natural environments, organisms living in chronically polluted sites are exposed to low concentrations of contaminants for long periods; in the other cases, the organism may be abruptly exposed to high levels contaminants upon the outfall of a pollutant in coastal waters (Torres et al. 2008). The adaptability of these species depends mainly on the type of pollutant, its concentration, duration of exposure and some physicochemical parameters of water (Pérez-Ruzafa et al. 2000, Lawson 2011).

Bioconcentration and Bioaccumulation Factor

Bioconcentration is the result of the direct uptake of a chemical by an organism only from water and the result of such a process is measured by the bioconcentration factor (BCF), which represents the ratio of steady state concentration of the respective chemical in the biota (mass of chemical per kg of organism dry weight) and the corresponding freely dissolved chemical concentration in the surrounding water (mass of chemical per unit of volume) (Geyer et al. 2000). Concerning the Biota-Sediment bioaccumulation factor (BAF or BSAF) is defined as the ratio between the contaminant concentration in the biota (mass of chemical per kg of biota dry weight) and that in the sediment (mass of chemical per kg of sediment dry weight) (Nenciu et al. 2014, Newman 2014). According to the definition given by the USEPA, a substance is considered bioaccumulative when it has a BCF ranging between

1000 and 5000 and very bioaccumulative if it has a BCF greater than 5000 (TSCA 2012). However, the European Chemicals Legislation (Registration, Evaluation, Authorization and Restriction of Chemicals, R.E.A.CH) stipulates that a substance fulfils the bioaccumulation criterion when the bioconcentration factor (BCF) is higher than 2000 (EC 2006). Regarding (BAF or BSAF) the different species of marine organisms can be classified into three groups, such as macroconcentrator (BSAF > 2), microconcentrator (1 < BSAF < 2) or deconcentrator (BSAF < 1) (Dallinger 1993). Several studies, generally *in situ*, have been carried out concerning the use of seagrasses and marine algae for the decontamination of marine and aquatic environment (Table 1). The bioconcentration factor (BCF) and the biota-sediment bioaccumulation factor (BAF) cited in Table 2, show that some species of seagrass and algae represent an enormous potential for removal of a variety of contaminants, including heavy metals, inorganic/organic pollutants by phytoremediation.

Phytoremediation

The use of plants directly or indirectly to remediate contaminated water is known as phytoremediation. This technology has emerged as a cost effective, noninvasive and publicly acceptable way to address the removal of environmental contaminants. Plants can be used to accumulate inorganic and organic contaminants, metabolize organic contaminants and encourage microbial degradation of organic contaminants in the root zone (Arthur et al. 2005). Different techniques can be distinguished within phytoremediation of polluted water according to their objectives, the characteristics of the environment, the contaminants and the used species: phytoextraction, phytodegradation (rhizodegradation), phytostabilization rhizofiltration (Brooks 1998, Salt et al. 1998, Kirkham 2006). Treatment of organic contaminants mainly involves phytostabilization, phytodegradation (rhizodegradation) and rhizofiltration. While phytostabilization, rhizofiltration and phytoaccumulation mechanisms are involved in the treatment of inorganic contaminants (Dhir 2013b). Rhizofiltration is a form of phytoremediation is a combination of phytoextraction and phytostabilisation, which refers to the approach of using hydroponically cultivated plant roots to remediate contaminated water through absorption, concentration and precipitation of pollutants. Due to their morphology, seagrass are more adapted for rhizofiltration (Salt et al. 1998, Alkorta et al. 2004). Phycoremediation is defined as the "use of algae to treat wastes or wastewaters". The algae comprise both the microalgae as well as the marine macroalgae, more commonly known as the seaweeds (Phang et al. 2015). The phytoremediation in marine ecosystem is affected by several factors including, species, biomass (the high-biomass possess higher contaminant removal potential) and physicochemical parameters of water and contaminant (Dhir 2013a).

Tolerance and Detoxification Mechanisms of Inorganic Contaminants

Plants have many endogenous genetic, biochemical and physiological properties to tolerate pollution, that make them ideal agents for soil and water remediation (Meagher 2000). In aquatic environments, the contaminant uptake from water

Table 2. Bioconcentration factors (BCFs) and bioaccumulation factors (BSAF) biota-sediment accumulation factor of contaminants in seagrasses and algae.

Species	Contaminant	BAF and BCF	References
Seagrasses			
Zostera japonica	As, Cd and Mn	$BAF_{As} = 3.43$; $BAF_{Mn} = 6.43$; $BAF_{Cd} = 30.95$	(Lin et al. 2016)
Halodule wrightii	Triazine	$BCF_{max} = 21634$ $BCF_{average} = 6681$	(Fernandez and Gardinali 2016)
Syringodium filiforme	Triazine	$BCF_{average} = 6809$	(Fernandez and Gardinali 2016)
Posidonia oceanica	Ni	$BCF_{Ni} = 59000$ $BAF_{Ni} = 2,7$	(Joksimovic and Stankovic 2012)
Mytilus galloprovincialis	As	$BCF_{As} = 88000$ $BAF_{As} = 1,7$	(Joksimovic and Stankovic 2012)
Holodule uninervis	PCDD	$BAF = 1.73$	(McLachlan et al. 2001)
Zostera marina	Total PAHs Total PCBs	$BAF_{root} = 3$ $BAF_{root} = 4.2$	(Huesemann et al. 2009)
Zostera marina	Triazine	$BCF = 25000$	(Scarlett et al. 1999)
Algae			
Phytoplankton (eight species)	Cr, Pb and Cu	$BCF_{Cr} = 4810$ $BCF_{Pb} = 24500$ $BCF_{Cu} = 1710$	(Tao et al. 2012)
Ulva spp. (macroalgae)	Pb	$BCF_{Pb} = 0,41–6,05$ $BAF_{pb} = 1,45–1,82$	(Jitar et al. 2015)
Ulva fasciata (macroalgae)	DDTs and PBDEs	$BCF_{DDTs} = 17,2.10^3$ $BCF_{PBDEs} = 6.10^3$ $BAF_{DDTs} = 0.25$ $BAF_{PBDEs} = 0.26$	(Qiu et al. 2017)
Phytoplankton species (*Chaetoceros lorenzianus, Nitzschia, Bacteriastrum hyalinum and Thalassionema nitzschioide*)	DDTs and PBDEs	$BCF_{DDTs} = 1642,7.10^3$ $BCF_{PBDEs} = 575,2.10^3$ $BAF_{DDTs} = 23.5$ $BAF_{PBDEs} = 24.6$	(Qiu et al. 2017) (Arias et al. 2017)
	PAHs: Phenanthrene	$BCF = 7.51$	
	Fluoranthene	$BCF = 12,89$	
Rhodomonas baltica	Pyrene	$BCF = 16.51$	

(directly or from adsorption/accumulation in food) is counteracted by endogenous enzymatic biotransformation and elimination processes. Hydrologic, geochemical and environmental conditions can also strongly affect bioaccumulation of a particular contaminant (either inorganic or organic) by interfering in its bioavailability in the aqueous phase (Guha 2004). According to Dhir (2013b), the presence of inorganic contaminants is very common in marine environment and polluted waters; these mainly include metals, ions/nutrients and radionuclides. Reactive oxygen species (ROS) generation derived from contaminant exposure leads to the inhibition or retardation of growth and cell division, impairment of photosynthetic activity and cell death (Di Toppi and Gabbrielli 1999). Bacteria and higher organisms have developed resistance mechanisms to toxic metals to make them innocuous. Organisms respond to heavy metal stress using different defence systems, such as exclusion, compartmentalization or complexation with binding proteins such as metallothioneins (MTs) or phytochelatins (PCs) and consequent translocation into vacuoles (Mejáre and Bülow 2001). Metal removal mechanism is a complex process that depends on the chemistry of metal ions, cell wall compositions, physiology of the organism as well as physicochemical factors like pH, temperature, time of exposure, ionic strength and metal concentration (Mishra et al. 2010). In this part, we paid a special attention to tolerance and detoxification mechanisms of heavy metals in seagrasses and marine algae.

Biosorption

Biosorption is a promising technology which pays attention to fabricate novel, cheap (low-cost) and highly-effective materials to apply in wastewater purification technology (Anastopoulos and Kyzas 2015). The term biosorption is employed to describe the range of processes by which biomass removes metals and other substances from solution, yet it can also be used in a stricter sense to mean uptake by dead or living biomass by purely physico-chemical processes such as adsorption or ion exchange (White et al. 1995). Biosorption is a method that can be used for the removal of pollutants from wastewater, especially those that are not easily biodegradable like heavy metals and industrial dyes (Mata et al. 2008). The use of a biosorbent such as dried algae and seagrass biomass in wastewater treatment often depends not only on the biosorptive capacity, but also on how well the biosorbent can be regenerated and recycled (Pennesi et al. 2015). The biosorbent ability of macrophytes (i.e., seaweeds and seagrasses) resides in the structure of the cell walls and cuticles of seagrasses and macroalgae (Flouty and Estephane 2012, Wang et al. 2013). Also micro-algal cell wall have a high binding affinity for metal cations via counterion interactions (Crist et al. 1981). The algal cell wall matrix contains different functional groups, such as, hydroxyl (OH), phosphoryl (PO_3O_2), amino (NH_2), carboxyl (COOH), sulphydryl (SH) and other charged groups, which are generated by complex heteropolysaccharides and lipid components which favor sequestration of positively charged molecules (Crist et al. 1981, Niu and Volesky 2000, Mohan et al. 2002, Daneshvar et al. 2007) (Table 3). The biosorption capacity of different chemical groups depends on several factors: the number of sites on the adsorbent material, accessibility, their chemical status (availability) and the affinity between site and metal and contaminants (bond strength) (Vieira and Volesky 2000). Proteins, lipids and occasionally, nucleic acids may also be present on the surface of the cell walls. These molecules, however, occur mainly in the plasma membrane and in the cytoplasm and therefore, are bound to metal ions through their functional groups (aminic, carboxylic, imidazolic, thiolic, thioesteric, nitrogen and oxygen in peptidic bonds) mostly within the cell (Majidi et al. 1990). The accumulation of heavy metals involves two processes: an initial rapid (passive) uptake (its a physical adsorption between the metal ions and the algae surface), followed by a much slower (active) uptake (chemical adsorption) (Bates et al. 1982, Garnham et al. 1992). During the passive uptake, metal ions adsorb onto the cell surface within a relatively short span of

Table 3. Binding groups for biosorption in some seaweeds and seagrasses (Pennesi et al. 2015).

Binding group	Chemical formula	Ligand atom	Occurrence in biomolecules	Phylum
Amine	$-NH_2$	Hydrogen	Protein Metallothionein	Chlorophyta Magnoliophyta
Carboxyl	$-COOH$	Oxygen	Cutin Alginic acid	Magnoliophyta Heterokontophyta
Carbonyl	$>C=O$	Oxygen	Alginic acid Protein Alginic acid	Heterokontophyta Chlorophyta Heterokontophyta
Hydroxyl	$-OH$	Oxygen	Polysaccharides Cutin Agarose (agar), carrageenan Fucoidan (sulfated polysaccharide) Agaropectin (agar), carrageenan	Chlorophyta Magnoliophyta Rhodophyta Heterokontophyta
Sulfonate	$-SO_2=O$	Oxygen	(Sulfated polysaccharides), porphyran, furcellaran (sulfated galactans)	Rhodophyta
Sulfhydryl (thiol)	$-SH$	Sulfur	Metallothionein	Magnoliophyta

time (few seconds or minutes), and the process is metabolism independent. However, the active uptake is metabolism-dependent, causing the transport of metal ions across the cell membrane into the cytoplasm (Gadd 1988). Biosorption of metals by algae may be affected by several factors, including concentration of metals and biomass, pH, temperature, the presence of competing ions and contact time (Dixit and Singh 2015). Feng and Aldrich (2004) indicated that marine alga *Ecklonia maxima* can be used as an efficient biosorbent material for the treatment of aqueous waste streams contaminated with Cu, Pb and Cd. Also, the *Sargassum* species showed a high efficiency of copper biosorption (Carsky and Mbhele 2013). *Cystoseira crinita* for Ni (II) and *Cystoseira barbata* for Cu (II) (Simeonova and Petkova 2007). The principal mechanism of metallic cation sequestration involved the formation of complexes between a metal ion and functional groups (carboxyl, carbonyl, amino, amido, sufonate, phosphate, etc.) present on the surface or inside the porous structure of the biological material (Fourest and Volesky 1997). It was shown that the presence of alginic acid in the cell wall played a major role in the complexation of Cd and Pb in *Sargassum fluatan* (Fourest and Volesky 1997), and also for Pb^{2+}, Cu^{2+}, Zn^{2+} and Cd^{2+} in *Sargassum filipendula* (Kleinübing et al. 2013). Effectively, the higher removal performance of those species is mainly explained by the presence of alginate (anionic polysaccharides) as one of the major components of brown algae (Kawai and Murata 2016). In red algae, the amorphous portion is formed by a number of sulfated galactans such as carrageenan, agar, furcellaran and porphyran. On the other hand, green algae may have an external capsule that is composed of protein or polysaccharides or both (Graham and Wilcox 2000). For seagrass, the surface of the leaves is covered by a layer of polymeric material called cuticle (cutin and cuticular waxes) (Smith et al. 2010). The thin cuticle facilitates uptake of ions and carbon; seagrasses are able to uptake nutrients and carbon directly through the leaves. In order to facilitate the exchange of solutes and gases, the cuticle of seagrasses is porous and perforated (Hemminga and Duarte 2000). The cuticle of seagrasses is also rich in carboxyl groups, which are involved in the bond with the protonated metal, while red and green algae have a lower capacity to immobilize protonated metals due to the presence of groups other than carboxylic ones (Table 2) (Pennesi et al. 2015). The recent research of Pennesi et al. (2013) suggest that biomass of *Posidonia oceanica* can be used as an efficient biosorbent for removal of vanadium (III) and molybdenum (V) from aqueous solutions. According to Pennesi et al. (2015), its assumed that the number of binding sites identified decreases in the order: brown algae > seagrasses > green algae > red algae. Biosorption is generally used for the treatment of heavy metal pollutants in wastewater; application of biosorption for organic and other pollutants could also be used for the treatment of wastewater (Tsezos and Bell 1989). An example of this is the bisorption of phenol by *Posidonia oceanica* fibres (Ncibi et al. 2006), and of 2–4 dichlorophenol onto *Posidonia oceanica* (Demirak et al. 2011). There are also reports of adsorption of phenol by *Sargassum* sp. and *Chaetomorpha* algae (Navarro et al. 2016) and the biosorption of methylene blue dye from wastewater by *Sargassum hemiphyllum* (Liang et al. 2017).

Phytochelatin and metallothionein

The consequences of xenobiotic bioaccumulation in a biological system are revealed at multiple hierarchical levels: from single organism effects to cross-linked trophic connections in the ecosystem as a whole (Newman and Unger 2003). Plants and algae have the ability to hyper accumulate various heavy metals by the action of phytochelatins and metallothioneins forming complexes with heavy metals and translocate them into vacuoles (Suresh and Ravishankar 2004). The mechanism of accumulation and adsorption of metals by algae involve adsorption onto the cell surface and binding to cytoplasmic ligands, phytochelatins and metallothioneins (Fig. 2) (Mehta and Gaur 2005). Phytochelatins are intracellular metal-binding peptides produced enzymatically by higher plants, fungi and algae in response to many metals (Ahner et al. 1995). Phytochelatins are polypeptides with the amino acid structure (y-glu-cys),-gly, where *n* ranges from 2 to 11 (Grill et al. 1985). They chelate metals through coordination with the reduced sulfur in cysteine (Ahner et al. 1995). Metallothioneins resemble phytochelatins in many ways structurally and functionally (Salt et al. 1995). However, metallothioneins are characterized by lower molecular weight (6–7 kDa) and result from mRNA translation, with a distinct enzymatic synthesis when compared with phytochelatins synthesis (Cobbett and Goldsbrough 2002, Romero-Isart and Vašák 2002). In plants, these molecules bind to free metal ions that carry them into vacuoles, where their toxicity is reduced, staying available for metabolic functions if the target metals are essential for growth (Murphy et al. 1997). Verbruggen et al. (2009) suggested that cellular vacuoles are the target compartments for sequestration phytochelatin-conjugated metals. Metal compartmentalization has been reported for several algae and seagrasses, for example, compartmentalization of Pb and Zn in *Chlorella saccharophila*, *Navicula incerta* and *Nitzschia closterium* (Jensen et al. 1982). Other chelation

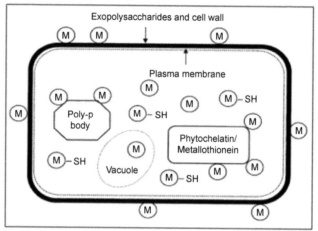

Fig. 2. Metal-binding sites of a typical algal cell. *The alphabet M represents the metal species (independent of its oxidation state)* (Mehta and Gaur 2005).

process is based in the ability of polyphosphate bodies to accumulate a number of heavy metals like Fe, Zn, Cd, Cu and Pb and may serve as protection of algal cells from metal toxicity (Pawlik-Skowronska and Skowrori 2001). The marine alga *Dunaliella tertiolecta* has been shown to have high phytochelatin content attributed to its capability to hyperaccumulate Zn and Cd (Tsuji et al. 2003). Ahner et al. (2002) reported that upon exposure to Cd or Cu, there was a significant increase in the intracellular concentration of phytochelatin in diatoms (*Emiliania huxleyi, Phaeodactylum tricornutum, Thalassiosira pseudonana, Thalassiosira weissflogii*), and green alga (*Dunaliella* sp.). Also in some seagrass species such as *Enhalus acoroides,* a strong correlation between phytochelatin and Pb tissue concentration was found in the root organ collected from sediment highly contaminated with Pb (Nguyen et al. 2017). The same could be verified in roots and leaves of *Thalassia hemprichii* treated with Pb (Tupan et al. 2014). Same findings were found in the seagrass *Posidonia oceanica* exposed to Cu and Cd, with the synthesis of metallothionein leading to a reduction of the oxidative stress caused by metals (Cozza et al. 2006). Another study showed that Cd may be transformed by microalgae to form sulphides, which have low solubilities and thus, lower toxicity due to its low bioavailability (Edwards et al. 2013).

Antioxidant response

The toxic effect of heavy metals appears to be related to production of reactive oxygen species (ROS) and the resulting unbalanced cellular redox status (Pinto et al. 2003b). It is known that heavy metals cause oxidative stress on the cell, which generates reactive oxygen species "ROS" ($O2^-$, H_2O_2....) that can react with cellular components and cause oxidation of lipids, carbohydrates, proteins, pigments and DNA. These results in growth inhibition and photosynthesis (Shamsi et al. 2008, Bouchama et al. 2016). These are specific but complex mechanisms involving morphological, physiological and biochemical adaptation. In particular, the ROS-combating antioxidant system, including diverse enzymes such as superoxide dismutase, catalase, glutathione peroxidase and ascorbate peroxidase, and non-enzymatic system consists of glutathione (GSH) and ascorbic acid (Pinto et al. 2003a, Gill and Tuteja 2010). An increasing activity of superoxide dismutase (SOD) and guaiacol peroxidase (POD) was founded in *Thalassia hemprichii* exposed to various concentrations of Zn^{2+}, Cd^{2+}, Pb^{2+} and Cu^{2+} (Li et al. 2012). Also all tissues of live sheaths and root/rhizomes experienced an increase in cysteine, glutathione (GSH), γ-glutamylcysteine (γ-EC) and phytochelatin-like peptides as a response to Cd exposure (Alvarez-Legorreta et al. 2008). An increase in catalase and GST activities was observed in *Posidonia oceanica* exposed to low concentrations of mercury chloride (Ferrat et al. 2002). The research of Lee and Shin (2003) showed an increase of guaiacol and ascorbate peroxidases and catalase activity whereas superoxide dismutase and glutathione reductase activity markedly decreased, in the marine alga *Nannochloropsis oculata* exposed to Cd. This study suggested that the ability of this marine algae to tolerate the harmful effects of cadmium is probably due to the modification of antioxidant enzyme levels.

Tolerance and Detoxification Mechanisms of Organic Contaminants

In recent years, various studies have demonstrated the bioaccumulation, biotransformation and biodegradation potential of several algal and seagrasses species for various organic contaminants, such as the bioaccumulation and biodegrading of two polycyclic aromatic hydrocarbons, phenanthrene (PHE) and fluoranthene (FLA) in diatoms *Skeletonema costatum* and *Nitzschia* sp. (Hong et al. 2008). Also the uptake of polychlorinated biphenyls (PCBs) by *Emilliana huxleyi, Skeletonema costatum, Thalassiosira nordenskioidii Bacillariophyceae Phaeodactylum tricornutum* (Gerofke et al. 2005). Selective phytoplankton, diatoms and microalgal species have shown the potential of biodegradation of organic contaminants, especially biotransformation of the complex organic compounds in lower carbon compounds (Singh et al. 2015). Akin to the biodegradation of endocrine disrupting chemicals including nonylphenols (NPs), bisphenol (BPA), 17 alpha-ethynylestradiol (EE2) and estradiol (E2) in marine diatom *Navicula incerta* (Liu et al. 2010). In another study, the bioaccumulation of PAHs and PCBs in roots and leaves of *Zostera marina* without major physiological disturbance (Huesemann et al. 2009). In a recent study, Fernandez and Gardinali (2016) have demonstrated the ability of *Halodule wrightii* to accumulate irgarol, a triazine herbicide. The biodegradation processes of PCBs, including dechlorination, can transform PCBs, altering their potential toxicity effectively (Borja et al. 2005). Although some of the organic pollutants may not be completely mineralised, they may be converted to less toxic or nontoxic compounds, which is an important bioremediation strategy (Subashchandrabose et al. 2013). It is clear that although the complete degradation of aromatic pollutants is rare, algae are capable of carrying out biotransformations on aromatic pollutants, such as the hydroxylation of naphthalene and benzo[*a*]pyrene to their hydroxylated intermediates. These initial biotransformation may be extremely important in the overall degradation of pollutants in the environment by other microorganisms like white rot fungi (Meulenberg et al. 1997). Subashchandrabose et al. (2013) reported that several factors such as algal size, density, morphology and metabolic activities are important in affecting the uptake and removal of the toxicants. According to Wolff (1976), the low decomposition rates of seagrass have an important role in the maintenance of the standing stock of organic matter in the coastal environment and even in deep sea environment.

Conclusion

Pollution problems greatly affect the marine ecosystem which is mainly sensitive to several types of contaminants; harmful organic and inorganic contaminants can be present at different levels in the marine environment. Increased consciousness of the necessity to safeguard those environments had motivated a search for alternative technologies to remove pollution from this sensible ecosystem. The present review emphasises the role of seagrass and algae in phytoremediation and phycoremediation technologies. Different techniques can be distinguished within phytoremediation of polluted water according to their objectives, the characteristics of the environment, the contaminants and the used species. Several species of seagrasses and algae have been reviewed in this chapter. Some of them represent an immense potential for removal of a variety of contaminants including heavy metals, inorganic/organic, by bioaccumulation, biodegradation, biotransformation and biosorption using the binding groups for biosorption, ROS-combating antioxidant system, phytochelatin and metallothionine.

References

Abdel-Shafy, H.I. and Mansour, M.S. 2016. A review on polycyclic aromatic hydrocarbons: source, environmental impact, effect on human health and remediation. Egyptian Journal of Petroleum 25: 107–123.

Adriano, D.C. 2001. Introduction. Trace elements in terrestrial environments. Springer.

Agency, U.S.E.P. 2002. USEPA document EPA-821-R-02-012. *In:* Agency, U.S.E.P. (ed.). Washington, DC: USEPA.

Ahner, B.A., Kong, S. and Morel, F.M. 1995. Phytochelatin production in marine algae. 1. An interspecies comparison. Limnology and Oceanography 40: 649–657.

Ahner, B.A., Wei, L., Oleson, J.R. and Ogura, N. 2002. Glutathione and other low molecular weight thiols in marine phytoplankton under metal stress. Marine Ecology Progress Series 232: 93–103.

Aioi, K., Komatsu, T. and Morita, K. 1998. The world's longest seagrass, Zostera caulescens from northeastern Japan. Aquatic Botany 61: 87–93.

Alkorta, I., Hernández-Allica, J., Becerril, J., Amezaga, I., Albizu, I. and Garbisu, C. 2004. Recent findings on the phytoremediation of soils contaminated with environmentally toxic heavy metals and metalloids such as zinc, cadmium, lead, and arsenic. Reviews in Environmental Science and Biotechnology 3: 71–90.

Almeda, R., Wambaugh, Z., Chai, C., Wang, Z., Liu, Z. and Buskey, E.J. 2013. Effects of crude oil exposure on bioaccumulation of polycyclic aromatic hydrocarbons and survival of adult and larval stages of gelatinous zooplankton. PLOS ONE 8: e74476.

Alvarez-Legorreta, T., Mendoza-Cozatl, D., Moreno-Sanchez, R. and Gold-Bouchot, G. 2008. Thiol peptides induction in the seagrass Thalassia testudinum (Banks ex König) in response to cadmium exposure. Aquatic Toxicology 86: 12–19.

Álvarez-Muñoz, D., Llorca, M., Blasco, J. and Barceló, D. 2016. Chapter 1—Contaminants in the Marine Environment. Marine Ecotoxicology. Academic Press.

Anastopoulos, I. and Kyzas, G.Z. 2015. Progress in batch biosorption of heavy metals onto algae. Journal of Molecular Liquids 209: 77–86.

Apostolopoulou, M.-V., Monteyne, E., Krikonis, K., Pavlopoulos, K., Roose, P. and Dehairs, F. 2014. Monitoring polycyclic aromatic hydrocarbons in the Northeast Aegean Sea using Posidonia oceanica seagrass and synthetic passive samplers. Marine Pollution Bulletin 87: 338–344.

Arias, A.H., Souissi, A., Glippa, O., Roussin, M., Dumoulin, D., Net, S., Ouddane, B. and Souissi, S. 2017. Removal and biodegradation of phenanthrene, fluoranthene and pyrene by the marine algae Rhodomonas baltica enriched from North Atlantic coasts. Bulletin of Environmental Contamination and Toxicology 98: 392–399.

Arthur, E.L., Rice, P.J., Rice, P.J., Anderson, T.A., Baladi, S.M., Henderson, K.L. and Coats, J.R. 2005. Phytoremediation—an overview. Critical Reviews in Plant Sciences 24: 109–122.

Ayangbenro, A.S. and Babalola, O.O. 2017. A new strategy for heavy metal polluted environments: a review of microbial biosorbents. International Journal of Environmental Research and Public Health 14: 94.

Bai, X. and Acharya, K. 2017. Algae-mediated removal of selected pharmaceutical and personal care products (PPCPs) from lake mead water. Science of the Total Environment 581-582: 734–740.

Bandekar, P.D. and Haragi, S.B. 2017. Physicochemical parameters in Karwar coastal water, central West coast of India.

Barka, S. 2012. Contribution of X-ray spectroscopy to marine ecotoxicology: Trace metal bioaccumulation and detoxification in marine invertebrates. Ecotoxicology. InTech.

Bates, S.S., Tessier, A., Campbell, P.G. and Buffle, J. 1982. Zinc adsorption and transport by Chlamydomonas varuiabilis and Scenedesmus subspicatus (Chlorophyceae) grown in semicontinuous culture. Journal of Phycology 18: 521–529.

Beek, B. 1999. Bioaccumulation New Aspects and Developments. Springer Science & Business Media.

Bhadja, P. and Kundu, R. 2012. Status of the seawater quality at few industrially important coasts of Gujarat (India) off Arabian Sea.

Binelli, A. and Provini, A. 2003. The PCB pollution of Lake Iseo (N. Italy) and the role of biomagnification in the pelagic food web. Chemosphere 53: 143–151.

Blackman, R. 1986. Oil in the sea: Inputs, fates, and effects: National Academy Press, Washington, DC. 1985. ISBN 0-309-03479-5. 601pp. Marine Pollution Bulletin 17: 45–46.

Boehm, P.D., Page, D.S., Brown, J.S., Neff, J.M. and Burns, W.A. 2004. Polycyclic aromatic hydrocarbon levels in mussels from Prince William Sound, Alaska, USA, document the return to baseline conditions. Environ. Toxicol. Chem. 23: 2916–2929.

Borja, J., Taleon, D.M., Auresenia, J. and Gallardo, S. 2005. Polychlorinated biphenyls and their biodegradation. Process Biochemistry 40: 1999–2013.

Bouchama, K., Rouabhi, R. and Djebar, M.R. 2016. Behavior of Phragmites australis (CAV.) Trin. Ex Steud used in phytoremediation of wastewater contaminated by cadmium. Desalination and Water Treatment 57: 5325–5330.

Bouchon, C., Lemoine, S., Dromard, C. and Bouchon-Navaro, Y. 2016. Level of contamination by metallic trace elements and organic molecules in the seagrass beds of Guadeloupe Island. Environmental Science and Pollution Research 23: 61–72.

Brooks, R. 1998. Plants that hyperaccumulate heavy metals CAB International Wallingford. UK Google Scholar.

Brown, J., Gjelsvik, R., Saxen, R. and Mattila, J. 2009. Knowledge gaps in relation to radionuclide levels and transfer to wild plants and animals, in the context of environmental impact assessments, and a strategy to fill them. Nordisk Kernesikkerhedsforskning.

Bryan, G.W. 1971. The effects of heavy metals (other than mercury) on marine and estuarine organisms. Proceedings of the Royal Society of London. Series B. Biological Sciences 177: 389–410.

Buesseler, K.O., Jayne, S.R., Fisher, N.S., Rypina, I.I., Baumann, H., Baumann, Z., Breier, C.F., Douglass, E.M., George, J., Macdonald, A.M., Miyamoto, H., Nishikawa, J., Pike, S.M. and Yoshida, S. 2012. Fukushima-derived radionuclides in the ocean and biota off Japan. Proceedings of the National Academy of Sciences 109: 5984–5988.

Burke, L., Reytar, K., Spalding, M. and Perry, A. 2011. Reefs at risk. World Resources Institute, Washington, DC, 124.

Carsky, M. and Mbhele, F. 2013. Adsorption of heavy metals using marine algae. South African Journal of Chemical Engineering 18: 40–51.

Center, I.M.K. 2012. International shipping facts and figures–information resources on trade. Safety, Security, Environment.

Chislock, M.F., Doster, E., Zitomer, R.A. and Wilson, A. 2013. Eutrophication: causes, consequences, and controls in aquatic ecosystems. Nature Education Knowledge 4: 10.

Cirelli, A.F., Ojeda, C., Castro, M.J. and Salgot, M. 2008. Surfactants in sludge-amended agricultural soils: a review. Environmental Chemistry Letters 6: 135–148.

Claessens, M., Vanhaecke, L., Wille, K. and Janssen, C.R. 2013. Emerging contaminants in Belgian marine waters: single toxicant and mixture risks of pharmaceuticals. Marine Pollution Bulletin 71: 41–50.

Clark, R.B., Frid, C. and Attrill, M. 1989. Marine Pollution, Clarendon Press Oxford.

Cobbett, C. and Goldsbrough, P. 2002. Phytochelatins and metallothioneins: roles in heavy metal detoxification and homeostasis. Annual Review of Plant Biology 53: 159–182.

Cowan-Ellsberry, C., Belanger, S., Dorn, P., Dyer, S., Mcavoy, D., Sanderson, H., Versteeg, D., Ferrer, D. and Stanton, K. 2014. Environmental safety of the use of major surfactant classes in North America. Critical Reviews in Environmental Science and Technology 44: 1893–1993.

Cozza, R., Pangaro, T., Maestrini, P., Giordani, T., Natali, L. and Cavallini, A. 2006. Isolation of putative type 2 metallothionein encoding sequences and spatial expression pattern in the seagrass Posidonia oceanica. Aquatic Botany 85: 317–323.

Creel, L. 2003. Ripple effects: population and coastal regions, Population Reference Bureau Washington, DC.

Crist, R.H., Oberholser, K., Shank, N. and Nguyen, M. 1981. Nature of bonding between metallic ions and algal cell walls. Environmental Science & Technology 15: 1212–1217.

Dahle, S., Savinov, V.M., Matishov, G.G., Evenset, A. and NÆS, K. 2003. Polycyclic aromatic hydrocarbons (PAHs) in bottom sediments of the Kara Sea shelf, Gulf of Ob and Yenisei Bay. Science of the Total Environment 306: 57–71.

Dallinger, R. 1993. Strategies of metal detoxification in terrestrial invertebrates. Ecotoxicology of Metals in Invertebrates, 245.

Daneshvar, N., Khataee, A., Rasoulifard, M. and Pourhassan, M. 2007. Biodegradation of dye solution containing Malachite Green: Optimization of effective parameters using Taguchi method. Journal of Hazardous Materials 143: 214–219.

Daughton, C.G. and Ternes, T.A. 1999. Pharmaceuticals and personal care products in the environment: agents of subtle change? Environmental Health Perspectives 107: 907.

Dean, T.A., Stekoll, M.S., Jewett, S.C., Smith, R.O. and Hose, J.E. 1998. Eelgrass (Zostera marina L.) in Prince William Sound, Alaska: effects of the Exxon Valdez oil spill. Marine Pollution Bulletin 36: 201–210.

Delorenzo, M.E. and Fleming, J. 2008. Individual and mixture effects of selected pharmaceuticals and personal care products on the marine phytoplankton species Dunaliella tertiolecta. Archives of Environmental Contamination and Toxicology 54: 203–210.

Demirak, A., Dalman, Ö., Tilkan, E., Yıldız, D., Yavuz, E. and Gökçe, C. 2011. Biosorption of 2,4 dichlorophenol (2,4-DCP) onto Posidonia oceanica (L.) seagrass in a batch system: Equilibrium and kinetic modeling. Microchemical Journal 99: 97–102.

Denny, M.W. and Gaines, S.D. 2007. Encyclopedia of tidepools and rocky shores, Univ. of California Press.

Dhir, B. 2013a. Introduction. Phytoremediation: Role of Aquatic Plants in Environmental Clean-Up. India: Springer India.

Dhir, B. 2013b. Phytoremediation: Role of aquatic plants in environmental clean-up, Springer.

Di toppi, L.S. and Gabbrielli, R. 1999. Response to cadmium in higher plants. Environmental and Experimental Botany 41: 105–130.

Dixit, S. and Singh, D. 2015. Phycoremediation: Future Perspective of Green Technology. Algae and Environmental Sustainability. Springer.

Duarte, C.M. 2002. The future of seagrass meadows. Environmental Conservation 29: 192–206.

E.C. 2006. Regulation Council of the European Union No. 1907/2006 of the European Parliament and of the Council of 18 December 2006 concerning the Registration, Evaluation, Authorisation and Restriction of Chemicals ("REACh").

Ebele, A.J., Abou-Elwafa Abdallah, M. and Harrad, S. 2017. Pharmaceuticals and personal care products (PPCPs) in the freshwater aquatic environment. Emerging Contaminants 3: 1–16.

Edwards, C.D., Beatty, J.C., Loiselle, J.B., Vlassov, K.A. and Lefebvre, D.D. 2013. Aerobic transformation of cadmium through metal sulfide biosynthesis in photosynthetic microorganisms. BMC Microbiology 13: 161.

El Gamal, A.A. 2010. Biological importance of marine algae. Saudi Pharmaceutical Journal 18: 1–25.

Eliana Gattullo, C., Bährs, H., Steinberg, C. and Loffredo, E. 2012. Removal of bisphenol A by the freshwater green alga Monoraphidium braunii and the role of natural organic matter.

Espiritu, J.A.A. and Paz-Alberto, A.M. 2018. Phytoremediation potential of seagrasses and seaweed species in the coastal resources of brgy. bolitoc, sta. cruz, zambales, philippines.

Feng, D. and Aldrich, C. 2004. Adsorption of heavy metals by biomaterials derived from the marine alga Ecklonia maxima. Hydrometallurgy 73: 1–10.

Fernandez, M.V. and Gardinali, P.R. 2016. Risk assessment of triazine herbicides in surface waters and bioaccumulation of Irgarol and M1 by submerged aquatic vegetation in Southeast Florida. Science of the Total Environment 541: 1556–1571.

Ferrat, L., Romeo, M., Gnassia-Barelli, M. and Pergent-Martini, C. 2002. Effects of mercury on antioxidant mechanisms in the marine phanerogam Posidonia oceanica. Diseases of Aquatic Organisms 50: 157–160.

Flouty, R. and Estephane, G. 2012. Bioaccumulation and biosorption of copper and lead by a unicellular algae Chlamydomonas reinhardtii in single and binary metal systems: a comparative study. Journal of Environmental Management 111: 106–114.

Fourest, E. and Volesky, B. 1997. Alginate properties and heavy metal biosorption by marine algae. Applied Biochemistry and Biotechnology 67: 215–226.

Gadd, G.M. 1988. Accumulation of metals by microorganisms and algae. Biotechnology: A Comprehensive Treatise 6: 401–433.

Gallardo, T. 2014. Marine Algae: General Aspects (Biology, Systematics, Field and Laboratory Techniques). Marine Algae: Biodiversity, Taxonomy, Environmental Assessment, and Biotechnology, 1.

Garnham, G.W., Codd, G.A. and Gadd, G.M. 1992. Accumulation of cobalt, zinc and manganese by the estuarine green microalga Chlorella salina immobilized in alginate microbeads. Environmental Science & Technology 26: 1764–1770.

Gayet, G., Baptist, F., Baraille, L., Caessteker, P., Clément, J.C., Gaillard, J., Gaucherand, S., Isselin-Nondedeu, F., Poinsot, C. and Quétier, F. 2016. Méthode nationale d'évaluation des fonctions des zones humides. Fondements théoriques, scientifiques et techniques. Onema, MNHN, 310.

Geissen, V., Mol, H., Klumpp, E., Umlauf, G., Nadal, M., Van der Ploeg, M., Van de Zee, S.E.A.T.M. and Ritsema, C.J. 2015. Emerging pollutants in the environment: A challenge for water resource management. International Soil and Water Conservation Research 3: 57–65.

Gerofke, A., Kömp, P. and Mclachlan, M.S. 2005. Bioconcentration of persistent organic pollutants in four species of marine phytoplankton. Environmental Toxicology and Chemistry 24: 2908–2917.

Geyer, H.J., Rimkus, G.G., Scheunert, I., Kaune, A., Schramm, K.-W., Kettrup, A., Zeeman, M., Muir, D.C., Hansen, L.G. and Mackay, D. 2000. Bioaccumulation and occurrence of endocrine-disrupting chemicals (EDCs), persistent organic pollutants (POPs), and other organic compounds in fish and other organisms including humans. Bioaccumulation–New Aspects and Developments. Springer.

150 *Ecotoxicology of Marine Organisms*

Ghosal, D., Ghosh, S., Dutta, T.K. and Ahn, Y. 2016. Current state of knowledge in microbial degradation of polycyclic aromatic hydrocarbons (PAHs): a review. Frontiers in Microbiology, 7.

Gill, S.S. and Tuteja, N. 2010. Reactive oxygen species and antioxidant machinery in abiotic stress tolerance in crop plants. Plant Physiology and Biochemistry 48: 909–930.

Gioia, R., Dachs, J., Nizzetto, L., Berrojalbiz, N., Galbán-malagón, C., Del vento, S., Méjanelle, L. and Jones, K. 2011. Sources, Transport and Fate of Organic Pollutants in the Oceanic Environment.

Graham, L. and Wilcox, L. 2000. Algae–Prentice Hall. Upper Saddle River, New Jersey.

Green, E.P. and Short, F.T. 2003. World Atlas of Seagrasses, Univ. of California Press.

Grill, E., Winnacker, E.-L. and Zenk, M.H. 1985. Phytochelatins: the principal heavy-metal complexing peptides of higher plants. Science 230: 674–677.

Guha, H. 2004. Biogeochemical influence on transport of chromium in manganese sediments: experimental and modeling approaches. Journal of Contaminant Hydrology 70: 1–36.

Guiry, M.D. 2012. How many species of algae are there? Journal of Phycology 48: 1057–1063.

Guo, W.-Q., Zheng, H.-S., Li, S., Du, J.-S., Feng, X.-C., Yin, R.-L., Wu, Q.-L., Ren, N.-Q. and Chang, J.-S. 2016. Removal of cephalosporin antibiotics 7-ACA from wastewater during the cultivation of lipid-accumulating microalgae. Bioresource Technology 221: 284–290.

Gupta, G., Kumar, V. and Pal, A. 2017. Microbial degradation of high molecular weight polycyclic aromatic hydrocarbons with emphasis on pyrene. Polycyclic Aromatic Compounds, 1–13.

Halling-Sørensen, B., Nyholm, N., Kusk, K.O. and Jacobsson, E. 2000. Influence of nitrogen status on the bioconcentration of hydrophobic organic compounds to Selenastrum capricornutum. Ecotoxicology and Environmental Safety 45: 33–42.

Hamza-Chaffai, A. 2014. Usefulness of bioindicators and biomarkers in pollution biomonitoring. International Journal of Biotechnology for Wellness Industries 3: 19–26.

Haynes, D. and Johnson, J.E. 2000. Organochlorine, heavy metal and polyaromatic hydrocarbon pollutant concentrations in the Great Barrier Reef (Australia) environment: a review. Marine Pollution Bulletin 41: 267–278.

Haynes, D., Müller, J. and Carter, S. 2000. Pesticide and herbicide residues in sediments and seagrasses from the great barrier reef world heritage area and queensland coast. Marine Pollution Bulletin 41: 279–287.

He, N., Sun, X., Zhong, Y., Sun, K., Liu, W. and Duan, S. 2016. Removal and biodegradation of nonylphenol by four freshwater microalgae. International Journal of Environmental Research and Public Health 13: 1239.

Hemminga, M.A. and Duarte, C.M. 2000. Seagrass Ecology, Cambridge University Press.

Hom-Diaz, A., Llorca, M., Rodríguez-Mozaz, S., Vicent, T., Barceló, D. and Blánquez, P. 2015. Microalgae cultivation on wastewater digestate: b-estradiol and 17a-ethynylestradiol degradation and transformation products identification. Journal of Environmental Management 155: e113.

Hong, Y.-W., Yuan, D.-X., Lin, Q.-M. and Yang, T.-L. 2008. Accumulation and biodegradation of phenanthrene and fluoranthene by the algae enriched from a mangrove aquatic ecosystem. Marine Pollution Bulletin 56: 1400–1405.

Hotchkiss, A.K., Rider, C.V., Blystone, C.R., Wilson, V.S., Hartig, P.C., Ankley, G.T., Foster, P. M., Gray, C.L. and Gray, L.E. 2008. Fifteen years after "Wingspread"—environmental endocrine disrupters and human and wildlife health: where we are today and where we need to go. Toxicological Sciences 105: 235–259.

Huesemann, M.H., Hausmann, T.S., Fortman, T.J., Thom, R.M. and Cullinan, V. 2009. *In situ* phytoremediation of PAH- and PCB-contaminated marine sediments with eelgrass (Zostera marina). Ecological Engineering 35: 1395–1404.

Hurtado, C., Cañameras, N., Domínguez, C., Price, G.W., Comas, J. and Bayona, J.M. 2017. Effect of soil biochar concentration on the mitigation of emerging organic contaminant uptake in lettuce. Journal of Hazardous Materials 323: 386–393.

Hylland, K. 2006. Polycyclic aromatic hydrocarbon (PAH) ecotoxicology in marine ecosystems. Part A. Journal of Toxicology and Environmental Health 69: 109–123.

I Batlle, J.V. and Vandenhove, H. 2014. Dynamic modelling of the radiological impact of the Fukushima accident on marine biota. Ann. Belg. Ver. Stralingsbescherming 38: 299–312.

IMO Centre (International Maritime Organization). 2012. International Shipping Facts and Figures - Information Resources on Trade, Safety, Security, Environment, Maritime Knowledge Center.

IPCS, W. 1995. Persistent Organic Pollutants: An Assessment Report on: DDT, Aldrin, Dieldrin, Endrin, Chlordane, Heptachlor, Hexachlorobenzene, Mirex, Toxaphene, Polychlorinated Biphenyls, Dioxins and Furans. Disponible à l'adresse Internet www. pops. int.

Jackson, M., Eadsforth, C., Schowanek, D., Delfosse, T., Riddle, A. and Budgen, N. 2015. Comprehensive review of several surfactants in marine environment: Fate and ecotoxicity. Environmental Toxicology and Chemistry.

Jaishankar, M., Tseten, T., Anbalagan, N., Mathew, B.B. and Beeregowda, K.N. 2014. Toxicity, mechanism and health effects of some heavy metals. Interdisciplinary Toxicology 7: 60–72.

Jensen, T.E., Rachlin, J.W., Jani, V. and Warkentine, B. 1982. An X-ray energy dispersive study of cellular compartmentalization of lead and zinc in Chlorella saccharophila (Chlorophyta), Navicula incerta and Nitzschia closterium (Bacillariophyta). Environmental and Experimental Botany 22: 319–328.

Jiang, J.-J., Lee, C.-L. and Fang, M.-D. 2014. Emerging organic contaminants in coastal waters: Anthropogenic impact, environmental release and ecological risk. Marine Pollution Bulletin 85: 391–399.

Jitar, O., Teodosiu, C., Oros, A., Plavan, G. and Nicoara, M. 2015. Bioaccumulation of heavy metals in marine organisms from the Romanian sector of the Black Sea. New Biotechnology 32: 369–378.

Johnston, E.L., Mayer-Pinto, M. and Crowe, T.P. 2015. Review: Chemical contaminant effects on marine ecosystem functioning. Journal of Applied Ecology 52: 140–149.

Joksimovic, D. and Stankovic, S. 2012. The trace metals accumulation in marine organisms of the southeastern Adriatic coast, Montenegro. Journal of the Serbian Chemical Society 77: 105–117.

Jones, K.C. and De Voogt, P. 1999. Persistent organic pollutants (POPs): state of the science. Environmental Pollution 100: 209–221.

Kachel, M.J. 2008. Threats to the Marine Environment: Pollution and Physical Damage. Particularly Sensitive Sea Areas: The IMO's Role in Protecting Vulnerable Marine Areas. Berlin, Heidelberg: Springer Berlin Heidelberg.

Karouna-Renier, N.K., Snyder, R.A., Allison, J.G., Wagner, M.G. and Rao, K.R. 2007. Accumulation of organic and inorganic contaminants in shellfish collected in estuarine waters near Pensacola, Florida: contamination profiles and risks to human consumers. Environmental Pollution 145: 474–488.

Kawai, S. and Murata, K. 2016. Biofuel production based on carbohydrates from both brown and red macroalgae: recent developments in key biotechnologies. International Journal of Molecular Sciences 17: 145.

Kenworthy, W., Durako, M.J., Fatemy, S., Valavi, H. and Thayer, G. 1993. Ecology of seagrasses in northeastern Saudi Arabia one year after the Gulf War oil spill. Marine Pollution Bulletin 27: 213–222.

Kirkham, M. 2006. Cadmium in plants on polluted soils: effects of soil factors, hyperaccumulation, and amendments. Geoderma. 137: 19–32.

Kirso, U. and Irha, N. 1998. Role of Algae in Fate of Carcinogenic Polycyclic Aromatic Hydrocarbons in the Aquatic Environment.

Kleinübing, S.J., Gai, F., Bertagnolli, C. and Silva, M.G.C.D. 2013. Extraction of alginate biopolymer present in marine alga sargassum filipendula and bioadsorption of metallic ions. Materials Research 16: 481–488.

Ko, F.-C., Chang, C.-W. and Cheng, J.-O. 2014. Comparative study of polycyclic aromatic hydrocarbons in coral tissues and the ambient sediments from Kenting National Park, Taiwan. Environmental Pollution 185: 35–43.

Lapworth, D.J., Baran, N., Stuart, M.E. and Ward, R.S. 2012. Emerging organic contaminants in groundwater: A review of sources, fate and occurrence. Environmental Pollution 163: 287–303.

Larkum, A.W., Drew, E.A. and Ralph, P.J. 2007. Photosynthesis and metabolism in seagrasses at the cellular level. Seagrasses: Biology, Ecology and Conservation. Springer.

Lawson, E. 2011. Physico-chemical parameters and heavy metal contents of water from the Mangrove Swamps of Lagos Lagoon, Lagos, Nigeria. Advances in Biological Research 5: 8–21.

Lee, M.Y. and Shin, H.W. 2003. Cadmium-induced changes in antioxidant enzymes from the marine alga Nanochloropsis oculata. Journal of Applied Phycology 15: 13–19.

Levinton, J.S. 1995. Marine Biology: Function, Biodiversity, Ecology, Oxford University Press New York.

Lewis, M.A., Dantin, D.D., Chancy, C.A., Abel, K.C. and Lewis, C.G. 2007. Florida seagrass habitat evaluation: A comparative survey for chemical quality. Environmental Pollution 146: 206–218.

Li, J., Dong, H., Zhang, D., Han, B., Zhu, C., Liu, S., Liu, X., Ma, Q. and Li, X. 2015a. Sources and ecological risk assessment of PAHs in surface sediments from Bohai Sea and northern part of the Yellow Sea, China. Marine Pollution Bulletin 96: 485–490.

Li, L., Huang, X., Borthakur, D. and Ni, H. 2012. Photosynthetic activity and antioxidative response of seagrass Thalassia hemprichii to trace metal stress. Acta Oceanologica. Sinica. 31: 98–108.

Li, P., Cao, J., Diao, X., Wang, B., Zhou, H., Han, Q., Zheng, P. and Li, Y. 2015b. Spatial distribution, sources and ecological risk assessment of polycyclic aromatic hydrocarbons in surface seawater from Yangpu Bay, China. Marine Pollution Bulletin 93: 53–60.

Liang, J., Xia, J. and Long, J. 2017. Biosorption of methylene blue by nonliving biomass of the brown macroalga Sargassum hemiphyllum. Water Science and Technology, wst2017343.

Lin, H., Sun, T., Xue, S. and Jiang, X. 2016. Heavy metal spatial variation, bioaccumulation, and risk assessment of Zostera japonica habitat in the Yellow River Estuary, China. Science of the Total Environment 541: 435–443.

Liu, Y., Guan, Y., Gao, Q., Tam, N.F.Y. and Zhu, W. 2010. Cellular responses, biodegradation and bioaccumulation of endocrine disrupting chemicals in marine diatom Navicula incerta. Chemosphere 80: 592–599.

Lobban, C.S. and Harrison, P.J. 1994. Seaweed Ecology and Physiology, Cambridge University Press.

Majidi, V., Laude Jr, D.A. and Holcombe, J.A. 1990. Investigation of the metal-algae binding site with cadmium-113 nuclear magnetic resonance. Environmental Science & Technology 24: 1309–1312.

Mata, Y., Blazquez, M., Ballester, A., Gonzalez, F. and Munoz, J. 2008. Characterization of the biosorption of cadmium, lead and copper with the brown alga Fucus vesiculosus. Journal of Hazardous Materials 158: 316–323.

Matamoros, V., Gutiérrez, R., Ferrer, I., García, J. and Bayona, J.M. 2015. Capability of microalgae-based wastewater treatment systems to remove emerging organic contaminants: a pilot-scale study. Journal of Hazardous Materials 288: 34–42.

Mckenzie, L., Yoshida, R., Mellors, J. and Coles, R. 2009. Seagrass-watch. Proceedings of a Workshop for Monitoring Seagrass Habitats in Indonesia. The Nature Concervancy, Coral Triangle Center, Sanur, Bali (ID), 9th Mei, 2009.

Mclachlan, M.S., Haynes, D. and Müller, J.F. 2001. PCDDs in the water/sediment–seagrass–dugong (Dugong dugon) food chain on the Great Barrier Reef (Australia). Environmental Pollution 113: 129–134.

Meador, J.P., Stein, J.E., Reichert, W.L. and Varanasi, U. 1995. Bioaccumulation of polycyclic aromatic hydrocarbons by marine organisms. In: Ware, G.W. (ed.) Reviews of Environmental Contamination and Toxicology: Continuation of Residue Reviews. New York, NY: Springer New York.

Meagher, R.B. 2000. Phytoremediation of toxic elemental and organic pollutants. Current Opinion in Plant Biology 3: 153–162.

Mehta, S. and Gaur, J. 2005. Use of algae for removing heavy metal ions from wastewater: progress and prospects. Critical Reviews in Biotechnology 25: 113–152.

Mejáre, M. and Bülow, L. 2001. Metal-binding proteins and peptides in bioremediation and phytoremediation of heavy metals. Trends in Biotechnology 19: 67–73.

Mellors, J. and Mckenzie, L. 2008. Seagrass-Watch.

Meulenberg, R., Rijnaarts, H.H., Doddema, H.J. and Field, J.A. 1997. Partially oxidized polycyclic aromatic hydrocarbons show an increased bioavailability and biodegradability. FEMS Microbiology Letters 152: 45–49.

Mille, G., Munoz, D., Jacquot, F., Rivet, L. and Bertrand, J.-C. 1998. The amoco cadiz oil spill: evolution of petroleum hydrocarbons in the ile grande salt marshes (Brittany) after a 13-year Period. Estuarine, Coastal and Shelf Science 47: 547–559.

Mishra, V., Balomajumder, C. and Agarwal, V.K. 2010. Zn(II) ion biosorption onto surface of eucalyptus leaf biomass: isotherm, kinetic, and mechanistic modeling. Clean–Soil, Air, Water 38: 1062–1073.

Mitra, A. and Zaman, S. 2015. Biodiversity of the blue zone. Blue Carbon Reservoir of the Blue Planet. Springer.

Mitra, G. 2017. Essential plant nutrients and recent concepts about their uptake. Essential Plant Nutrients. Springer.

Mohan, S.V., Rao, N.C., Prasad, K.K. and Karthikeyan, J. 2002. Treatment of simulated Reactive Yellow 22 (Azo) dye effluents using Spirogyra species. Waste Management 22: 575–582.

Moss, C.E. 1973. Health Physics 25: 197–198.

Mueller, J.G., Cerniglia, C.E. and Pritchard, P.H. 1996. Bioremediation of environments contaminated by polycyclic aromatic hydrocarbons. Biotechnology Research Series 6: 125–194.

Murphy, A., Zhou, J., Goldsbrough, P.B. and Taiz, L. 1997. Purification and immunological identification of metallothioneins 1 and 2 from Arabidopsis thaliana. Plant Physiology 113: 1293–1301.

Nagelkerken, I. 2009. Evaluation of nursery function of mangroves and seagrass beds for tropical decapods and reef fishes: patterns and underlying mechanisms. Ecological Connectivity among Tropical Coastal Ecosystems. Springer.

Naik, M.M. and Dubey, S.K. 2017. Marine Pollution and Microbial Remediation, Springer.

Navarro, A.E., Hernandez-vega, A., Masud, M.E., Roberson, L.M. and Diaz-vázquez, L.M. 2016. Bioremoval of phenol from aqueous solutions using native caribbean seaweed. Environments 4: 1.

Ncibi, M.C., Mahjoub, B. and Seffen, M. 2006. Biosorption of phenol onto Posidonia oceanica (L.) seagrass in batch system: Equilibrium and kinetic modelling. The Canadian Journal of Chemical Engineering 84: 495–500.

Ncibi, M.C., Mahjoub, B. and Seffen, M. 2008. Adsorptive removal of anionic and non-ionic surfactants from aqueous phase using Posidonia oceanica (L.) marine biomass. Journal of Chemical Technology and Biotechnology 83: 77–83.

Nenciu, M., Rosioru, D., Oros, A., Galatchi, M. and Rosoiu, N. 2014. Bioaccumulation of Heavy Metals in Seahorse Tissue at the Romanian Black Sea Coast. 14th SGEM GeoConference on Water Resources. Forest, Marine And Ocean Ecosystems, www. sgem. org, SGEM2014 Conference Proceedings, ISBN, 2014. 978–619.

Newman, M. and Unger, M. 2003. Fundamentals of Ecotoxicology. Lewis Publishers. Boca Raton, Florida, 458.

Newman, M.C. 2014. Fundamentals of Ecotoxicology—The Science of Pollution CRC Press.

Nguyen, X.-V., Le-ho, K.-H. and Papenbrock, J. 2017. Phytochelatin 2 accumulates in roots of the seagrass Enhalus acoroides collected from sediment highly contaminated with lead. BioMetals 30: 249–260.

Niu, H. and Volesky, B. 2000. Gold-cyanide biosorption with L-cysteine. Journal of Chemical Technology and Biotechnology 75: 436–442.

Noyes, P.D., Mcelwee, M.K., Miller, H.D., Clark, B.W., Van tiem, L.A., Walcott, K.C., Erwin, K.N. and Levin, E.D. 2009. The toxicology of climate change: environmental contaminants in a warming world. Environment International 35: 971–986.

Pampanin, D.M. and Sydnes, M.O. 2013. Polycyclic Aromatic Hydrocarbons a Constituent of Petroleum: Presence and Influence in the Aquatic Environment. Hydrocarbon. Rijeka: InTech.

Pawlik-Skowronska, B. and Skowroriski, T. 2001. Freshwater algae. Metals in the Environment. CRC Press.

Peng, F.-Q., Ying, G.-G., Yang, B., Liu, S., Lai, H.-J., Liu, Y.-S., Chen, Z.-F. and Zhou, G.-J. 2014. Biotransformation of progesterone and norgestrel by two freshwater microalgae (Scenedesmus obliquus and Chlorella pyrenoidosa): Transformation kinetics and products identification. Chemosphere 95: 581–588.

Pennesi, C., Totti, C. and Beolchini, F. 2013. Removal of vanadium(III) and molybdenum(V) from wastewater using Posidonia oceanica (Tracheophyta) biomass. PLOS ONE 8: e76870.

Pennesi, C., Rindi, F., Totti, C. and Beolchini, F. 2015. Marine Macrophytes: Biosorbents. Springer Handbook of Marine Biotechnology. Springer.

Pérez-ruzafa, A., Navarro, S., Barba, A., Marcos, C., Camara, M., Salas, F. and Gutierrez, J. 2000. Presence of pesticides throughout trophic compartments of the food web in the Mar Menor Lagoon (SE Spain). Marine Pollution Bulletin 40: 140–151.

Pergent, G., Labbe, C., Lafabrie, C., Kantin, R. and Pergent-Martini, C. 2011. Organic and inorganic human-induced contamination of Posidonia oceanica meadows. Ecological Engineering 37: 999–1002.

Phang, S.-M., Chu, W.-L. and Rabiei, R. 2015. Phycoremediation. *In:* Sahoo, D. and Seckbach, J. (eds.). The Algae World. Dordrecht: Springer Netherlands.

Pidwirny, M. 2017. Chapter 28: Biogeochemical Cycling and Ecosystem Productivity: Single chapter from the eBook Understanding Physical Geography, Our Planet Earth Publishing.

Pintado-Herrera, M.G., Wang, C., Lu, J., Chang, Y.-P., Chen, W., Li, X. and Lara-Martín, P.A. 2017. Distribution, mass inventories, and ecological risk assessment of legacy and emerging contaminants in sediments from the Pearl River Estuary in China. Journal of Hazardous Materials 323: 128–138.

Pinto, E., Sigaud-Kutner, T., Leitao, M.A., Okamoto, O.K., Morse, D. and Colepicolo, P. 2003a. Heavy metal-induced oxidative stress in algae. Journal of Phycology.

Pinto, E., Sigaud-Kutner, T., Leitao, M.A., Okamoto, O.K., Morse, D. and Colepicolo, P. 2003b. Heavy metal-induced oxidative stress in algae. Journal of Phycology 39: 1008–1018.

Prasad, M.N.V. 2001. Metals in the Environment: Analysis by Biodiversity, CRC Press.

Prichard, E. and Granek, E.F. 2016. Effects of pharmaceuticals and personal care products on marine organisms: from single-species studies to an ecosystem-based approach. Environmental Science and Pollution Research 23: 22365–22384.

Qiu, Y.-W., Zeng, E.Y., Qiu, H., Yu, K. and Cai, S. 2017. Bioconcentration of polybrominated diphenyl ethers and organochlorine pesticides in algae is an important contaminant route to higher trophic levels. Science of the Total Environment 579: 1885–1893.

Richardson, S.D. and Ternes, T.A. 2011. Water analysis: emerging contaminants and current issues. Analytical Chemistry 83: 4614–4648.

Romero-Isart, N. and Vašák, M. 2002. Advances in the structure and chemistry of metallothioneins. Journal of Inorganic Biochemistry 88: 388–396.

Roose, P., Albaigés, J., Bebianno, M.J., Camphuysen, C.J., Cronin, M., De leeuw, J., Gabrielsen, G., Hutchinson, T., Hylland, K., Jansson, B., Jenssen, B.M., Schulz-Bull, D., Szefer, P., Webster, L., Bakke, T., Janssen, C., Calewaert, J.-B. and Mcdonough, N. 2011. Monitoring Chemical Pollution in Europe's Seas: Programmes, practices and priorities for research.

Rubio-Clemente, A., Torres-Palma, R.A. and Peñuela, G.A. 2014. Removal of polycyclic aromatic hydrocarbons in aqueous environment by chemical treatments: A review. Science of the Total Environment 478: 201–225.

Salt, D.E., Blaylock, M., Kumar, N.P., Dushenkov, V., Ensley, B.D., Chet, I. and Raskin, I. 1995. Phytoremediation: a novel strategy for the removal of toxic metals from the environment using plants. Nature Biotechnology 13: 468–474.

Salt, D.E., Smith, R. and Raskin, I. 1998. Phytoremediation. Annual Review of Plant Biology 49: 643–668.

Sánchez-Avila, J., Tauler, R. and Lacorte, S. 2012. Organic micropollutants in coastal waters from NW Mediterranean Sea: sources distribution and potential risk. Environment International 46: 50–62.

Sánchez-Rodrıguez, I., Huerta-Diaz, M., Choumiline, E., Holguın-Quinones, O. and Zertuche-González, J. 2001. Elemental concentrations in different species of seaweeds from Loreto Bay, Baja California Sur, Mexico: implications for the geochemical control of metals in algal tissue. Environmental Pollution 114: 145–160.

Sauvé, S. and Desrosiers, M. 2014. A review of what is an emerging contaminant. Chemistry Central Journal 8: 1–7.

Scarlett, A., Donkin, P., Fileman, T., Evans, S. and Donkin, M. 1999. Risk posed by the antifouling agent Irgarol 1051 to the seagrass, Zostera marina. Aquatic Toxicology 45: 159–170.

Schwarzbauer, J. 2006. Organic Contaminants in Riverine and Groundwater Systems_ Aspects of the Anthropogenic Contribution. *In:* SPRINGER (ed.). Verlag Berlin Heidelberg.

Shamsi, I., Wei, K., Zhang, G., Jilani, G. and Hassan, M. 2008. Interactive effects of cadmium and aluminum on growth and antioxidative enzymes in soybean. Biologia Plantarum 52: 165–169.

Simeonova, A. and Petkova, S. 2007. Biosorption of heavy metals by marine algae Ulva rigida, Cystoseira barbata and C. crinita. Вісник Дніпропетровського університету. Біологія, екологія, 1.

Singh, B., Bauddh, K. and Bux, F. 2015. Algae and Environmental Sustainability, Springer.

Singh, B.K., Singh, V.P. and Singh, M.N. 2002. Bioremediation of contaminated water bodies. *In:* Ved pal, S. and Raymond, D.S. (eds.). Progress in Industrial Microbiology. Elsevier.

Smith, A., Coupland, G., Dolan, L., Harberd, N., Jones, J., Martin, C., Sablowski, R. and Amey, A. 2010. Plant Biology. Garland Science. Taylor and Francis Group, LLC.

Solé, A. and Matamoros, V. 2016. Removal of endocrine disrupting compounds from wastewater by microalgae co-immobilized in alginate beads. Chemosphere 164: 516–523.

Song, J. 2011. Biogeochemical processes of biogenic elements in China marginal seas, Springer Science & Business Media.

Stevenson, F.J.C. and Michael, A. 1999. Cycles of Soils: Carbon, Nitrogen, Phosphorus, Sulfur, Micronutrients, John Wiley & Sons.

Strasdeit, H., Duhme, A.-K., Kneer, R., Zenk, M.H., Hermes, C. and Nolting, H.-F. 1991. Evidence for discrete Cd (SCys) 4 units in cadmium phytochelatin complexes from EXAFS spectroscopy. Journal of the Chemical Society, Chemical Communications, 1129–1130.

Subashchandrabose, S.R., Ramakrishnan, B., Megharaj, M., Venkateswarlu, K. and Naidu, R. 2013. Mixotrophic cyanobacteria and microalgae as distinctive biological agents for organic pollutant degradation. Environment International 51: 59–72.

Sudharsan, S., Seedevi, P., Ramasamy, P., Subhapradha, N., Vairamani, S. and Shanmugam, A. 2012. Heavy metal accumulation in seaweeds and sea grasses along southeast coast of India. J. Chem. Pharm. Res. 4: 4240–4244.

Sumpter, J., Jobling, S. and Tyler, C. 1996. Oestrogenic substances in the aquatic environment and their potential impact on animals, particularly fish. Seminar Series-Society for Experimental Biology, 1996. Cambridge University Press, 205–224.

Suresh, B. and Ravishankar, G. 2004. Phytoremediation—a novel and promising approach for environmental clean-up. Critical Reviews in Biotechnology 24: 97–124.

Talbot, F.F. and Wilkinson, C.C. 2001. Coral reefs, mangroves and seagrasses: A sourcebook for managers, Australian Institute of Marine Science (AIMS).

Tanabe, S. 2002. Contamination and toxic effects of persistent endocrine disrupters in marine mammals and birds. Marine Pollution Bulletin 45: 69–77.

Tao, Y., Yuan, Z., Xiaona, H. and Wei, M. 2012. Distribution and bioaccumulation of heavy metals in aquatic organisms of different trophic levels and potential health risk assessment from Taihu lake, China. Ecotoxicology and Environmental Safety 81: 55–64.

Thomaidi, V.S., Stasinakis, A.S., Borova, V.L. and Thomaidis, N.S. 2015. Is there a risk for the aquatic environment due to the existence of emerging organic contaminants in treated domestic wastewater? Greece as a case-study. Journal of Hazardous Materials 283: 740–747.

Todd, P.A., Ong, X. and Chou, L.M. 2010. Impacts of pollution on marine life in Southeast Asia. Biodiversity and Conservation 19: 1063–1082.

Tornero, V. and Hanke, G. 2016. Chemical contaminants entering the marine environment from sea-based sources: A review with a focus on European seas. Marine Pollution Bulletin 112: 17–38.

Torres, M.A., Barros, M.P., Campos, S.C., Pinto, E., Rajamani, S., Sayre, R.T. and Colepicolo, P. 2008. Biochemical biomarkers in algae and marine pollution: a review. Ecotoxicology and Environmental Safety 71: 1–15.

TSCA. 2012. TSCA Work Plan Chemicals: Methods Document. Office of Pollution Prevention and Toxics, U.S. Environmental Protection Agency.

Tsezos, M. and Bell, J. 1989. Comparison of the biosorption and desorption of hazardous organic pollutants by live and dead biomass. Water Research 23: 561–568.

Tsuji, N., Hirayanagi, N., Iwabe, O., Namba, T., Tagawa, M., Miyamoto, S., Miyasaka, H., Takagi, M., Hirata, K. and Miyamoto, K. 2003. Regulation of phytochelatin synthesis by zinc and cadmium in marine green alga, Dunaliella tertiolecta. Phytochemistry 62: 453–459.

Tupan, C.I., Herawati, E. and Arfiati, D. 2014. Detection of phytochelatin and glutathione in seagrass Thalassia hemprichii as a detoxification mechanism due to lead heavy metal exposure. Aquatic Science and Technology 2: 67–78.

Van Brummelen Tc, V.H.B., Crommentuijn, T. and Kalf, D. 1998. Bioavailability and ecotoxicity of PAHs. pp. 203–263. In: Neilson, A. and Hutzinger, O. (eds.). PAHs and Related Compounds. The Handbook of Environmental Chemistry.

Vasanthi, D., Karuppasamy, P., Santhanam, P., Kumar, S.D. and Malarvannan, G. 2015. Phytoremediation to remove nutrients and textile dye effluent using seagrass (Cymodocea rotundata). Advances in Biological Research 9: 405–412.

Verbruggen, N., Hermans, C. and Schat, H. 2009. Molecular mechanisms of metal hyperaccumulation in plants. New Phytologist 181: 759–776.

Vieira, R.H. and Volesky, B. 2000. Biosorption: a solution to pollution? International Microbiology 3: 17–24.

Vymazal, J. and Kröpfelová, L. 2008. Horizontal Flow Constructed Wetlands, Springer.

Wang, J., Evangelou, B.P., Nielsen, M.T. and Wagner, G.J. 1992. Computer, simulated evaluation of possible mechanisms for sequestering metal ion activity in plant vacuoles II. Zinc. Plant Physiology 99: 621–626.

Wang, N.-X., Li, Y., Deng, X.-H., Miao, A.-J., Ji, R. and Yang, L.-Y. 2013. Toxicity and bioaccumulation kinetics of arsenate in two freshwater green algae under different phosphate regimes. Water Research 47: 2497–2506.

Wang, Y., Liu, J., Kang, D., Wu, C. and Wu, Y. 2017. Removal of pharmaceuticals and personal care products from wastewater using algae-based technologies: a review. Reviews in Environmental Science and Bio/Technology 16: 717–735.

Waycott, M., Duarte, C.M., Carruthers, T.J., Orth, R.J., Dennison, W.C., Olyarnik, S., Calladine, A., Fourqurean, J.W., Heck, K.L. and Hughes, A.R. 2009. Accelerating loss of seagrasses across the globe threatens coastal ecosystems. Proceedings of the National Academy of Sciences 106: 12377–12381.

Wetzel, D.L. and Van Vleet, E.S. 2004. Accumulation and distribution of petroleum hydrocarbons found in mussels (Mytilus galloprovincialis) in the canals of Venice, Italy. Marine Pollution Bulletin 48: 927–936.

White, C., Wilkinson, S.C. and Gadd, G.M. 1995. The role of microorganisms in biosorption of toxic metals and radionuclides. International Biodeterioration & Biodegradation 35: 17–40.

WHO. 2013. State of the science of endocrine disrupting chemicals—2012—An assessment of the state of the science of endocrine disruptors prepared by a group of experts for the United Nations Environment Programme (UNEP) and WHO. 296.

WHO (World Health Organization WHO) . 2009. Children's Health and the Environment. WHO Training Package for the Health Sector. World Health Organization.

Wolff, T. 1976. Utilization of seagrass in the deep sea. Aquatic Botany 2: 161–174.

Xie, P. 2017. Research Progress on Degradation of PPCPs by Micro-Algae.

Ying, G.-G. 2006. Fate, behavior and effects of surfactants and their degradation products in the environment. Environment International 32: 417–431.

Zhang, Y., Habteselassie, M.Y., Resurreccion, E.P., Mantripragada, V., Peng, S., Bauer, S. and Colosi, L.M. 2014. Evaluating removal of steroid estrogens by a model alga as a possible sustainability benefit of hypothetical integrated algae cultivation and wastewater treatment systems. ACS Sustainable Chemistry & Engineering 2: 2544–2553.

Zieman, J.C., Orth, R., Phillips, R.C., Thayer, G. and Thorhaug, A. 1984. Effects of Oil on Seagrass Ecosystems. Restoration of Habitats Impacted by Oil Spills, Butterworth Boston. 1984. Edited by John Cairns, Jr. and Arthur L. Buikema, Jr., pp. 37–64, 5 Fig, 1 Tab, 82 Ref.

7
Ecotoxicology of Pharmaceuticals in Coastal and Marine Organisms

Vanessa F. Fonseca,[1,*] and *Patrick Reis-Santos*[1,2,*]

INTRODUCTION

For as long as they have been produced, pharmaceutical compounds have been released in the environment. And although these compounds and other personal care products are classified as contaminants of emerging concern, this term does not necessarily imply their occurrence in the environment as recent. It rather alludes to contaminants from multiple sources (domestic, industrial or agricultural) that escaped prior notice and classically were not monitored in spite of their potential to cause adverse effects to the environment; or to compounds for which only recently have environmental concerns been fully raised (Glassmeyer et al. 2007, Sauvé and Desrosiers 2014). In the end, the use of the term 'emerging contaminants' has the intention to highlight the largely unregulated nature of the presence in the environment of substances such as pharmaceutical compounds, but also others such as cosmetics, UV blocker agents (sunscreens) or fragrances (Daughton 2016). Furthermore, the continuous and rapid technological development in highly sensitive analytical instrumentation has enabled the discovery and quantification of numerous compounds and substances in the aquatic environment, and from complex matrices, which had previously been undetected (Pérez and Barceló 2007, Sanderson and Thomsen 2009, Klosterhaus et al. 2013).

Pharmaceuticals have come under particular scrutiny regarding their occurrence and effects on aquatic environments due to a few key features. Firstly, both

[1] MARE – Marine and Environmental Sciences Centre, Faculdade de Ciências, Universidade de Lisboa, Campo Grande, 1749-016 Lisboa, Portugal.
[2] Southern Seas Ecology Laboratories, School of Biological Sciences, The University of Adelaide, South Australia 5005, Australia.
* Corresponding authors: vffonseca@fc.ul.pt; pnsantos@fc.ul.pt

human and veterinary pharmaceutical compounds are continuously released into the environment worldwide, resulting in their ubiquitous and persistent presence. Moreover, their concentrations in aquatic ecosystems are projected to continue to rise, with mounting environmental concerns, due to an expected increase in both the access and the widespread use of medication by a growing global population (Kuster and Adler 2014). Additionally, unlike several chemical contaminants, pharmaceutical compounds are biologically active and target particular metabolic pathways that in many cases are evolutionary conserved (Gunnarsson et al. 2008, Furuhagen et al. 2014), eliciting effects at very low environmental concentrations (e.g., ng/L), and shown to specifically affect multiple algae and animal functions (e.g., Franzelitti et al. 2013, Aguirre-Martínez et al. 2015, Minguez et al. 2016). However, it is important to notice that the term pharmaceuticals does not refer to a specific or unambiguous class of molecules sharing an *a priori* defined set of chemical, physical or biological similarities, but to a varied group of therapeutic compounds used for human or veterinary treatment encompassing a wide range of kinetics, metabolism, modes of action (MOA), and ultimately, an array of potential underlying effects to the environment (Taylor and Senac 2014).

In this context, over the last couple of decades, growing attention has been given to monitoring and evaluating the presence and the ecotoxicology of pharmaceutical compounds in the aquatic environment (Daughton 2016). Yet, in comparison to freshwater systems, where studies on the occurrence and potential effects of pharmaceuticals are manifold, transition and coastal marine environments have been comparatively overlooked or poorly investigated. In part, this is likely due to the assumption that dispersion and dilution processes, including from freshwater sources to estuarine and coastal environments, would be suffice to lessen or cancel any potential effects. Only recently has this trend begun to be reversed, with research gradually focusing towards coastal areas and showcasing that pharmaceuticals are present throughout transition and marine environments at levels potentially or effectively adverse to different levels of biological complexity (e.g., Fatta-Kassinos et al. 2011, Klosterhaus et al. 2013, Gaw et al. 2014, Aminot et al. 2016, Arpin-Pont et al. 2016, Du et al. 2016, Fabbri and Franzellitti 2016). Moreover, it is important to highlight that presumed impacts on transition and coastal environments are expected to continue to increase allied to population growth and coastal settlement, as well as from accessory human activities such as aquaculture (Burridge et al. 2010, Gaw et al. 2014, Tornero and Hanke 2016). Overall, the increase in research and literature since 2014 regarding the occurrence, fate and ecotoxicology of pharmaceuticals in coastal and marine environments may, at least in part, be attributed to a review by Gaw et al. (2014), and the call for research prioritization. At that time, Gaw et al. (2014) found 49 studies from the year 2000 onwards reporting concentrations of pharmaceuticals in marine and coastal environments. A number that has since raised considerably, and we were able to compile information from 124 studies (since the year 2000) focusing on the occurrence and effects of pharmaceuticals in transition and coastal marine environments [Web of Science search in February 2017 with the terms: Marine AND pharmaceutical AND (occurrence OR effect* OR toxicity)].

In the present chapter, we aim to provide a brief overview of the most recent advances in the literature regarding the occurrence and ecotoxicology of pharmaceuticals in coastal and marine environments. We critically assess recent research and provide an integrative analysis focusing on the sources of major therapeutic classes of pharmaceuticals to transition and coastal marine environments, their pathways and ecotoxicology to different levels of biological complexity, highlighting reported adverse effects of pharmaceuticals exposure in coastal and marine organisms. In the interest of a focused approach, the scope of the current chapter has been restricted to major therapeutic pharmaceutical compounds, excluding natural and synthesized hormones. Overall, we will prioritize *in situ* evaluations of effects of environmentally relevant concentrations, highlight knowledge gaps and present-day challenges, and provide an outline of key areas and opportunities where future research should be prioritized to underpin the delineation of effective management options. Ultimately, understanding the effects of pharmaceuticals on the marine environment and unraveling their ecotoxicology, MOA and bioaccumulation rates, together with research on their occurrence and fate, is key to safeguarding potential threats to environmental and human health and supporting effective risk management strategies.

Sources and Occurrence of Pharmaceuticals in Coastal and Marine Environments

In this section, we outline the major sources of pharmaceutical contaminants in marine environments as well as their contamination pathways, highlighting ranges of concentrations found, and then briefly refer the physical and chemical processes that may influence the environmental concentrations of these contaminants in both water and sediments.

The sources and pathways of pharmaceutical contaminants in coastal and marine environments are manifold (Kummerer 2009c, Gaw et al. 2014). However, estuarine and coastal areas receive a complex mixture of pharmaceutical contaminants from a set of overarching key origins, namely: (i) human household use; (ii) hospital use; (iii) veterinary applications, via aquaculture or from the terrestrial environment, including livestock production or household pets' care; (iv) and industrial and commercial activities linked to the production of pharmaceuticals (Fig. 1). All these produce large amounts of waste that via a multitude of entwined pathways result in the presence of pharmaceutical compounds, their metabolites and transformation by-products, both directly or via diffuse routes. Assumptions that pharmaceuticals would be negligible due to hydrodynamics or dilution processes in coastal and marine environments are by large currently refuted (Gaw et al. 2014, Fabbri and Franzellitti 2016). In fact, pharmaceutical contaminants have been detected in marine environments at distances that exceed tens and even hundreds of kilometers from what would be their anticipated sources (e.g., WWTP marine outfalls, coastal areas) (Wille et al. 2010, Zhang et al. 2013, Alygizakis et al. 2016). Even when contaminants are not found in water or sediments, or detected only sporadically, pharmaceuticals are still detected in marine organisms such as bivalves and fish, showcasing their potential for bioconcentration (Wille et al. 2011, Maruya et al. 2012, Klosterhaus et al. 2013).

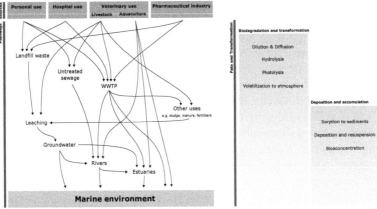

Fig. 1. Major sources and pathways of pharmaceutical contamination into coastal and marine environments. Also shown, the main fate and transformation processes that affect the presence and concentration of pharmaceutical compounds.

The major route of entry for pharmaceuticals and their by-products in natural aquatic environments are point source wastewater discharges of treated [i.e., outflow of waste water treatment plants (WWTP) and septic tanks] and untreated sewage (Glassmeyer et al. 2007, Fatta-Kassinos et al. 2011). Estuarine and coastal marine environments, particularly those near urban clusters, receive large volumes of these effluent discharges both directly via coastal or offshore underwater outfalls (e.g., Togola and Budzinski 2008, Alygizakis et al. 2016), and indirectly via loadings from streams and rivers where wastewater discharges have taken place (Xu et al. 2013, Cantwell et al. 2016). For instance, the annual loads of pharmaceuticals flushed out to sea from the Yangtze estuary are estimated to surpass 150 metric tons, as a result of the discharge of c. 50×10^6 m^3 of sewage (Qi et al. 2014). In another study in South-west France, an assessment of 53 compounds produced an estimated influx of c. 10 kg per day of pharmaceuticals to the Garonne estuary (Aminot et al. 2016). It is important to highlight that even in peak conditions, treatment plants are unable to remove all pharmaceutical contaminants from wastewaters, and that performance efficiency and WWTP removal rates of pharmaceuticals vary significantly from 100 percent to < 1 percent, depending on the type of treatment, operating conditions, chemical loads and the specific physico-chemical properties of the different pharmaceutical compounds (Kim et al. 2007, Gros et al. 2010, Luo et al. 2014, Silva et al. 2014). As a result, over the years, different human and veterinary pharmaceuticals have been found in coastal and marine waters over a wide range of concentrations: for example, 0.01 ng/L (e.g., roxithromycin—antibiotic; Yan et al. 2013) to 6800 ng/L (e.g., norfloxacin—antibiotic; Zou et al. 2011) and even above 200000 ng/L in areas closely affected by WWTP effluents (e.g., paracetamol—analgesic; Togola and Budzinski 2008); as well as in sediments (e.g., from 0.01 to c. 17 ng/g dry weight metoprolol—β-blocker; Cantwell et al. 2016) (see also Gaw et al. 2014, Arpin-Pont et al. 2016, Fabbri and Franzellitti 2016). It is worth highlighting that the contamination and persistence of pharmaceuticals in some transition and coastal environments such as bays, inlets, estuaries and coastal lagoons where water residency and flushing times are reduced, or with periodic connections to the sea, may be of added concern. Many of these coastal areas are favored human settlement or seasonal holiday hubs, and in addition to direct sewage discharges and other local loadings (e.g., river input, groundwater contamination) there is an increased potential risk hazard associated to the confined nature and the distinctive physico-chemical properties of these systems, where dilution and dispersion of contaminants is likely reduced and changes in sorption kinetics will affect the accumulation of pharmaceuticals over time (Dougherty et al. 2010, Liu et al. 2013, Moreno-González et al. 2015, Aminot et al. 2016).

Other sea based human activities such as shipping, particularly cruise and large passenger ships may have a significant impact on specific coastal and marine areas as a result of wastewater discharges (Alygizakis et al. 2016, Westhof et al. 2016). Over 20 million passengers board cruise ships every year (Cruise Line International Association Industry Outlook), with individual liners that hold passenger and crew numbers above those of small townships regularly visiting highly sought confined or sensitive areas, and though wastewater discharges are regulated (Annex IV of the MARPOL convention on pollution prevention), treatment performance still lacks effective administrative regulation or monitoring, so the potential for contamination is substantial. For instance, Westhof et al. (2016) estimated annual loads of ibuprofen exceed 3.3 Kg for a ship with 4000 persons on board.

Wastewaters from healthcare and pharmaceutical production facilities are other key sources of pharmaceutical compounds to estuarine and coastal marine environments (Fig. 1). By their very nature, hospital activities generate a sizable quantity of contaminated effluents. These are dependent on numerous factors which include, but are not limited to, bed density, number of patients or medical specialties, with several studies characterizing pharmaceutical residues in hospital wastewaters in different regions worldwide (e.g., Santos et al. 2013, Herrmann et al. 2015, Oliveira et al. 2015, Azuma et al. 2016). Compiling information on hospital and healthcare facility effluents, Oliveira et al. (2017) showcased that many pharmaceuticals were present at concentrations below 10 µg/L, though for several of the most common active ingredients, values were significantly higher (e.g., paracetamol 1368 µg/L, ciprofloxacin 125 µg/L). Overall, though healthcare facilities have been pointed out as key contributors, everyday household discharges are still generally acknowledged as the main contributor of human use pharmaceuticals to the environment (see Kummerer 2009c, Le Corre et al. 2012, Herrmann et al. 2015). In part, this is due to the sheer number of users and the amount of pharmaceutical consumption that takes place in the domestic context, with many outpatients also continuing treatment or receiving palliative care outside hospital facilities.

Regarding drug manufacturing, a number of studies have also reported environmental contamination as well as the damaging effects of exposure to effluents from pharmaceutical production sites, with several evidences of high concentrations in effluents, with contamination values reaching tens of mg/L (Fick et al. 2009, Cardoso et al. 2014, Larsson 2014). Remarkably, and for purposes of management and supervision, it is possible to reconstruct exposure pathways and disentangle factory source contamination from human use by evaluating the ratio of pharmaceutical precursor and of its human metabolites (e.g., Prasse et al. 2010). Overall, drug factory discharges, environmental risk and contamination patterns are not specifically linked to use patterns or seasonality, and will mainly affect coastal and marine environments via their localization, or via loading of rivers and streams with subsequent contamination downstream. Though pharmaceutical industries are mostly located in south east Asia (e.g., Bangladesh, China, India and Pakistan), production sites elsewhere (e.g., Europe, US) are also identified as significant contamination sources (see Cardoso et al. 2014, Larsson 2014, Rehman et al. 2015).

Veterinary applications of pharmaceuticals in both aquaculture and land based animal husbandry or livestock productions are also known contributors of pharmaceuticals to natural environments (Fig. 1). In response to the rising demand in seafood products worldwide, aquaculture has been seeing a continued boost in both the number of farms as well as production yield, and this is in part associated to the availability of an array of pharmaceutical compounds that enhance productivity (Sapkota et al. 2008, Tornero and Hanke 2016). The range of veterinary pharmaceuticals accessible to fish farmers include antibiotics, analgesics and antiparasitics, among others, some of them of generic human use, with many compounds applied prophylactically (Cabello 2006, Burridge et al. 2010, Tornero and Hanke 2016). Thus, any sea (coastal or estuarine) based aquaculture activities are direct entry points of pharmaceuticals into the marine environment. By large, pharmaceuticals are incorporated into feed, though other administration routes such as dilution and immersion in baths are available. In any case, these pharmaceutical compounds, as well as their excreted metabolites and transformation products will fuel environmental contamination and elicit impacts on non-target organisms (see Sapkota et al. 2008, Burridge et al. 2010, Chen et al. 2015). Other key pathways for pharmaceuticals to enter estuarine and coastal environments are wastewater discharges from land based aquaculture activities (Le and Munekage 2004, Zou et al. 2011). Though best practices vary worldwide, pond or tank based aquaculture of multiples species of crustaceans and fish is extensive throughout estuarine and coastal environments, with discharges made directly to these areas. Leakage from ponds is also an acknowledged pathway for veterinary pharmaceuticals to reach estuarine and coastal waters. Over the years, the range of concentrations in wastewater discharges or water and sediments surrounding aquaculture activities has been found to vary widely from a few ng/L to 2.5 mg/L (i.e., oxolinic acid) (Le and Munekage 2004, Chen et al. 2015, Kim et al. 2017). In some cases, the effect of released pharmaceutical loads may be aggravated by a combination of low flow conditions and the local abundance of juveniles of many species, as aquaculture farms are established in areas (e.g., estuarine habitats, mangroves) renowned to have a key nursery role (Beck et al. 2001) where potential effects on juvenile biota may be subsequently exported to adult populations (Rochette et al. 2010, Vasconcelos et al. 2011, Fonseca et al. 2015).

Discharges of both treated and untreated wastewaters from land based agricultural and livestock productions may also contribute to the presence of pharmaceuticals in estuaries and coastal environments (e.g., Lim et al. 2013, Awad et al. 2014, Paíga et al. 2016) (Fig. 1). Furthermore, fecal excretion of metabolites over pastoral lands can contaminate the groundwater via leaching or run-off, as can the use of waste products as fertilizers (e.g., manure). Overall, the presence of pharmaceuticals in groundwater can originate from many sources, including sewage contamination (e.g., via septic tanks or sewer leakage), contaminated leachate from landfill (e.g., animal carcasses, pharmaceutical waste products), use of contaminated sludge as fertilizer, as well as by the use of grey waters for irrigation (see review by Sui et al. 2015). The latter can all be relevant sources of pharmaceutical contamination to coastal marine environments, due to discharge and connectivity between groundwater and coastal systems (Dougherty et al. 2010, Sui et al. 2015) (Fig. 1). Overall, due to the added risk of contamination to public drinking water, evaluating pathways of groundwater contamination is also paramount (e.g., Fick et al. 2009).

Upon pharmaceutical intake, significant fractions of parent compound are excreted unprocessed as well as in the form of metabolites and transformation by-products. Irrespective of their sources, the fate and persistence of pharmaceuticals in coastal environments, as well as their subsequent potential to affect biota or bioconcentrate, are linked to key physico-chemical processes: transport, biodegradation, transformation and sequestration (Fig. 1). Overall, pathways and routes of exposure rely on both dissolved and particle transport (upon sorption to particulate matter or sediments), with the fate of individual pharmaceuticals influenced by environmental conditions (e.g., salinity, suspended particulate matter, hydrodynamics, water column mixing, pH, turbidity or light penetration) as well as by their own physical and chemical properties (Glassmeyer et al. 2007). Thus, information collated for degradation or sorption of pharmaceuticals in freshwater environments may not be directly applicable in estuarine and marine contexts (see for instance Fenet et al. 2014, Gaw et al. 2014, Zhao et al. 2015, Fabbri and Franzellitti 2016). Nonetheless, hydrolysis, photolysis, biodegradation and adsorption are recognized as the most likely to alter pharmaceutical compounds. Understanding the complex interactions among sorption kinetics, potential resuspension and transport is crucial; all these processes play a key role in the fate of pharmaceuticals along the interface between estuarine and coastal marine environments, and will determine the persistence of these compounds in the environment, their availability for bioaccumulation and exchange between environments (Kummerer 2009c, Liu et al. 2013). For instance, sorption to colloids can represent a sink for pharmaceuticals, increasing persistence but decreasing bioavailability (pending resuspension) whilst pH and salinity variations can also affect the ionization and solubility of different compounds. In transition areas, changing environmental conditions as well as major seasonal variations in both environmental conditions and contaminant inputs (e.g., in recreational and holiday areas) may have significant repercussions in the occurrence of pharmaceuticals, bioavailability and transfer to the marine environment (Liu et al. 2013, McEneff et al. 2014, Moreno-González et al. 2015, Zhao et al. 2015).

Ecotoxicological Effects of Pharmaceutical Exposure on Coastal and Marine Organisms

Pharmaceuticals are designed to elicit biological effects at low doses, targeting specific metabolic and physiological pathways to achieve the desired therapeutic effects in human and veterinary medicine. In addition to high specificity at low concentrations, the evolutionary conservation of most molecular targets across taxa implies that environmental concentrations of pharmaceuticals have the potential to chronically impact non-target aquatic organisms, with several adverse effects of pharmaceutical exposure reported in organisms at different levels of biological organisation (e.g., Huerta et al. 2012).

The majority of ecotoxicological data on pharmaceutical compounds pertains to the freshwater environment (reviews by Crane et al. 2006, Fent et al. 2006), yet scientific contributions on occurrence and effects of pharmaceuticals on coastal and marine biota are increasing. For the current chapter, we found 124 studies focusing on ecotoxicology of pharmaceuticals in coastal and marine biota, as well as on bioaccumulation in wild coastal and marine organisms. Six major therapeutic classes, namely analgesic non-steroid anti-inflammatory drugs (NSAIDs), antidepressants, antibiotics, anticonvulsants, antihypertensives and lipid regulators, clearly stood out and encompassed c. 91 percent of all studies. This is likely due to their frequent detection in the marine environment as well as to their higher sales and consumption. Nonetheless, we cannot exclude a bias towards better-known or more-established compounds and MOA, with researchers favoring the possibility of data comparison with available information.

Concerning major taxonomic groups, mollusks are the most frequent group of organisms in pharmaceutical accumulation and toxicity studies (69 studies), followed by crustaceans (32 studies) and fish (27 studies) (Fig. 2). Research with marine microorganisms, specifically microalgae (11 studies) and bacteria (six studies), were predominantly standard toxicity tests. Overall, research on the effects of the major therapeutic classes is well distributed among taxonomic groups. Though there is less data available for antihypertensives and lipid regulators on mollusks, in comparison to other therapeutic classes (Fig. 2). In terms of study type, most of the data relate to molecular changes (71 studies), which include gene and protein expression as well as other biochemical changes (e.g., biomarkers of oxidative stress and xenobiotics biotransformation) (Fig. 2). Molecular endpoints are ubiquitous to all therapeutic classes and are the primary endpoint in analgesic and NSAIDs toxicity studies, whereas behavior endpoints are particularly associated with antidepressant exposures. Effects on development, mortality and reproduction of marine biota are also important endpoints in pharmaceutical toxicity assessments (27, 21 and 17 studies, respectively). Twenty-two studies reported the accumulation of pharmaceuticals in finfish, crustaceans and shellfish tissues (Fig. 2). This is a noteworthy increase from the 14 studies identified in Gaw et al. (2014). Overall, antibiotics are the main therapeutic class investigated in bioaccumulation studies of marine and coastal organisms, with many studies linked to major aquaculture production. For instance, quinolones, sulfonamides and macrolides were detected in wild mollusk species collected along the coast of the Bohai Sea in China, with the highest concentrations ranging from 36 to 1575 µg/kg dw (Li et al. 2012). Evaluating bioaccumulation and biomagnification of several antibiotic agents in a marine trophic web in Laizhou Bay (China), trimethoprim, nine sulfonamide, five fluoroquinolone and four macrolide antibiotics were all detected in marine invertebrates and fish (Liu et al. 2017). Additionally, sulfonamides and trimethoprim were found to biomagnify along the food web, whilst fluoroquinolones and macrolides were biodiluted. Nonetheless, local seafood consumption was considered unlikely to pose a major human health risk, regarding antibiotic concentrations. Other studies have focused

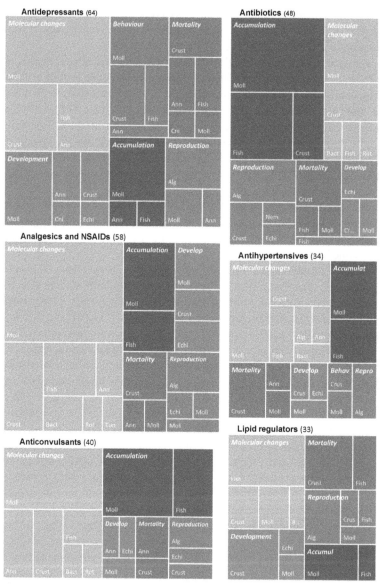

Fig. 2. Tree map representation of studies on the effects of pharmaceutical exposure in coastal and marine organisms per therapeutic class, biological endpoints and major taxonomic groups. Therapeutic classes are antidepressants, analgesics and non-steroid anti-inflammatories (NSAIDs), anticonvulsants, antibiotics, antihypertensives and lipid regulators. Biological endpoints and respective abbreviations are molecular changes, accumulation (accumul), development (develop), mortality, reproduction (repro) and behavior (behav). Major taxonomic groups and respective abbreviations are fish, tunicates (tun), echinoderms (echi), mollusks (moll), crustaceans (crust), rotifers (rot), annelids (ann), nematods (nem), cnidarians (cni), algae (alg) and bacteria (bact). Individual box sizes are proportional to number of entries, and total number of entries per therapeutic class is shown (*n*). Note that a single study may have multiple entrances per therapeutic class (total number of studies 124).

Color version at the end of the book

on field monitoring of several pharmaceutical compounds in coastal waters via long term caging experiences with marine bivalves. Wille et al. (2011) detected five pharmaceuticals in caged mussels along the Belgium coast, namely salicylic acid, paracetamol, propranolol, ofloxacin and carbamazepine (highest concentrations ranging from 11 to 490 ng/g dw). McEneff et al. (2014) quantified carbamazepine, mefenamic acid and trimethoprim (peak concentrations of 7.28 to 9.22 ng/g dw) in *Mytilus* spp. following a one year experiment in the Irish coast. Yet, knowledge on the fate, biotransformation and bioaccumulation of pharmaceutical compounds in the marine environment is still insufficient, as evidenced by the lack of concordance between field-derived bioaccumulation factors for ribbed horse mussels (*Geukensia demissa*) and model-predicted bioconcentration factors (Klosterhaus et al. 2013), as well as the lack of correlations between accumulation and observed molecular effects of pharmaceuticals (Mezzelani et al. 2016b).

Analgesics and NSAIDs reduce pain and inflammation and are amongst the highest consumed pharmaceuticals worldwide (Fent et al. 2006). Representative compounds include acetaminophen, diclofenac, ibuprofen, ketoprofen and indomethacin. Biological targets are cyclooxygenase isoforms (Cox1 and Cox2) and these drugs act by non-specific inhibition of the synthesis of various prostaglandins from arachidonic acid (Vane and Botting 1998). Besides being involved in inflammation and pain responses, prostaglandins also play important roles in various physiological functions, including reproduction processes, reducing hypertension, fatty acids metabolism and synthesis of the protective gastric mucosa (Jones 1972). In fish and marine invertebrates, prostaglandins have been related with reproduction, ion regulation and immune responses (Sorbera et al. 2001, Rowley et al. 2005). Accordingly, the marine clam, *Ruditapes philippinarum*, exhibited significant immunological alterations following a seven day exposure to ibuprofen, particularly at the highest concentrations tested (500 and 1000 ug/L) (Matozzo et al. 2012). In the crustacean *Carcinus maenas*, osmoregulatory capacity was impaired with environmentally relevant concentrations of diclofenac over 7 days (10 ng/L and 100 ng/L), although no stress-related effects were observed in these individuals (Eades and Waring 2010). The potential of NSAIDs for endocrine disruption has been suggested in mussels *Mytilus galloprovincialis* exposed to 250 ng/L of ibuprofen or diclofenac for two weeks, with males and females presenting elevated levels of gonad vitellogenin-like proteins (Gonzalez-Rey and Bebianno 2012, 2014). However, a shorter time frame study with diclofenac (1 to 1000 ug/L, 96 h) and bivalves *Mytilus* spp. did not show differences in the expression of these vitellogenin-like proteins (Schmidt et al. 2011). Akin to endocrine disruption, variable neurotoxic responses have been reported in mussels in response to analgesics and NSAIDs, namely via tissue specific inhibition and the increased activity of acetylcholinesterase (AChE) (e.g., Milan et al. 2013, Mezzelani et al. 2016a, 2016b). Multibiomaker approaches highlighted changes in immunological responses, lipid metabolism and DNA integrity in *M. galloprovincialis* exposed separately to various analgesics and NSAIDs (acetaminophen, diclofenac, ibuprofen, ketoprofen or nimesulide) at 25 μg/L and 0.5 μg/L concentrations for 14 days (Mezzelani et al. 2016a, 2016b). Subsequent gene transcription analysis, via DNA microarrays, corroborated biomarker responses, highlighting the similarities on proposed MOA of NSAIDs between bivalves and vertebrate species (Mezzelani et al. 2016b). However, the lack of a significant change in oxidative stress biomarkers (catalase, glutathione peroxidase and glutathione reductase activities, total glutathione, and total oxyradical scavenging capacity) or recovery of the antioxidant system indicate that prooxidant response is not a key target in the pharmacology of these compounds (Gonzalez-Rey and Bebianno 2014, Mezzelani et al. 2016a, 2016b).

To date, the effects of analgesics and NSAIDs in coastal and marine fish species have only been evaluated *in vitro*, and to test their effects on the activities of several enzymes related to xenobiotic and steroid metabolism. Ribalta and Solé (2014) reported that diclofenac significantly interfered in the CYP1A and CYP3A systems of Mediterranean fishes, particularly in the middle slope gadiform *Trachyrincus scabrous*. Ibuprofen exposure (100 µM concentration) also inhibited the activity of the CYP3A4 enzyme, benzyloxy-4-[trifluoromethyl]-coumarin-O-debenzyloxylase (BFCOD) in the liver microsomal fraction of *Solea solea*, whilst acetaminophen had no effects on measured enzyme activities (Crespo and Solé 2016).

Antidepressants are neuroactive drugs for the treatment of depression and related psychiatric disorders (e.g., anxiety, obsessive-compulsive disorder, post-traumatic stress disorder). Selective serotonin reuptake inhibitors (SSRIs, such as fluoxetine, sertraline, citalopram) and serotonin and norepinephrine reuptake inhibitors (SNRIs, such as venlafaxine) are some of the most prescribed antidepressants. Their highly specific MOA is based on the modulation of neurotransmission in the human brain, targeting and blocking serotonin and norepinephrine reuptake proteins which leads to increased levels of these neurotransmitters in the synaptic cleft (Hiemke and Härtter 2000). Serotonin is also present in lower vertebrates and invertebrates, and as in humans, this biogenic monoamine appears to be involved in various physiological functions and behaviors interacting with reproduction and neuroendocrine processes (e.g., Winberg and Nilsson 1993, Winberg et al. 1997, Fong 1998). A review of the effects of antidepressant exposure on mollusks and crustaceans outlined the impacts of antidepressants on the organisms' metabolism, growth, reproduction, feeding, locomotion and behavior, yet the bulk of information was related to freshwater invertebrates (Fong and Ford 2014). Noteworthy, changes to spawning and larval release in bivalves as well as impaired locomotion and fecundity in snails occurred at environmentally relevant concentrations of antidepressants; although the occurrence of non-monotonic dose response curves were also reported with significant biological effects at lower but not at higher concentrations (Fong and Ford 2014). Regarding toxicity to marine invertebrates, altered cognitive capacities (learning and memory retention) and less efficient cryptic behaviors were observed in cuttlefish *Sepia officinalis* following fluoxetine exposure at hatchling stages (1 ng/L to 100 ng/L) (Di Poi et al. 2013, 2014). Fluoxetine at concentrations ranging from 43 µg/L to 4.34 mg/L, also induced foot detachment from the substrate in five species of marine snails from different habitats (Fong and Molnar 2013). This potentially lethal outcome was also observed in two marine snail species exposed to venlafaxine, albeit different locomotion behaviors at the onset of foot detachment suggest that venlafaxine and fluoxetine have different physiological mechanisms of action (Fong et al. 2015). In another study, long-term exposure to low concentrations of fluoxetine (0.3 ng/L to 300 ng/L) diminished algal clearance rates, growth and gonadosomatic index in California mussel *M. californianus* (Peters and Granek 2016). Pharmacological effects of fluoxetine and trait-based sensitivity have also been described for the marine worm *Hediste diversicolor*, based on increased serotonin levels in the coelomic fluid and tissues. Fluoxetine effects on *H. diversicolor* included weight loss (up to 2 percent at 500 µg/L), decreased feeding rates (68 percent at 500 µg/L) and increased oxygen consumption and ammonia excretion (from 10 µg/L), but only limited influence on predator avoidance behaviors (Hird et al. 2016). Regarding sertraline, an early life-stage bioassay with sea urchin embryos found it to be highly toxic given the development of significant abnormalities at ng/L range concentrations (Ribeiro et al. 2015).

As for invertebrates, information on the toxicity of antidepressants to fish stem mainly from freshwater species. These include deleterious effects on physiology, reproduction, behavior (e.g., reproductive, predator avoidance, territorial and defensive behaviors) and the potential of SSRIs as endocrine disruption compounds in various fish species (e.g., Mennigen et al. 2010a, 2010b, Schultz et al. 2011, Weinberger and Klaper 2014). Few studies have evaluated the effects of antidepressants on coastal and marine fish species. Two *in vitro* studies reported species specific responses when assessing the impact of multiple pharmaceuticals on various enzyme activities in coastal and deep-sea fishes: (1) In Solé and Sanchez-Hernandez (2015), fluoxetine had no effect on carboxylesterase (CbE). The latter is involved in metabolism and activation of numerous exogenous and endogenous compounds in humans and has been associated with pesticide detoxification in fish (Wheelock et al. 2008). In Ribalta and Solé (2014), fluoxetine inhibited cytochrome P450, which are enzymes linked to phase I of xenobiotic metabolism as well as to the metabolism of endogenous compounds (e.g., steroids). Moreover, high concentrations of fluoxetine, administered intraperitoneally to the gulf toadfish *Opsanus beta*, affected the branchial urea excretion and intestinal osmoregulation and resulted in a severe stress response with high levels of plasma cortisol (Morando et al. 2009). Additionally, fluoxetine has also been shown to affect marine fish behavior by reducing locomotor activity (EC50 155 µg/L at 32 h of exposure) (Winder et al. 2012).

Beta-adrenergic receptor antagonists or β-blockers are antihypertensive drugs, commonly used to treat high blood pressure, angina, arrhythmias and other cardiac conditions. The MOA of blockers such as propranolol, atenolol and metoprolol consists in their specific binding to adrenoreceptors, competing with β-adrenergic agonists, decreasing resting heart rate, cardiac output and cardiac muscles contractibility, among others (Bourne 1981). A comparative physiology review described the similarity in beta-adrenergic receptors between mammals and fish, highlighting the diversity of physiological processes mediated by these receptors, and proposed biomarkers for β-blockers exposure included cardiovascular dysfunction, with subsequent potential negative effects on fish growth and fecundity (Owen et al. 2007). Only recently have the effects of β-blockers been evaluated in marine fish. Both studies were *in vitro* 100 µM propranolol exposures and described decreased CbE and BFCOD activities in the microsomal fraction of liver of coastal and deep-sea fishes (Solé and Sanchez-Hernandez 2015, Crespo and Solé 2016).

Sublethal toxicology of propranolol on marine invertebrates include molecular, physiological and behavior changes. Motor activity of amphipod *Gammarus* sp. has been shown to decrease even in the presence of predator cues, with respiration rate and feeding rate increasing with propranolol concentrations (100 µg/L to 5 mg/L), probably to compensate for higher energy requirements (Wiklund et al. 2011). However, another study documented a decreased feeding rate with associated oxidative damage and neurotoxicity in mussels (147 µg/L propranolol) (Solé et al. 2010). Exposure has also been linked with lower scope for growth, byssus strength and byssus abundance, potentially reducing substrate fixation ability in blue mussels, albeit at remarkably high propranolol concentrations (1 to 10 mg/L) (Ericson et al. 2010). A series of complementary experiments with mussels *M. galloprovincialis*, evaluated the MOA, molecular targets and associated endpoints, as well as unspecific effects of exposure to pharmaceuticals interacting with the cAMP-dependent pathway. cAMP cell signaling influences various physiological functions of mussels, namely their reproduction, metabolic regulation, and filtering efficiency (Fabbri and Capuzzo 2010). Overall, exposure to environmentally-relevant concentrations of propranolol revealed differences on cAMP-related endpoints, suggesting differential expression of molecular targets in digestive glands, mantle/gonads and gill tissues (Franzellitti et al. 2011). Furthermore, coexposure to fluoxetine and propranolol suggested adrenergic regulation in the digestive gland, whereas serotonergic prevailed in the mantle/gonads of exposed mussels (Franzellitti et al. 2013). A multibiomarker approach revealed altered lysosomal parameters in mussels exposed to low propranolol concentration (0.3 ng/L), but other oxidative stress responses were only observed in the combined fluoxetine and propranolol treatment (Franzellitti et al. 2015). Furthermore, transcriptional and functional regulation of genes (e.g., ABCB) and transporters (e.g., P-glycoprotein) related to the multixenobiotic resistance (MXR) system highlighted the potential of propranolol to impair immunotoxic response in mussels, thus potentially affecting the ability to extrude contaminants and cope with environmental stressors in general (Franzellitti and Fabbri 2013, Franzellitti et al. 2016).

Anticonvulsants, also termed antiseizure or antiepileptics, are neuroactive drugs that interact with the central nervous system to treat epilepsy, bipolar disorder and are increasingly used as mood-stabilizers. Several compounds lead to decreased neuronal activity through different MOA. For example, benzodiazepines (such as diazepam or lorazepam) enhance the γ-aminobutyric acid (GABA) neurotransmitter affinity for its receptor increasing chloride channel opening frequency, whilst carbamazepine acts via the blockage of sodium voltage-dependent channels of excitatory neurons inhibiting their sustained firing. Both result in lower cell excitation (Rang et al. 1999). A high degree of evolutionary conservation in GABA receptors (e.g., in fish) whose functions are related with reducing neuronal excitability and muscle tension has been reported (Carr and Chambers 2001). Even if carbamazepine's MOA is not fully understood, molecular targets appear to be conserved in mussels *M. galloprovincialis* following *in vivo* exposure, with reduction of the second messenger cyclic AMP and cAMP-dependent protein kinase (PKA), akin to responses in mammals (Martin-Diaz et al. 2009). Follow-up studies have described transcriptional and functional impairment of the MXR system in this species, highlighting the potential of carbamazepine, and others (i.e., fluoxetine and propranolol), in inducing immunotoxicological effects in marine bivalves at environmental relevant concentrations (Franzellitti et al. 2010, 2014, 2016). Other recent studies have focused on the effects of carbamazepine exposure on biomarker responses in several marine invertebrate species. Biomarkers of cellular health (e.g., lysosomal membrane stability, LMS), xenobiotic metabolism (e.g., EROD, GST), oxidative stress (e.g., CAT, SOD, LPO), neurotoxicity (AChE) and genotoxicity (DNAd) have all been induced by varying exposure concentrations of carbamazepine in crab *C. maenas* (Aguirre-Martínez et al. 2013a, 2013c), clams *R. philippinarum* (Aguirre-Martínez et al. 2013b, 2016, Almeida et al. 2014) and *Scrobicularia plana* (Freitas et al. 2015), and in the polychaetes *H. diversicolor* (Pires et al. 2016) and *Diopatra neapolitana* (Freitas et al. 2015). Toxicity of anticonvulsants in coastal and marine fish has seldom been reported. Reduced oxidative stress response, increased swimming lethargy and abnormal posture were observed in the euryhaline fish *Gambusia holbokrii* following acute diazepam exposure (in mg/L range) (Nunes et al. 2008), with acute toxicity LC50 estimated at 12.7 mg/L (Nunes et al. 2005). *In vitro* assays confirmed inhibitory action of carbamazepine on CbE and BFCOD activity in coastal and deep-sea fish species (Solé and Sanchez-Hernandez 2015, Crespo and Solé 2016).

Lipid regulators or antilipidemic drugs include two major groups of lipid lowering agents: statins (e.g., simvastatin) and fibrates (e.g., bezafibrate, gemfibrozil). Their therapeutic role is to decrease the concentration of cholesterol and triglycerides (fibrates only) in blood plasma. Statins, such as simvastatin and atorvastatin, inhibit the activity of the enzyme HMG-CoA (3-hydroxymethylglutaryl coenzyme A reductase), which is responsible for feedback control of cholesterol synthesis. As a result of decreased intracellular cholesterol concentration, there is an over expression of low-density lipoprotein (LDL) receptors in hepatocyte membranes which leads to the resorption of circulating LDL cholesterol. Recent findings suggest that the MOA of statins is highly conserved, and hypothesized that all metazoan taxa might be susceptible to the effects of statins at environmentally relevant concentrations (Santos et al. 2016). Fibrates are peroxisomal proliferators whose MOA is not yet fully described. Their action is mediated through changes in the expression of the genes involved in lipoprotein metabolism. Fibrates bind to nuclear transcription factors of peroxisome proliferator activated receptors (PPARs), which then interacts with various cellular pathways determining hepatic lipid uptake and the metabolism of free fatty acids (Rang et al. 1999).

Antilipidemic toxicity data in marine organisms is limited, nonetheless recent studies have reported a variety of effects on the development and reproduction of invertebrates, whereas in fish, responses have been mainly assessed through molecular and biochemical changes. Chronic exposure to low levels of simvastatin (64 ng/L to 8 µg/L) in the marine amphipod *G. locusta* ensured severe impacts on growth, gonad maturation and fecundity, the latter at relevant environmental concentrations (Neuparth et al. 2014). In sea urchin *Paracentrotus lividus*, Ribeiro et al. (2015) described delayed embryo development and increased percentage of embryo abnormalities when exposed to simvastatin (5 and 2 mg/L, respectively). Accordingly, another study considering a range of realistic environmental concentrations of simvastatin (0.16 and 1.6 µg/L), reported a decrease in development time and a concomitant increase in body length and growth rate of copepods *Nitokra spinipes* (Dahl et al. 2006). Regarding gemfibrozil, exposure at 1 mg/L induced vitellogenin-like proteins (ALP) in *Mytilus* spp., which may be indicative of the potential for endocrine disruption by this fibrate (Schmidt et al. 2011). Concerning lipid regulators' toxicity to fish, gemfibrozil exposure (150 µg/L) upregulated PPAR-related genes transcription in juvenile *Sparus aurata*, albeit no concomitant activation of PPAR pathways was observed (Teles et al. 2016). Activation of immune responses was also suggested following increased mRNA levels of genes linked with pro-inflammatory processes at 15 ug/L gemfibrozil. Increase in cortisol, as evidence of stress related effects from gemfibrozil exposure were also observed, even if only at a concentration of 1.5 mg/L (Teles et al. 2016). Gemfibrozil (injected at 1 mg/kg body weight in *Solea senegalensis*) also induced the activity of CYP-related and phase II (UDPGT) biotransformation enzymes, whilst inhibiting antioxidant defenses (Solé et al. 2014). Furthermore, simvastatin and fenofibrate have been shown to inhibit CbE activity in various coastal and deep-sea fishes (Solé and Sanchez-Hernandez 2015), with simvastatin exposure also decreasing AChE levels in estuarine *Fundulus heteroclitus*, (1.25 mg/L and LC50 of 2.68 mg/L) (Key et al. 2009).

Antibiotics are used in both human and veterinary medicine to treat bacterial infections, but may also be used as animal growth promoters. This group encompasses compounds derived from natural products (e.g., secondary metabolites of bacterial origin), semi-synthetic derivatives or completely synthetic compounds which act through various mechanisms, such as suppression of bacterial cell wall or of protein synthesis and growth (Kummerer 2009a). Penicillins (e.g., penicillin and amoxicillin), macrolides (e.g., erythromycin), quinolones (e.g., ciprofloxacin) and tetracyclins (e.g., tetracycline) are amongst the most common types of antibiotics.

As antibiotics are designed to target microorganisms, their toxicity on bacteria and microalgae is commonly 2 to 3 orders of magnitude above effect levels reported for higher trophic groups (Kummerer 2009a). Accordingly, exposure to clarithromycin and clindamycin induced significant growth inhibition in the marine diatom *Skeletonema marinoi* at very low concentrations (EC50 of 156 and 154 ng/L, respectively) (Minguez et al. 2016). In contrast, Aguirre-Martínez et al. (2015) reported an EC50 of 400 mg/L for inhibition of bacterial luminescence in *Vibrio fischeri*, after 15 min of exposure to the antibiotic novobiocin, and an IC50 of 72.8 mg/L for growth inhibition in the algae *Isochysis galbana* (96 h exposure period). This study also reported effects for other pharmaceuticals in the mg/L range, yet novobiocin showed the highest toxicity for microorganisms when compared with IC50 values determined for carbamazepine, ibuprofen and caffeine. Growth of marine microalgae (*I. galbana* and *Tetraselmis chui*) was inhibited by three different antibiotics not usually found in the environment (chloramphenicol, florfenicol and thiamphenicol) with EC50 values ranging from 1.3 to 158 mg/L. Concerning other phototrophs, one study reported that sulfathiazole exposure, in concentrations commonly used in aquaculture (25 to 50 mg/L), induced growth inhibition on macroalgae *Ulva lactuca* (Leston et al. 2014).

In marine bivalves, exposure to trimethophin (300 to 900 ng/L) and to amoxicillin (100 to 400 µg/L) affected haemocyte parameters in both *R. philippinarum* and

M. galloprovincialis (Matozzo et al. 2015, Matozzo et al. 2016). The genotoxicity of amoxicillin was also confirmed via increased micronucleus frequency in both species' haemolymph (Matozzo et al. 2016). Similarly, exposure to environmental concentrations of oxytetracycline resulted in decreased lysosomal membrane stability in mussels (Banni et al. 2015). Regarding crustaceans, Han et al. (2016) described several toxicity effects of trimethophin exposure (in the mg/L range) in copepod *Tigriopus japonicas*, including increased ROS levels, upregulation of antioxidant and xenobiotic detoxication-related genes, delayed development time and impaired reproduction. Antibiotic toxicity in marine fish, encompasses thus far, feeding behavior and biomarker responses in juveniles of the common goby *Pomatoschistus microps* exposed to cefalexin (from 1.3 to 10 mg/L) (Fonte et al. 2016). At 20°C and over four days exposure, predation performance was significantly impaired (> 5 mg/L) and lipid peroxidation levels increased (at 10 mg/L). At 25°C, the cefalexin toxicity increased with a decrease of predation performance at 2.5 mg/L (Fonte et al. 2016).

Antibiotics could also have relevant ecosystem level effects through changes to microbial communities and their functions (e.g., denitrification, organic matter decomposition), compromising ecosystem health (Kummerer 2009a, Caracciolo et al. 2015). Furthermore, constant environmental exposure could promote the development of antibiotic resistance (Kummerer 2009b), which is a public health issue if resistance is transferred to human pathogens (Baran et al. 2011).

Despite the limited number of studies, in comparison to freshwater systems, the information currently available on the ecotoxicity of pharmaceuticals to coastal and marine species can already be taken into consideration for management and regulation purposes. The examples highlighted in this chapter clearly demonstrate that multiple pharmaceutical compounds have adverse effects on coastal and marine organisms at environmentally relevant concentrations (e.g., Franzellitti et al. 2016, Minguez et al. 2016). However, current legislation is still mostly based on freshwater toxicity data, even though the marine environment may be more sensitive to pharmaceutical residues than freshwater (Minguez et al. 2016—based on a comparative toxicity analyses of 48 pharmaceuticals in both marine and freshwater microalgae and crustacean species). Ultimately, there are still multiple shortcomings in the evaluation of pharmaceutical contamination in coastal and marine environments that we should aim to resolve.

Knowledge Gaps, Current Challenges and Futures Perspectives

There is still a lack of information regarding concentrations, fate and ecotoxicology of pharmaceuticals in coastal and marine environments (Brausch et al. 2012, Fabbri and Franzellitti 2016). Additionally, there is also a clear disparity of information among regions worldwide which we should tackle. For developing regions, where population increase, higher standards of living and improved access to pharmaceuticals will likely contribute to increased environmental contamination; this could be a key moment to start early monitoring schemes to evaluate environmental accumulation, and to develop associated strategies to minimize detrimental impacts both from household use and commercial enterprise (e.g., aquaculture, industry). In developed countries, mitigation plans are necessary as the environmental pressure exerted by pharmaceuticals will continue to rise linked to population ageing and prevalence of chronic diseases. However, up to now most approaches are limited to spatial or temporal isolated data, lacking long-term aims, rather than encompassing large regional and temporal coverage. The latter is particularly important in coastal and transition systems, where variations in loadings are associated to natural fluctuations in physical and chemical conditions (e.g., salinity, river flow, temperature, water chemistry), which may imply significant changes to the fate of pharmaceuticals in the environment (Glassmeyer et al. 2007, Zhao et al. 2015).

In the long run, management strategies for contamination by pharmaceuticals should aim to act in advance of ensuing adverse effects, promote the development of a suit of best practices to reduce their occurrence in the environment, and drive the improvement of systems that constrain potential contamination sources or increase the effectiveness of the removal and degradation of these compounds from the environment. The first line of action to reduce the potential entry of pharmaceuticals in the environment are WWTP, with continued research on the behavior, degradation and varying removal efficiencies of different WWTP treatments for multiple therapeutic classes still required. Developing novel methodologies that enhance the efficacy of WWTPs tertiary treatment to specifically remove or degrade pharmaceutical compounds is an acknowledged path for reducing the potential impact of pharmaceuticals (Margot et al. 2013, Calisto et al. 2017). In fact, Directive 2013/39 EU (European Parliament 2013) underlines the importance of finding new ways of tackling water pollution by pharmaceuticals, and unravelling the physico-chemical processes that determine degradation and transformation of pharmaceutical compounds, their metabolites and by-products that will further contribute to resolving these issues.

Different pharmaceuticals have been shown to bioaccumulate (Klosterhaus et al. 2013) and even biomagnify (Liu et al. 2017), yet in general, there is insufficient information on bioaccumulation and the impacts of pharmaceutical residues across the trophic web, namely for top-predators (Gaw et al. 2014). Likewise, given the effects of pharmaceuticals on bacteria and algae (Backhaus et al. 2011, Minguez et al. 2016) and the high degree of homology between chloroplasts and bacteria as well as among other metabolic pathways across multiple phyla (Brain et al. 2008), the lack of research on higher marine phototrophs (e.g., halophytes, plants) is conspicuous.

Compiling information on bioaccumulation, effects and understanding MOA and adverse outcomes of pharmaceuticals are critical for the effective management of pharmaceutical contamination and to safeguard coastal and marine biota. Thousands of different active pharmaceutical ingredients are available for human and veterinary use, which impedes assessing the full spectrum of contaminants in any given monitoring scheme. Furthermore, the consumed amount and toxicity of individual drugs varies greatly; thus it is key to prioritize research directives, monitoring and regulation. Several options have been forwarded over the years (e.g., Schreiber et al. 2011, Caldwell et al. 2014, Rudd et al. 2014), though three main aspects to take into consideration are generally consumption levels, ecotoxicological risk and persistence in the environment. Rather than in isolation, these facets should be evaluated simultaneously, as directing resources to higher risk but low use or persistence pharmaceuticals may not prove a good investment of time and resources. Approaches based on MOA take into consideration the evolutionary and functional conservation of molecular targets of pharmaceuticals (e.g., receptors, enzymes), and cellular and physiological processes across species, which enables the identification of relevant endpoints and experimental conditions to determine drug toxicity (Christen et al. 2010, Fabbri and Franzellitti 2016, Santos et al. 2016). Furthermore, chronic exposure assessments at environmentally significant concentrations are central to evaluate the risk posed by pharmaceutical substances (Fabbri and Franzellitti 2016). Acute testing has several limitations that can compromise resulting environmental regulation. Yet, contamination thresholds are still mostly based on acute standard toxicity tests. Even though, they are less sensitive than other endpoints in non-model species (e.g., Aguirre-Martínez et al. 2015), and neglect potential long-term effects from chronic exposures, which are more representative of the persistent contamination organisms experience in their natural environment (Crane et al. 2006, Fent et al. 2006).

Ecotoxicological assessments should strive to fill the gap between sub-cellular endpoints and adverse individual or population level effects. This is a major challenge and requires the development of frameworks that synthesize data at many levels of biological organization. The adverse outcome pathways (AOP) is a good example of this, and several studies have illustrated the potential of AOP for population-modelling and predictive ecotoxicology (Ankley et al. 2010, Franzellitti et al. 2014, Hird et al. 2016). Furthermore, the utility of the AOP approach has been demonstrated for cross species extrapolation and integrating life-history theory (Groh et al. 2015). One of the key issues is ensuring baseline toxicity studies produce robust and accurate quantitative data that can be subsequently integrated in population modelling approaches. Ideally dose-response or concentration-response relationships for both lethal and sub-lethal effects should be defined allowing response curves, effect-thresholds and the probability of effects occurring at different levels of biological organization to be estimated (Kramer et al. 2011).

Ultimately, monitoring of prioritized pharmaceuticals, metabolites and by-products in coastal environments should complement risk assessment and is integral to current European policy. The EU dynamic watch list for emerging contaminants under the Water Framework Directive (WFD—Directive 2000/60/EC) includes pharmaceutical diclofenac, and two hormones 17-beta-estradiol (E2), and 17-alpha-ethinylestradiol (EE2), with three additional antibiotics proposed for inclusion (erythromycin, clarithromycin and azithromycin). In addition to consistent water collections and analysis, monitoring strategies can build upon the success of programs such as Mussel Watch (Goldberg and Bertine 2000), which would allow for both bioaccumulation (Wille et al. 2011, McEneff et al. 2014) and monitoring of effects and ecotoxicology via a standardized set of biomarkers (Franzellitti et al. 2015, Mezzelani et al. 2016a). Other prospective monitoring tools for baseline concentration data include the use of passive sampling devices (Martínez Bueno et al. 2016) or the use of unmanned automated sampling devices in ships and marine platforms of opportunity (Brumovsky et al. 2016).

Acknowledgments

This work had the support of the Fundação para a Ciência e a Tecnologia (FCT) via UID/MAR/04292/2013 and project grant PTDC/MAR-EST/3048/2014. PRS was funded with FCT postdoctoral grant SFRH/BPD/95784/2013.

References

Aguirre-Martínez, G.V., Buratti, S., Fabbri, E., Del Valls, T.A. and Martin-Diaz, M.L. 2013a. Stability of lysosomal membrane in *Carcinus maenas* acts as a biomarker of exposure to pharmaceuticals. Environ. Monit. Assess. 185: 3783–3793.

Aguirre-Martínez, G.V., Buratti, S., Fabbri, E., Del Valls, T.A. and Martín-Díaz, M.L. 2013b. Using lysosomal membrane stability of haemocytes in *Ruditapes philippinarum* as a biomarker of cellular stress to assess contamination by caffeine, ibuprofen, carbamazepine and novobiocin. J. Environ. Sci.-China 25: 1408–1418.

Aguirre-Martínez, G.V., Del Valls, T.A. and Martin-Diaz, M.L. 2013c. Early responses measured in the brachyuran crab *Carcinus maenas* exposed to carbamazepine and novobiocin: application of a 2-tier approach. Ecotox. Environ. Saf. 97: 47–58.

Aguirre-Martínez, G.V., Owuor, M.A., Garrido-Perez, C., Salamanca, M.J., Del Valls, T.A. and Martin-Diaz, M.L. 2015. Are standard tests sensitive enough to evaluate effects of human pharmaceuticals in aquatic biota? Facing changes in research approaches when performing risk assessment of drugs. Chemosphere 120: 75–85.

Aguirre-Martínez, G.V., Del Valls, T.A. and Martín-Díaz, M.L. 2016. General stress, detoxification pathways, neurotoxicity and genotoxicity evaluated in *Ruditapes philippinarum* exposed to human pharmaceuticals. Ecotox. Environ. Saf. 124: 18–31.

Almeida, A., Calisto, V., Esteves, V.I., Schneider, R.J., Soares, A., Figueira, E. and Freitas, R. 2014. Presence of the pharmaceutical drug carbamazepine in coastal systems: Effects on bivalves. Aquat. Toxicol. 156: 74–87.

Alygizakis, N.A., Gago-Ferrero, P., Borova, V.L., Pavlidou, A., Hatzianestis, I. and Thomaidis, N.S. 2016. Occurrence and spatial distribution of 158 pharmaceuticals, drugs of abuse and related metabolites in offshore seawater. Sci. Total Environ. 541: 1097–1105.

Aminot, Y., Le Menach, K., Pardon, P., Etcheber, H. and Budzinski, H. 2016. Inputs and seasonal removal of pharmaceuticals in the estuarine Garonne River. Mar. Chem. 185: 3–11.

Ankley, G.T., Bennett, R.S., Erickson, R.J., Hoff, D.J., Hornung, M.W., Johnson, R.D., Mount, D.R., Nichols, J.W., Russom, C.L., Schmieder, P.K., Serrrano, J.A., Tietge, J.E. and Villeneuve, D.L. 2010. Adverse outcome pathways: a conceptual framework to support ecotoxicology research and risk assessment. Environ. Toxicol. Chem. 29: 730–741.

Arpin-Pont, L., Bueno, M.J.M., Gomez, E. and Fenet, H. 2016. Occurrence of PPCPs in the marine environment: a review. Environ. Sci. Pollut. Res. 23: 4978–4991.

Awad, Y.M., Kim, S.-C., Abd El-Azeem, S.A.M., Kim, K.-H., Kim, K.-R., Kim, K., Jeon, C., Lee, S.S. and Ok, Y.S. 2014. Veterinary antibiotics contamination in water, sediment, and soil near a swine manure composting facility. Environ. Earth Sci. 71: 1433–1440.

Azuma, T., Arima, N., Tsukada, A., Hirami, S., Matsuoka, R., Moriwake, R., Ishiuchi, H., Inoyama, T., Teranishi, Y., Yamaoka, M., Mino, Y., Hayashi, T., Fujita, Y. and Masada, M. 2016. Detection of pharmaceuticals and phytochemicals together with their metabolites in hospital effluents in Japan, and their contribution to sewage treatment plant influents. Sci. Total Environ. 548-549: 189–197.

Backhaus, T., Porsbring, T., Arrhenius, Å., Brosche, S., Johansson, P. and Blanck, H. 2011. Single-substance and mixture toxicity of five pharmaceuticals and personal care products to marine periphyton communities. Environ. Toxicol. Chem. 30: 2030–2040.

Banni, M., Sforzini, S., Franzellitti, S., Oliveri, C., Viarengo, A. and Fabbri, E. 2015. Molecular and cellular effects induced in *Mytilus galloprovincialis* treated with oxytetracycline at different temperatures. PLOS ONE 10: e0128468.

Baran, W., Adamek, E., Ziemiańska, J. and Sobczak, A. 2011. Effects of the presence of sulfonamides in the environment and their influence on human health. J. Hazard Mater. 196: 1–15.

Beck, M.W., Heck Jr., K.L., Able, K.W., Childers, D.L., Eggleston, D.B., Gillanders, B.M., Halpern, B., Hays, C.G., Hoshino, K., Minello, T.J., Orth, R.J., Sheridan, P.F. and Weinstein, M.P. 2001. The identification, conservation, and management of estuarine and marine nurseries for fish and invertebrates. BioScience 51: 633–641.

Bourne, G.R. 1981. The metabolism of β-adrenoreceptor blocking drugs. *In*: Bridges, J.W. and Chasseaud, L.F. (eds.). Progress in Drug Metabolism, Book 6. John Wiley and Sons, London.

Brain, R.A., Hanson, M.L., Solomon, K.R. and Brooks, B.W. 2008. Aquatic plants exposed to pharmaceuticals: effects and risks. *In*: Whitacre, D.M. (ed.). Rev. Environ. Contam. Toxicol. Book 192. Springer, New York, NY.

Brausch, J.M., Connors, K.A., Brooks, B.W.R. and Rand, G.M. 2012. Human pharmaceuticals in the aquatic environment: a review of recent toxicological studies and considerations for toxicity testing. Rev. Environ. Contam. Toxicol. 218: 1–99.

Brumovsky, M., Becanova, J., Kohoutek, J., Thomas, H., Petersen, W., Sorensen, K., Sanka, O. and Nizzetto, L. 2016. Exploring the occurrence and distribution of contaminants of emerging concern through unmanned sampling from ships of opportunity in the North Sea. J. Marine Syst. 162: 47–56.

Burridge, L., Weis, J.S., Cabello, F., Pizarro, J. and Bostick, K. 2010. Chemical use in salmon aquaculture: A review of current practices and possible environmental effects. Aquaculture 306: 7–23.

Cabello, F.C. 2006. Heavy use of prophylactic antibiotics in aquaculture: a growing problem for human and animal health and for the environment. Environmental Microbiol. 8: 1137–1144.

Caldwell, D.J., Mastrocco, F., Margiotta-Casaluci, L. and Brooks, B.W. 2014. An integrated approach for prioritizing pharmaceuticals found in the environment for risk assessment, monitoring and advanced research. Chemosphere 115: 4–12.

Calisto, V., Jaria, G., Silva, C.P., Ferreira, C.I.A., Otero, M. and Esteves, V.I. 2017. Single and multi-component adsorption of psychiatric pharmaceuticals onto alternative and commercial carbons. J. Environ. Manage. 192: 15–24.

Cantwell, M.G., Katz, D.R., Sullivan, J.C., Ho, K., Burgess, R.M. and Cashman, M. 2016. Selected pharmaceuticals entering an estuary: Concentrations, temporal trends, partitioning, and fluxes. Environ. Toxicol. Chem. 35: 2665–2673.

Caracciolo, A.B., Topp, E. and Grenni, P. 2015. Pharmaceuticals in the environment: Biodegradation and effects on natural microbial communities. A review. J. Pharm. Biomed. Anal. 106: 25–36.

Cardoso, O., Porcher, J.-M. and Sanchez, W. 2014. Factory-discharged pharmaceuticals could be a relevant source of aquatic environment contamination: Review of evidence and need for knowledge. Chemosphere 115: 20–30.

Carr, R.L. and Chambers, J.E. 2001. Toxic responses of the nervous system. *In*: Schlenk, E. and Benson, W.H. (eds.). Target Organ Toxicity in Marine and Freshwater Teleosts, Book 2. Taylor and Francis, London.

Chen, H., Liu, S., Xu, X.-R., Liu, S.-S., Zhou, G.-J., Sun, K.-F., Zhao, J.-L. and Ying, G.-G. 2015. Antibiotics in typical marine aquaculture farms surrounding Hailing Island, South China: Occurrence, bioaccumulation and human dietary exposure. Mar. Pollut. Bull. 90: 181–187.

Christen, V., Hickmann, S., Rechenberg, B. and Fent, K. 2010. Highly active human pharmaceuticals in aquatic systems: A concept for their identification based on their mode of action. Aquat. Toxicol. 96: 167–181.

Crane, M., Watts, C. and Boucard, T. 2006. Chronic aquatic environmental risks from exposure to human pharmaceuticals. Sci. Total Environ. 367: 23–41.

Crespo, M. and Solé, M. 2016. The use of juvenile *Solea solea* as sentinel in the marine platform of the Ebre Delta: *in vitro* interaction of emerging contaminants with the liver detoxification system. Environ. Sci. Pollut. R 23: 19229–19236.

Dahl, U., Gorokhova, E. and Breitholtz, M. 2006. Application of growth-related sublethal endpoints in ecotoxicological assessments using a harpacticoid copepod. Aquat. Toxicol. 77: 433–438.

Daughton, C.G. 2016. Pharmaceuticals and the Environment (PiE): Evolution and impact of the published literature revealed by bibliometric analysis. Sci. Total Environ. 562: 391–426.

Di Poi, C., Darmaillacq, A.S., Dickel, L., Boulouard, M. and Bellanger, C. 2013. Effects of perinatal exposure to waterborne fluoxetine on memory processing in the cuttlefish *Sepia officinalis*. Aquat. Toxicol. 132: 84–91.

Di Poi, C., Bidel, F., Dickel, L. and Bellanger, C. 2014. Cryptic and biochemical responses of young cuttlefish *Sepia officinalis* exposed to environmentally relevant concentrations of fluoxetine. Aquat. Toxicol. 151: 36–45.

Dougherty, J.A., Swarzenski, P.W., Dinicola, R.S. and Reinhard, M. 2010. Occurrence of herbicides and pharmaceutical and personal care products in surface water and groundwater around Liberty Bay, Puget Sound, Washington. J. Environ. Qual. 39: 1173–1180.

Du, B., Haddad, S.P., Luek, A., Scott, W.C., Saari, G.N., Burket, S.R., Breed, C.S., Kelly, M., Broach, L., Rasmussen, J.B., Chambliss, C.K. and Brooks, B.W. 2016. Bioaccumulation of human pharmaceuticals in fish across habitats of a tidally influenced urban bayou. Environ. Toxicol. Chem. 35: 966–974.

Eades, C. and Waring, C.P. 2010. The effects of diclofenac on the physiology of the green shore crab *Carcinus maenas*. Mar. Environ. Res. 69: S46–S48.

Ericson, H., Thorsen, G. and Kumblad, L. 2010. Physiological effects of diclofenac, ibuprofen and propranolol on Baltic Sea blue mussels. Aquat. Toxicol. 99: 223–231.

European Parliament. 2013. Directive 2013/39/EU of the European Parliament and of the Council of 12 August 2013 amending Directives 2000/60/EC and 2008/105/EC as regards priority substances in the field of water policy.

Fabbri, E. and Capuzzo, A. 2010. Cyclic AMP signaling in bivalve molluscs: an overview. Part A J. Exp. Zool. 313A: 179–200.

Fabbri, E. and Franzellitti, S. 2016. Human pharmaceuticals in the marine environment: Focus on exposure and biological effects in animal species. Environ. Toxicol. Chem. 35: 799–812.

Fatta-Kassinos, D., Meric, S. and Nikolaou, A. 2011. Pharmaceutical residues in environmental waters and wastewater: current state of knowledge and future research. Anal. Bioanal. Chem. 399: 251–275.

Fenet, H., Arpin-Pont, L., Vanhoutte-Brunier, A., Munaron, D., Fiandrino, A., Martínez Bueno, M.-J., Boillot, C., Casellas, C., Mathieu, O. and Gomez, E. 2014. Reducing PEC uncertainty in coastal zones: A case study on carbamazepine, oxcarbazepine and their metabolites. Environ. Int. 68: 177–184.

Fent, K., Weston, A.A. and Caminada, D. 2006. Ecotoxicology of human pharmaceuticals. Aquat. Toxicol. 76: 122–159.

Fick, J., Söderström, H., Lindberg, R.H., Phan, C., Tysklind, M. and Larsson, D.G.J. 2009. Contamination of surface, ground, and drinking water from pharmaceutical production. Environ. Toxicol. Chem. 28: 2522–2527.

Fong, P.P. 1998. Zebra mussel spawning is induced in low concentrations of putative serotonin reuptake inhibitors. Bio. Bull. 194: 143–149.

Fong, P.P. and Molnar, N. 2013. Antidepressants cause foot detachment from substrate in five species of marine snail. Mar. Environ. Res. 84: 24–30.

Fong, P.P. and Ford, A.T. 2014. The biological effects of antidepressants on the molluscs and crustaceans: A review. Aquat. Toxicol. 151: 4–13.

Fong, P.P., Bury, T.B., Dworkin-Brodsky, A.D., Jasion, C.M. and Kell, R.C. 2015. The antidepressants venlafaxine ("Effexor") and fluoxetine ("Prozac") produce different effects on locomotion in two species of marine snail, the oyster drill (*Urosalpinx cinerea*) and the starsnail (*Lithopoma americanum*). Mar. Environ. Res. 103: 89–94.

Fonseca, V.F., Vasconcelos, R.P., Tanner, S.E., França, S., Serafim, A., Lopes, B., Company, R., Bebianno, M.J., Costa, M.J. and Cabral, H.N. 2015. Habitat quality of estuarine nursery grounds: Integrating non-biological indicators and multilevel biological responses in *Solea senegalensis*. Ecol. Ind. 58: 335–345.

Fonte, E., Ferreira, P. and Guilhermino, L. 2016. Temperature rise and microplastics interact with the toxicity of the antibiotic cefalexin to juveniles of the common goby (*Pomatoschistus microps*): Post-exposure predatory behaviour, acetylcholinesterase activity and lipid peroxidation. Aquat. Toxicol. 180: 173–185.

Franzellitti, S., Buratti, S., Donnini, F. and Fabbri, E. 2010. Exposure of mussels to a polluted environment: Insights into the stress syndrome development. Comp. Biochem. Phys. C 152: 24–33.

Franzellitti, S., Buratti, S., Valbonesi, P., Capuzzo, A. and Fabbri, E. 2011. The beta-blocker propranolol affects cAMP-dependent signaling and induces the stress response in Mediterranean mussels, *Mytilus galloprovincialis*. Aquat. Toxicol. 101: 299–308.

Franzellitti, S., Buratti, S., Valbonesi, P. and Fabbri, E. 2013. The mode of action (MOA) approach reveals interactive effects of environmental pharmaceuticals on *Mytilus galloprovincialis*. Aquat. Toxicol. 140-141: 249–256.

Franzellitti, S. and Fabbri, E. 2013. Cyclic-AMP mediated regulation of ABCB mRNA expression in mussel haemocytes. PLOS ONE 8: e61634.

Franzellitti, S., Buratti, S., Capolupo, M., Du, B., Haddad, S.P., Chambliss, C.K., Brooks, B.W. and Fabbri, E. 2014. An exploratory investigation of various modes of action and potential adverse outcomes of fluoxetine in marine mussels. Aquat. Toxicol. 151: 14–26.

Franzellitti, S., Buratti, S., Du, B., Haddad, S.P., Chambliss, C.K., Brooks, B.W. and Fabbri, E. 2015. A multibiomarker approach to explore interactive effects of propranolol and fluoxetine in marine mussels. Environ. Pollut. 205: 60–69.

Franzellitti, S., Striano, T., Valbonesi, P. and Fabbri, E. 2016. Insights into the regulation of the MXR response in haemocytes of the Mediterranean mussel (*Mytilus galloprovincialis*). Fish Shellfish Immun. 58: 349–358.

Freitas, R., Almeida, A., Pires, A., Velez, C., Calisto, V., Schneider, R.J., Esteves, V.I., Wrona, F.J., Figueira, E. and Soares, A.M. 2015. The effects of carbamazepine on macroinvertebrate species: Comparing bivalves and polychaetes biochemical responses. Water Res. 85: 137–147.

Furuhagen, S., Fuchs, A., Lundström Belleza, E., Breitholtz, M. and Gorokhova, E. 2014. Are pharmaceuticals with evolutionary conserved molecular drug targets more potent to cause toxic effects in non-target organisms? PLOS ONE 9: e105028.

Gaw, S., Thomas, K.V. and Hutchinson, T.H. 2014. Sources, impacts and trends of pharmaceuticals in the marine and coastal environment. Philos. Trans. R Soc. Lond. B Biol. Sci. 369: 20130572.

Glassmeyer, S.T., Kolpin, D.W.T.F.E. and Focazio, J.T. 2007. Environmental presence and persistence of pharmaceuticals: An overview. *In*: Aga, D.S. (ed.). Fate of Pharmaceuticals in the Environment and in Water Treatment Systems. CRC Press, USA.

Goldberg, E.D. and Bertine, K.K. 2000. Beyond the mussel watch—new directions for monitoring marine pollution. Sci. Total Environ. 247: 165–174.

Gonzalez-Rey, M. and Bebianno, M.J. 2012. Does non-steroidal anti-inflammatory (NSAID) ibuprofen induce antioxidant stress and endocrine disruption in mussel *Mytilus galloprovincialis*? Environ. Toxicol. Phar. 33: 361–371.

Gonzalez-Rey, M. and Bebianno, M.J. 2014. Effects of non-steroidal anti-inflammatory drug (NSAID) diclofenac exposure in mussel *Mytilus galloprovincialis*. Aquat. Toxicol. 148: 221–230.

Groh, K.J., Carvalho, R.N., Chipman, J.K., Denslow, N.D., Halder, M., Murphy, C.A., Roelofs, D., Rolaki, A., Schirmer, K. and Watanabe, K.H. 2015. Development and application of the adverse outcome pathway framework for understanding and predicting chronic toxicity: I. Challenges and research needs in ecotoxicology. Chemosphere 120: 764–777.

Gros, M., Petrović, M., Ginebreda, A. and Barceló, D. 2010. Removal of pharmaceuticals during wastewater treatment and environmental risk assessment using hazard indexes. Environ. Int. 36: 15–26.

Gunnarsson, L., Jauhiainen, A., Kristiansson, E., Nerman, O. and Larsson, D.G.J. 2008. Evolutionary conservation of human drug targets in organisms used for environmental risk assessments. Environ. Sci. Technol. 42: 5807–5813.

Han, J., Lee, M.C., Kim, D.H., Lee, Y.H., Park, J.C. and Lee, J.S. 2016. Effects of trimethoprim on life history parameters, oxidative stress, and the expression of cytochrome P450 genes in the copepod *Tigriopus japonicus*. Chemosphere 159: 159–165.

Herrmann, M., Olsson, O., Fiehn, R., Herrel, M. and Kümmerer, K. 2015. The significance of different health institutions and their respective contributions of active pharmaceutical ingredients to wastewater. Environ. Int. 85: 61–76.

Hiemke, C. and Härtter, S. 2000. Pharmacokinetics of selective serotonin reuptake inhibitors. Pharmacol. Therapeut. 85: 11–28.

Hird, C.M., Urbina, M.A., Lewis, C.N., Snape, J.R. and Galloway, T.S. 2016. Fluoxetine exhibits pharmacological effects and trait-based sensitivity in a marine worm. Environl. Scie. Technol. 50: 8344–8352.

Huerta, B., Rodriguez-Mozaz, S. and Barcelo, D. 2012. Pharmaceuticals in biota in the aquatic environment: analytical methods and environmental implications. Anal. Bioanal. Chem. 404: 2611–2624.

Jones, R.L. 1972. Functions of prostaglandins. Pathobiol. Ann. 2: 359–380.

Key, P.B., Hoguet, J., Chung, K.W., Venturella, J.J., Pennington, P.L. and Fulton, M.H. 2009. Lethal and sublethal effects of simvastatin, irgarol, and PBDE-47 on the estuarine fish, *Fundulus heteroclitus*. J. Environ. Sci. Health B 44: 379–382.

Kim, H.-Y., Lee, I.-S. and Oh, J.-E. 2017. Human and veterinary pharmaceuticals in the marine environment including fish farms in Korea. Sci. Total Environ. 579: 940–949.

Kim, S., Weber, A.S., Batt, A. and Aga, D.S. 2007. Removal of pharmaceuticals in biological wastewater treatment plants. *In*: Aga, D.S. (ed.). Fate of Pharmaceuticals in the Environment and in Water Treatment Systems. CRC Press, USA.

Klosterhaus, S.L., Grace, R., Hamilton, M.C. and Yee, D. 2013. Method validation and reconnaissance of pharmaceuticals, personal care products, and alkylphenols in surface waters, sediments, and mussels in an urban estuary. Environ. Int. 54: 92–99.

Kramer, V.J., Etterson, M.A., Hecker, M., Murphy, C.A., Roesijadi, G., Spade, D.J., Spromberg, J.A., Wang, M. and Ankley, G.T. 2011. Adverse outcome pathways and ecological risk assessment: bridging to population-level effects. Environ. Toxicol. Chem. 30: 64–76.

Kummerer, K. 2009a. Antibiotics in the aquatic environment—a review—part I. Chemosphere 75: 417–434.

Kummerer, K. 2009b. Antibiotics in the aquatic environment—a review—part II. Chemosphere 75: 435–441.

Kummerer, K. 2009c. The presence of pharmaceuticals in the environment due to human use-present knowledge and future challenges. J. Environ. Manage. 90: 2354–2366.

Kuster, A. and Adler, N. 2014. Pharmaceuticals in the environment: scientific evidence of risks and its regulation. Philos. Trans. R Soc. Lond. B Biol. Sci. 369: 20130587.

Larsson, D.G. 2014. Pollution from drug manufacturing: review and perspectives. Philos. Trans. R Soc. Lond. B Biol. Sci. Biological Sciences 369: 20130571.

Le Corre, K.S., Ort, C., Kateley, D., Allen, B., Escher, B.I. and Keller, J. 2012. Consumption-based approach for assessing the contribution of hospitals towards the load of pharmaceutical residues in municipal wastewater. Environ. Int. 45: 99–111.

Le, T.X. and Munekage, Y. 2004. Residues of selected antibiotics in water and mud from shrimp ponds in mangrove areas in Viet. Nam. Mar. Pollut. Bull. 49: 922–929.

Leston, S., Nunes, M., Viegas, I., Nebot, C., Cepeda, A., Pardal, M.A. and Ramos, F. 2014. The influence of sulfathiazole on the macroalgae *Ulva lactuca*. Chemosphere 100: 105–110.

Lim, W., Shi, Y., Gao, L., Liu, J. and Cai, Y. 2012. Investigation of antibiotics in mollusks from coastal waters in the Bohai Sea of China. Environ. Pollut. 162: 56–62.

Lim, S.J., Seo, C.-K., Kim, T.-H. and Myung, S.-W. 2013. Occurrence and ecological hazard assessment of selected veterinary medicines in livestock wastewater treatment plants. J. Environ. Sci. Heal. B 48: 658–670.

Liu, D., Lung, W.-S. and Colosi, L.M. 2013. Effects of sorption kinetics on the fate and transport of pharmaceuticals in estuaries. Chemosphere 92: 1001–1009.

Liu, S., Zhao, H., Lehmler, H.J., Cai, X. and Chen, J. 2017. Antibiotic pollution in marine food webs in Laizhou Bay, North China: trophodynamics and human exposure implication. Environ. Sci. Technol. 51: 2392–2400.

Luo, Y., Guo, W., Ngo, H.H., Nghiem, L.D., Hai, F.I., Zhang, J., Liang, S. and Wang, X.C. 2014. A review on the occurrence of micropollutants in the aquatic environment and their fate and removal during wastewater treatment. Sci. Total Environ. 473-474: 619–641.

Margot, J., Kienle, C., Magnet, A., Weil, M., Rossi, L., de Alencastro, L.F., Abegglen, C., Thonney, D., Chèvre, N., Schärer, M. and Barry, D.A. 2013. Treatment of micropollutants in municipal wastewater: Ozone or powdered activated carbon? Sci. Total Environ. 461-462: 480–498.

Martin-Diaz, L., Franzellitti, S., Buratti, S., Valbonesi, P., Capuzzo, A. and Fabbri, E. 2009. Effects of environmental concentrations of the antiepilectic drug carbamazepine on biomarkers and cAMP-mediated cell signaling in the mussel *Mytilus galloprovincialis*. Aquat. Toxicol. 94: 177–185.

Martínez Bueno, M.J., Herrera, S., Munaron, D., Boillot, C., Fenet, H., Chiron, S. and Gómez, E. 2016. POCIS passive samplers as a monitoring tool for pharmaceutical residues and their transformation products in marine environment. Environ. Sci. Pollut. R 23: 5019–5029.

Maruya, K.A., Vidal-Dorsch, D.E., Bay, S.M., Kwon, J.W., Xia, K. and Armbrust, K.L. 2012. Organic contaminants of emerging concern in sediments and flatfish collected near outfalls discharging treated wastewater effluent to the Southern California Bight. Environ. Toxicol. Chem. 31: 2683–2688.

Matozzo, V., Rova, S. and Marin, M.G. 2012. The nonsteroidal anti-inflammatory drug, ibuprofen, affects the immune parameters in the clam *Ruditapes philippinarum*. Mar. Environ. Res. 79: 116–121.

Matozzo, V., De Notaris, C., Finos, L., Filippini, R. and Piovan, A. 2015. Environmentally realistic concentrations of the antibiotic Trimethoprim affect haemocyte parameters but not antioxidant enzyme activities in the clam *Ruditapes philippinarum*. Environ. Pollut. 206: 567–574.

Matozzo, V., Bertin, V., Battistara, M., Guidolin, A., Masiero, L., Marisa, I. and Orsetti, A. 2016. Does the antibiotic amoxicillin affect haemocyte parameters in non-target aquatic invertebrates? The clam *Ruditapes philippinarum* and the mussel *Mytilus galloprovincialis* as model organisms. Mar. Environ. Res. 119: 51–58.

McEneff, G., Barron, L., Kelleher, B., Paull, B. and Quinn, B. 2014. A year-long study of the spatial occurrence and relative distribution of pharmaceutical residues in sewage effluent, receiving marine waters and marine bivalves. Sci. Total Environ. 476-477: 317–326.

Mennigen, J.A., Lado, W.E., Zamora, J.M., Duarte-Guterman, P., Langlois, V.S., Metcalfe, C.D., Chang, J.P., Moon, T.W. and Trudeau, V.L. 2010a. Waterborne fluoxetine disrupts the reproductive axis in sexually mature male goldfish, *Carassius auratus*. Aquat. Toxicol. 100: 354–364.

Mennigen, J.A., Sassine, J., Trudeau, V.L. and Moon, T.W. 2010b. Waterborne fluoxetine disrupts feeding and energy metabolism in the goldfish *Carassius auratus*. Aquat. Toxicol. 100: 128–137.

Mezzelani, M., Gorbi, S., Da Ros, Z., Fattorini, D., d'Errico, G., Milan, M., Bargelloni, L. and Regoli, F. 2016a. Ecotoxicological potential of non-steroidal anti-inflammatory drugs (NSAIDs) in marine organisms: Bioavailability, biomarkers and natural occurrence in *Mytilus galloprovincialis*. Mar. Environ. Res. 121: 31–39.

Mezzelani, M., Gorbi, S., Fattorini, D., d'Errico, G., Benedetti, M., Milan, M., Bargelloni, L. and Regoli, F. 2016b. Transcriptional and cellular effects of non-steroidal anti-inflammatory drugs (NSAIDs) in experimentally exposed mussels, *Mytilus galloprovincialis*. Aquat. Toxicol. 180: 306–319.

Milan, M., Pauletto, M., Patarnello, T., Bargelloni, L., Marin, M.G. and Matozzo, V. 2013. Gene transcription and biomarker responses in the clam *Ruditapes philippinarum* after exposure to ibuprofen. Aquat. Toxicol. 126: 17–29.

Minguez, L., Pedelucq, J., Farcy, E., Ballandonne, C., Budzinski, H. and Halm-Lemeille, M.P. 2016. Toxicities of 48 pharmaceuticals and their freshwater and marine environmental assessment in northwestern France. Environ. Sci. Pollut. Res. 23: 4992–5001.

Morando, M.B., Medeiros, L.R. and McDonald, M.D. 2009. Fluoxetine treatment affects nitrogen waste excretion and osmoregulation in a marine teleost fish. Aquat. Toxicol. 95: 164–171.

Moreno-González, R., Rodriguez-Mozaz, S., Gros, M., Barceló, D. and León, V.M. 2015. Seasonal distribution of pharmaceuticals in marine water and sediment from a mediterranean coastal lagoon (SE Spain). Environ. Res. 138: 326–344.

Neuparth, T., Martins, C., de los Santos, C.B., Costa, M.H., Martins, I., Costa, P.M. and Santos, M.M. 2014. Hypocholesterolaemic pharmaceutical simvastatin disrupts reproduction and population growth of the amphipod *Gammarus locusta* at the ng/L range. Aquat. Toxicol. 155: 337–347.

Nunes, B., Carvalho, F. and Guilhermino, L. 2005. Acute toxicity of widely used pharmaceuticals in aquatic species: *Gambusia holbrooki*, *Artemia parthenogenetica* and *Tetraselmis chuii*. Ecotox. Environ. Saf. 61: 413–419.

Nunes, B., Gaio, A.R., Carvalho, F. and Guilhermino, L. 2008. Behaviour and biomarkers of oxidative stress in *Gambusia holbrooki* after acute exposure to widely used pharmaceuticals and a detergent. Ecotox. Environ. Saf. 71: 341–354.

Oliveira, T.S., Murphy, M., Mendola, N., Wong, V., Carlson, D. and Waring, L. 2015. Characterization of pharmaceuticals and personal care products in hospital effluent and waste water influent/effluent by direct-injection LC-MS-MS. Sci. Total Environ. 518-519: 459–478.

Oliveira, T.S., Al Aukidy, M. and Verlicchi, P. 2017. Occurrence of Common Pollutants and Pharmaceuticals in Hospital Effluents. Springer Berlin Heidelberg, Berlin, Heidelberg.

Owen, S.F., Giltrow, E., Huggett, D.B., Hutchinson, T.H., Saye, J., Winter, M.J. and Sumpter, J.P. 2007. Comparative physiology, pharmacology and toxicology of beta-blockers: mammals versus fish. Aquat. Toxicol. 82: 145–162.

Paíga, P., Santos, L.H.M.L.M., Ramos, S., Jorge, S., Silva, J.G. and Delerue-Matos, C. 2016. Presence of pharmaceuticals in the Lis river (Portugal): Sources, fate and seasonal variation. Sci. Total Environ. 573: 164–177.

Pérez, S. and Barceló, D. 2007. Advances in the analysis of pharmaceuticals in the aquatic environment. *In*: Aga, D.S. (ed.). Environmental Presence and Persistence of Pharmaceuticals: An Overview. CRC Press, USA.

Peters, J.R. and Granek, E.F. 2016. Long-term exposure to fluoxetine reduces growth and reproductive potential in the dominant rocky intertidal mussel, *Mytilus californianus*. Sci. Total Environ. 545: 621–628.

Pires, A., Almeida, A., Calisto, V., Schneider, R.J., Esteves, V.I., Wrona, F.J., Soares, A.M.V.M., Figueira, E. and Freitas, R. 2016. Hediste diversicolor as bioindicator of pharmaceutical pollution: Results from single and combined exposure to carbamazepine and caffeine. Comp. Biochem. Phys. C 188: 30–38.

Prasse, C., Schlüsener, M.P., Schulz, R. and Ternes, T.A. 2010. Antiviral drugs in wastewater and surface waters: a new pharmaceutical class of environmental relevance? Environ. Sci. Technol. 44: 1728–1735.

Qi, W., Müller, B., Pernet-Coudrier, B., Singer, H., Liu, H., Qu, J. and Berg, M. 2014. Organic micropollutants in the Yangtze River: Seasonal occurrence and annual loads. Sci. Total Environ. 472: 789–799.

Rang, H.P., Dale, M.M. and Ritter, J.M. 1999. Pharmacology. Churchill Livingstone, Edinburgh.

Rehman, M.S., Rashid, N., Ashfaq, M., Saif, A., Ahmad, N. and Han, J.I. 2015. Global risk of pharmaceutical contamination from highly populated developing countries. Chemosphere 138: 1045–1055.

Ribalta, C. and Solé, M. 2014. *In vitro* interaction of emerging contaminants with the cytochrome p450 system of mediterranean deep-sea fish. Environl. Scie. Technol. 48: 12327–12335.

Ribeiro, S., Torres, T., Martins, R. and Santos, M.M. 2015. Toxicity screening of diclofenac, propranolol, sertraline and simvastatin using *Danio rerio* and *Paracentrotus lividus* embryo bioassays. Ecotox. Environ. Saf. 114: 67–74.

Rochette, S., Rivot, E., Morin, J., Mackinson, S., Riou, P. and Le Pape, O. 2010. Effect of nursery habitat degradation on flatfish population: Application to *Solea solea* in the Eastern Channel (Western Europe). J. Sea Res. 64: 34–44.

Rowley, A.F., Vogan, C.L., Taylor, G.W. and Clare, A.S. 2005. Prostaglandins in non-insect an invertebrates: recent insights and unsolved problems. J. Exp. Biol. 208: 3–14.

Rudd, M.A., Ankley, G.T., Boxall, A.B.A. and Brooks, B.W. 2014. International scientists' priorities for research on pharmaceutical and personal care products in the environment. Integr. Environ. Assess. 10: 576–587.

Sanderson, H. and Thomsen, M. 2009. Comparative analysis of pharmaceuticals versus industrial chemicals acute aquatic toxicity classification according to the United Nations classification system for chemicals. Assessment of the (Q)SAR predictability of pharmaceuticals acute aquatic toxicity and their predominant acute toxic mode-of-action. Toxicol. Lett. 187: 84–93.

Santos, L.H., Gros, M., Rodriguez-Mozaz, S., Delerue-Matos, C., Pena, A., Barceló, D. and Montenegro, M.C. 2013. Contribution of hospital effluents to the load of pharmaceuticals in urban wastewaters: identification of ecologically relevant pharmaceuticals. Sci. Total Environ. 461-462: 302–316.

Santos, M.M., Ruivo, R., Lopes-Marques, M., Torres, T., de los Santos, C.B., Castro, L.F. and Neuparth, T. 2016. Statins: An undesirable class of aquatic contaminants? Aquat. Toxicol. 174: 1–9.

Sapkota, A., Sapkota, A.R., Kucharski, M., Burke, J., McKenzie, S., Walker, P. and Lawrence, R. 2008. Aquaculture practices and potential human health risks: Current knowledge and future priorities. Environ. Int. 34: 1215–1226.

Sauvé, S. and Desrosiers, M. 2014. A review of what is an emerging contaminant. Chem. Cent. J. 8: 15.

Schmidt, W., O'Rourke, K., Hernan. R. and Quinn, B. 2011. Effects of the pharmaceuticals gemfibrozil and diclofenac on the marine mussel (*Mytilus* spp.) and their comparison with standardized toxicity tests. Mar. Pollut. Bull. 62: 1389–1395.

Schreiber, R., Gündel, U., Franz, S., Küster, A., Rechenberg, B. and Altenburger, R. 2011. Using the fish plasma model for comparative hazard identification for pharmaceuticals in the environment by extrapolation from human therapeutic data. Regul. Toxicol. Pharm. 61: 261–275.

Schultz, M.M., Painter, M.M., Bartell, S.E., Logue, A., Furlong, E.T., Werner, S.L. and Schoenfuss, H.L. 2011. Selective uptake and biological consequences of environmentally relevant antidepressant pharmaceutical exposures on male fathead minnows. Aquat. Toxicol. 104: 38–47.

Silva, L.J.G., Pereira, A.M.P.T., Meisel, L.M., Lino, C.M. and Pena, A. 2014. A one-year follow-up analysis of antidepressants in Portuguese wastewaters: Occurrence and fate, seasonal influence, and risk assessment. Sci. Total Environ. 490: 279–287.

Solé, M., Shaw, J.P., Frickers, P.E., Readman, J.W. and Hutchinson, T.H. 2010. Effects on feeding rate and biomarker responses of marine mussels experimentally exposed to propranolol and acetaminophen. Anal. Bioanal. Chem. 396: 649–656.

Solé, M., Fortuny, A. and Mananos, E. 2014. Effects of selected xenobiotics on hepatic and plasmatic biomarkers in juveniles of *Solea senegalensis*. Environ. Res. 135: 227–235.

Solé, M. and Sanchez-Hernandez, J.C. 2015. An *in vitro* screening with emerging contaminants reveals inhibition of carboxylesterase activity in aquatic organisms. Aquat. Toxicol. 169: 215–222.

Sorbera, L.A., Asturiano, J.F., Carrillo, M. and Zanuy, S. 2001. Effects of polyunsaturated fatty acids and prostaglandins on oocyte maturation in a marine teleost, the european sea bass (*Dicentrarchus labrax*). Biol. Reprod. 64: 382–389.

Sui, Q., Cao, X., Lu, S., Zhao, W., Qiu, Z. and Yu, G. 2015. Occurrence, sources and fate of pharmaceuticals and personal care products in the groundwater: A review. Emerging Contaminants 1: 14–24.

Taylor, D. and Senac, T. 2014. Human pharmaceutical products in the environment—the "problem" in perspective. Chemosphere 115: 95–99.

Teles, M., Fierro-Castro, C., Na-Phatthalung, P., Tvarijonaviciute, A., Soares, A., Tort, L. and Oliveira, M. 2016. Evaluation of gemfibrozil effects on a marine fish (*Sparus aurata*) combining gene expression with conventional endocrine and biochemical endpoints. J. Hazard Mater. 318: 600–607.

Togola, A. and Budzinski, H. 2008. Multi-residue analysis of pharmaceutical compounds in aqueous samples. J. Chromatogr. A 1177: 150–158.

Tornero, V. and Hanke, G. 2016. Chemical contaminants entering the marine environment from sea-based sources: A review with a focus on European seas. Mar. Pollut. Bull. 112: 17–38.

Vane, J.R. and Botting, R.M. 1998. Mechanism of action of anti-inflammatory drugs. Int. J. Tissue React. 20: 3–15.

Vasconcelos, R.P., Reis-Santos, P., Costa, M.J. and Cabral, H.N. 2011. Connectivity between estuaries and marine environment: Integrating metrics to assess estuarine nursery function. Ecol. Ind. 11: 1123–1133.

Weinberger, J. 2nd and Klaper, R. 2014. Environmental concentrations of the selective serotonin reuptake inhibitor fluoxetine impact specific behaviors involved in reproduction, feeding and predator avoidance in the fish *Pimephales promelas* (fathead minnow). Aquat. Toxicol. 151: 77–83.

Westhof, L., Köster, S. and Reich, M. 2016. Occurrence of micropollutants in the wastewater streams of cruise ships. Emerging Contaminants 2: 178–184.

Wheelock, C.E., Phillips, B.M., Anderson, B.S., Miller, J.L., Miller, M.J. and Hammock, B.D. 2008. Applications of carboxylesterase activity in environmental monitoring and toxicity identification evaluations (TIEs). Rev. Environ. Contam. Toxicol. 195: 117–178.

Wiklund, A.K., Oskarsson, H., Thorsén, G. and Krumblad, L. 2011. Behavioural and physiological responses to pharmaceutical exposure in macroalgae and grazers from a Baltic Sea littoral community. Aquat. Biol. 14: 29–39.

Wille, K., Noppe, H., Verheyden, K., Vanden Bussche, J., De Wulf, E., Van Caeter, P., Janssen, C.R., De Brabander, H.F. and Vanhaecke, L. 2010. Validation and application of an LC-MS/MS method for the simultaneous quantification of 13 pharmaceuticals in seawater. Anal. Bioanal. Chem. 397: 1797–1808.

Wille, K., Kiebooms, J.A.L., Claessens, M., Rappé, K., Vanden Bussche, J., Noppe, H., Van Praet, N., De Wulf, E., Van Caeter, P., Janssen, C.R., De Brabander, H.F. and Vanhaecke, L. 2011. Development of analytical strategies using U-HPLC-MS/MS and LC-ToF-MS for the quantification of micropollutants in marine organisms. Anal. Bioanal. Chem. 400: 1459–1472.

Winberg, S. and Nilsson, G.E. 1993. Roles of brain monoamine neurotransmitters in agonistic behaviour and stress reactions, with particular reference to fish. Comp. Biochem. Phys. C 106: 597–614.

Winberg, S., Nilsson, A., Hylland, P., Söderström, V. and Nilsson, G.E. 1997. Serotonin as a regulator of hypothalamic-pituitary-interrenal activity in teleost fish. Neurosci. Lett. 230: 113–116.

Winder, V.L., Pennington, P.L., Hurd, M.W. and Wirth, E.F. 2012. Fluoxetine effects on sheepshead minnow (*Cyprinodon variegatus*) locomotor activity. J. Environ. Sci. Health B 47: 51–58.

Xu, W., Yan, W., Li, X., Zou, Y., Chen, X., Huang, W., Miao, L., Zhang, R., Zhang, G. and Zou, S. 2013. Antibiotics in riverine runoff of the pearl river delta and pearl river estuary, China: concentrations, mass loading and ecological risks. Environ. Pollut. 182: 402–407.

Yan, C., Yang, Y., Zhou, J., Liu, M., Nie, M., Shi, H. and Gu, L. 2013. Antibiotics in the surface water of the Yangtze Estuary: Occurrence, distribution and risk assessment. Environ. Pollut. 175: 22–29.

Zhang, R., Tang, J., Li, J., Zheng, Q., Liu, D., Chen, Y., Zou, Y., Chen, X., Luo, C. and Zhang, G. 2013. Antibiotics in the offshore waters of the Bohai Sea and the Yellow Sea in China: Occurrence, distribution and ecological risks. Environ. Pollut. 174: 71–77.

Zhao, H., Zhou, J.L. and Zhang, J. 2015. Tidal impact on the dynamic behavior of dissolved pharmaceuticals in the Yangtze Estuary, China. Sci. Total Environ. 536: 946–954.

Zou, S., Xu, W., Zhang, R., Tang, J., Chen, Y. and Zhang, G. 2011. Occurrence and distribution of antibiotics in coastal water of the Bohai Bay, China: Impacts of river discharge and aquaculture activities. Environ. Pollut. 159: 2913–2920.

8

Mercury Exposure, Fish Consumption and Provisional Tolerable Weekly Intake
An Overview

Vieira, H.C., Soares, A.M.V.M., Morgado, F. and Abreu, S.N.*

INTRODUCTION

Mercury (Hg) appears at the top of the priority list of hazardous substances published by the US Agency for toxic substances and disease registry (ATSDR 2013) due to its persistence, ability to enter into biological systems (Wolkin et al. 2012, Liao et al. 2016) and its high toxicity even at low concentrations (Hajeb et al. 2009).

Originating from natural sources (Siegel and Siegel 1984, Gustin et al. 2000, Coolbaugh et al. 2002) such as volcanic emissions, geothermal releases, and biomass burning (Renzoni et al. 1998, Pirrone et al. 2010), and anthropogenic sources (Pacyna et al. 2001, Dommergue et al. 2002) as mining, chloro-alkali production and fossil fuels combustion (Renzoni et al. 1998, Pirrone et al. 2010), Hg can be found either in aquatic or terrestrial environments (Gochfeld 2003) in three oxidation states (Hg^0, Hg^{+1} and Hg^{+2}) as well in various forms (Ullrich et al. 2001, Rasmussen et al. 2005). All forms of Hg are toxic, however, the level of toxicity is strongly related to the chemical properties of each form (Lindqvist and Rodhe 1985, Gochfeld 2003, Harris et al. 2003). For example, inorganic forms may be transformed into organic forms (methylated species). Methylmercury (MeHg), the most toxic form of Hg (due to the neurotoxic effects), can be processed through bacteria, like sulfate-reducing bacteria (Compeau and Bartha 1985), and accumulate in the food chain, resulting in a process of bioaccumulation and biomagnification, leading to high concentrations

Department of Biology & CESAM, University of Aveiro, 3810-193 Aveiro, Portugal.
* Corresponding author; hugovieira@ua.pt

of Hg in fish. Ultimately, fish is consumed by humans, resulting in an increased risk of adverse effects to human health. Human exposure to MeHg is almost exclusively a result of the consumption of fish and shellfish (Mergler et al. 2007, Hajeb et al. 2009). In predatory marine fish, about 90 percent of the Hg exists in the methylated form (WHO 2008).

Apart from being pointed as the main route of MeHg to humans, fish consumption has also been recognised as an important component of a healthy diet (Malm et al. 1995, Mergler et al. 2007). Due its high nutritional value, fish is rich in protein with essential amino acids, macroelements (calcium and phosphorus), microelements (iodine, selenium and zinc), vitamins and unsaturated fatty acids. On the other hand, fish is considered an important source of n-3 fatty acids such as docosahexaenoic acid (22:6, n-3, DHA) and eicosapentaenoic acid (20:5, n-3, EPA) (Egeland and Middaugh 1997, FAO 2012, WHO 2003). All these elements play an important role preventing the development of some diseases, especially regarding cardiac and circulatory disorders, and reducing mortality in patients with coronary diseases (Kris-Etherton et al. 2002).

The concern of Hg pollution arises from a human health problem observed in various parts of the world, resultant from Hg exposure through the consumption of marine and freshwater seafood (Lin and Pehkonen 1999, Baeyens et al. 2003, Carrasco et al. 2011, You et al. 2014). The effects of Hg in human health have gained more importance essentially since the Hg poisoning incident occurred in the 1950s at Minamata Bay in Japan (Ullrich et al. 2001, Rasmussen et al. 2005). Following this episode, some international agencies such as United States Environmental Protection Agency (USEPA) and World Health Organization (WHO), have established reference doses (RfD) to the mercury tolerable intake levels, aiming to prevent health risks. Currently, Food and Agriculture Organization (FAO)/WHO Joint Expert Committee on Food Additives (JECFA) established the RfD "provisional tolerable weekly intake" (PTWI) for MeHg at 0.23 mg kg bw^{-1} day^{-1} (EFSA 2012) and USEPA fixed the RfD in 0.1 mg MeHg kg bw^{-1} day^{-1} (USEPA 1997a). Recently, PTWI suggested by JECFA for MeHg was revised by the European Food Safety Authority (EFSA) to 0.19 µg MeHg kg bw^{-1} day^{-1} (EFSA 2012).

Hair has been used in several studies as a bioindicator of Hg exposure for human populations (Dorea et al. 2003, Agusa et al. 2005). Hair grows approximately 1 cm per month, during hair formation, Hg present in the circulating blood is incorporated into the hair follicles, where it becomes stable, registering and providing the accumulation pattern and history of exposure (Dolbec et al. 2001, Agusa et al. 2007, Díez et al. 2008, Freire et al. 2010).

There are several reasons to use hair as bioindicator for mercury exposure: (1) it is easy to collect in a noninvasive manner, (2) it captures temporal exposure history as Hg is incorporated into growing hair and (3) it does not require special facilities for transport or storage (Cizdziel and Gerstenberger 2004). PTWI and RfD correspond to a hair Hg concentration of 2.2 and 1.0 µg g^{-1}, respectively. WHO, through analysis of neurotoxicological data, considered Hg concentration of 50 µg g^{-1} in human hair as a "no observed adverse effect level" (NOAEL) value for MeHg (WHO 1990).

Interspecific Hg Variation in Fish

Having the highest seafood consumption per capita in Europe per year, Portugal also occupies the top rank in countries with the highest seafood consumption *per capita* in the world (Failler et al. 2007, FAO 2010). The Azores archipelago is the Portuguese region with the highest consumption rate *per capita* of fishery products, where each Azorean consumes about 80 kg of fish per year (Megapesca 2007).

In the Portuguese Mainland, anthropogenic sources, mainly chlor-alkali plants (Mil-Homens et al. 2008) are the highest Hg contributors. One important example is the case of Laranjo basin in Ria de Aveiro, that received a highly contaminated effluent discharge from a chlor-alkali plant Hg cell located in the Estarreja industrial complex, from the 1950s until 1994 (Pereira et al. 2009). On the other hand, the volcanic Portuguese archipelagos, particularly the Azores, are mostly exposed to Hg from natural sources (e.g., volcanic activity) linked with geochemical anomalies and hot springs (Loppi 2001), being a remote area and away from large anthropogenic sources, the Azores have no significant discharge of Hg from industrial sources (Rodrigues et al. 2004).

The interspecific Hg variation in fish is the result of its intrinsic trophic position, growth rate, specimen age and food web complexity (Magalhães et al. 2007, Koenig et al. 2013). Thus, different fish may contain very different Hg content. Hg concentrations presented in Table 1 were based on seven studies that evaluated concentrations found in 28 fish species caught near the Azores archipelago.

Sparisoma cretense was the fish species with the lowest Hg concentration, on the other hand, the species *Mora moro* presented the highest Hg concentration.

Hg concentration of each species was compared to permissible limits set forth by Commission Regulation (EC) no. 1881/2006 of December 2006. This regulation establishes 0.5 µg g^{-1} as the maximum Hg concentration for most fish species, with an "exception list" of species whose maximum Hg concentration allowed increases to 1 µg g^{-1}. Regarding the permissible limits for fish consumption, only *Mora moro* exhibited higher values than those allowed (0.81 µg g^{-1} vs. 0.5 µg g^{-1}). Higher values for Hg concentrations in *Mora moro* were also observed by Koening et al. (2013) in a study on deep-sea organisms from the NW Mediterranean.

Fish Consumption and Hg Exposure Assessment Tools

Food frequency questionnaires (FFQ) are tools designed to evaluate the usual diet of a population. FFQ have the advantages of being easy and quick to apply, and practical to identify usual food consumption, in addition to its low cost (Ocké et al. 1997). Over the years, the use of FFQ have been increasingly used in studies where human exposure to Hg is assessed through fish consumption (MacIntosh et al. 1997, Björnberg et al. 2003, Passos et al. 2003, Oken et al. 2016). On the other hand, it is possible to reconstitute recent past exposure history (Cernichiari et al. 1995) by cutting the hair into centimeter-long samples and analyzing the Hg content of each centimeter independently (Passos et al. 2003).

The information related to the fish consumption (number of meals per week) was assessed during hair sampling, using a FFQ where each individual was asked to

Table 1. Hg concentrations ($\mu g\ g^{-1}$) in 28 fish species caught near the Azores archipelago and maximum Hg level set forth by Commission Regulation (EC) no. 1881/2006 of December 2006 for fish consumption.

Scientific name	[Hg]($\mu g\ g^{-1}$)	Maximum level	References
Aphanopus carbo	0.89	**1.0**	Afonso (2007)
Aphanopus carbo	0.71	**1.0**	Costa (2009)
Capros aper	0.05	0.5	Monteiro (1996)
Ceratoscopelus maderensis	0.12	0.5	Monteiro (1996)
Chelon labrosus	0.20	0.5	Andersen (1997)
Conger conger	0.25	0.5	Andersen (1997)
Conger conger	0.41	0.5	Magalhães (2007)
Diplodus sargus cadenati	3.41	0.5	Andersen (1997)
Diplodus sargus cadenati	0.30	0.5	Andersen (1997)
Electrona rissoi	0.10	0.5	Monteiro (1996)
Helicolenus dactylopterus	0.29	0.5	Monteiro (1991)
Helicolenus dactylopterus	0.26	0.5	Andersen (1997)
Katsuwonus pelamis	0.19	**1.0**	Andersen (1997)
Katsuwonus pelamis	0.04	**1.0**	Torres (2016)
Lepidopus caudatus	0.28	**1.0**	Andersen (1997)
Lepidopus caudatus	0.32	**1.0**	Magalhães (2007)
Macrorhamphosus scolopax	0.02	0.5	Monteiro (1996)
Maurolicus muelleri	0.11	0.5	Monteiro (1996)
Mora moro	0.81	0.5	Magalhães (2007)
Mullus surmuletus	0.12	**1.0**	Andersen (1997)
Muraena helena	0.05	0.5	Andersen (1997)
Myctophum punctatum	0.10	0.5	Monteiro (1996)
Pagellus acarne	0.20	**1.0**	Magalhães (2007)
Pagellus bogaraveo	0.26	**1.0**	Andersen (1997)
Pagrus pagrus	0.47	0.5	Andersen (1997)
Phycis blennoides	0.15	0.5	Magalhães (2007)
Phycis phycis	0.10	0.5	Andersen (1997)
Phycis phycis	0.13	0.5	Magalhães (2007)
Polyprion americanus	0.27	0.5	Magalhães (2007)
Pontinus kuhlii	0.16	0.5	Monteiro (1991)
Scomber japonicus	0.03	0.5	Monteiro (1996)
Thunnus alalunga	0.37	**1.0**	Andersen (1997)
Thunnus obesus	0.14	**1.0**	Torres (2016)
Trachurus picturatus	0.05	0.5	Monteiro (1996)
Trachurus picturatus	0.04	0.5	Andersen (1997)
Trachurus picturatus	0.16	0.5	Magalhães (2007)

complete a questionnaire detailing age, gender, body weight, height, smoking habits and frequency of fish consumption.

In an initial study, the hair samples were obtained randomly from 29 males and 81 females in Terceira Island, Azores (Fig. 1) who had been fully informed about the purpose of the study through a descriptive document.

In Terceira Island, the average of Hg concentration (n = 110) was 0.86 ± 0.05 $\mu g\ g^{-1}$. These Hg concentrations can be considered relatively low, when compared with the normal values established by international agencies for fish consumption population, once the USEPA established as normal levels the concentration of 1 $\mu g\ g^{-1}$ of Hg in hair (USEPA 1997a), or can be assumed to be even lower when related with the 2 $\mu g\ g^{-1}$ assumed by WHO as normal level of Hg in hair (WHO 1990). Having stated that, about 33 percent of the population in Terceira Island exceeded the RfD indicated by USEPA, whereas 5.9 percent of the same population exceeded the normal level set forth by WHO (Fig. 2).

The volunteers were then grouped according to the fish consumption habits and levels of consumption. In the initial approach, the volunteers were divided into two groups according to their fish-eating habits (consume and not consume). The highest Hg level 0.89 ± 0.05 $\mu g\ g^{-1}$ was obtained in the first group that admitted eating fish (Consume), as opposed to 0.39 ± 0.08 $\mu g\ g^{-1}$ of the second group that said they did not eat fish (Not Consume) (Fig. 3).

In the second approach, the volunteers were grouped into four categories: never (0 fish meals per week), 1–2 meals per week, 3–4 meals per week and 5 or more meals per week according the frequency of fish consumption (Fig. 4). A significant increase (p < 0.001) was found between the volunteers who claimed that they do not eat fish and the volunteers who admitted having five or more meals of fish per week. A significant increase was also observed between all groups of fish consumption except when comparing Hg concentration between the group who admitted eating fish 3–4

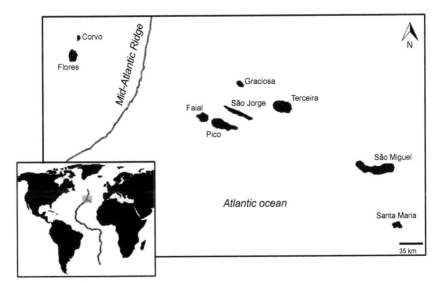

Fig. 1. Map of study area.

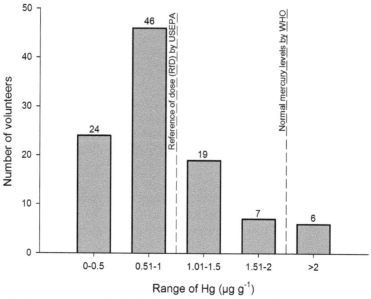

Fig. 2. Distribution of Hg concentration in hair of the 110 inhabitants of Terceira Island.

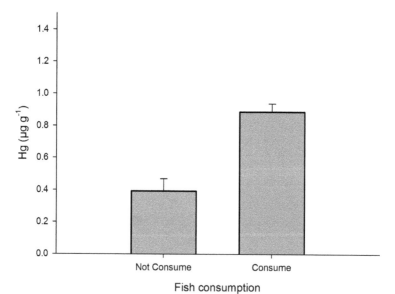

Fig. 3. Hg concentration in the hair of the inhabitants of Terceira Island in relation to fish consumption.

meals per week and the group that consumed 3 or 4 fish meals per week. Similar results have been reported by Al-Majed and Preston (2000) and Díez et al. (2008).

Comparing Hg in the hair of the residents and their habits of fish consumption it was found a positive statistical correlation in the first approach where the study group was divided into two groups according to their fish-eating habits (R = 0.267,

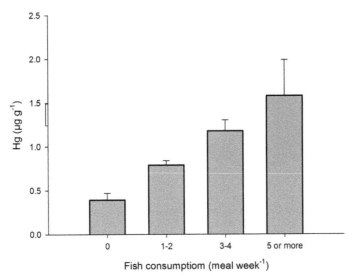

Fig. 4. Hg concentration in the hair of the inhabitants of Terceira Island in relation to fish consumption (meal week^{-1}).

$p < 0.05$) and also in the second approach where the volunteers were grouped into four categories (R = 0.415, $p < 0.001$). Other authors like Holsbeek et al. (1996) and Hajeb et al. (2008), have also associated the consumption of fish as a potential route of uptake of Hg.

In a second study, hair samples were obtained randomly from 157 young volunteers of two different high schools, 84 inhabitants (46 females and 41 males) from Angra do Heroísmo (Azores) and 73 inhabitants (41 females and 32 males) from Murtosa (Mainland), with ages between 14 and 18 years. Volunteers were invited to participate in the study after a detailed explanation of the main objectives.

Considering the Hg concentration of 1 µg g^{-1} in hair established as RfD by USEPA (USEPA 1997c), 24.4 percent of the Azorean volunteers and 52.9 percent of the sampled population in the Mainland exceeded this safety standard (Fig. 5). These results suggest higher Hg levels in the Mainland when compared to data from the Azores.

Hg hair concentration in the Azores ranged from 0.03 to 2.13 µg g^{-1} and the average was 0.82 ± 0.06 µg g^{-1}, on the other hand, the average in the Mainland was 1.13 ± 0.07 µg g^{-1} varying from 0.03 µg g^{-1} to 2.60 µg g^{-1}. Significant differences in average of Hg concentrations present in hair between sampling site were found. Fish consumption can be the key to these differences, since there is a significant difference ($p < 0.05$) when we compare the number of fish meals per week of the volunteers from the Mainland and the Azorean volunteers, 3.31 ± 0.22 meals week^{-1} and 2.08 ± 0.18 meals week^{-1}, respectively.

Hg accumulation in the hair was also evaluated separately for both sampling sites (Azores and Mainland), and the volunteers of both sites were grouped, in the initial approach, into two categories (Fig. 6a) according to their fish-eating habits

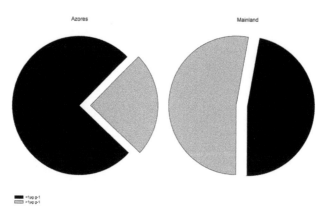

Azores   Mainland

■ <1µg g-1
▨ >1µg g-1

Fig. 5. Distribution of Hg concentration in the hair obtained in both sampling sites (Azores and Mainland).

(consume and not consume). In the second approach, the group was divided into four categories: never (0 meals per week), 1–2 meals per week, 3–4 meals per week and 5 or more meals per week (Fig. 6b).

Volunteers from both sites that admitted consuming fish present higher Hg concentrations than volunteers who did not consume fish (Fig. 6a). This difference is statistically significant for both sites ($p < 0.05$).

In the second approach, where volunteers were grouped into four categories of fish consumption rates (Fig. 6B), the volunteers who assume that they do not have fish meals were 6.4 percent in the Azores and 4.3 percent of the Mainland. On the other hand, 66.7 percent assumed to have a maximum of two meals of fish per week in the Azores, making the category of 1–2 meals per week the most representative for the Azorean volunteers; however, this category represents only 31.4 percent of the Mainland volunteers. In contrast to what was previously observed, the category of 3–4 meals per week, Mainland was the most representative, where 40 percent of the individuals assumed to eat fish 3 or 4 times per week in opposition to the 21.8 percent found in the Azores. In the category with higher fish consumption (5 or more meals of fish per week), 5.1 percent of the Azores volunteers were represented as opposed to 24.3 percent in the Mainland. The range of Hg concentration for each category was $0.03 - 0.49$ µg g^{-1} (never), $0.13 - 2.12$ µg g^{-1} (1–2 meal per week), $0.28 - 2.07$ µg g^{-1} (3–4 meals per week) and $0.95 - 2.00$ µg g^{-1} (≥ 5 meals per week) in the Azores while in the Mainland it ranged between $0.03 - 0.32$ µg g^{-1} (never), $0.27 - 1.77$ µg g^{-1}, (1–2 meal per week) $0.41 - 2.47$ µg g^{-1} (3–4 meals per week) and $0.93 - 2.60$ µg g^{-1} (≥ 5 meals per week).

The Hg concentration in both sampling sites (Azores and the Mainland) showed an increment with the increase of fish meals and are in accordance with Al-Majed and Preston (2000), Díez et al (2008), Vieira et al. (2013) and Shao et al. (2013).

The levels of Hg concentration indicated a significant increase in the case of the individuals who claimed to never eat fish in relation to those who admitted to consume fish 3–4 meals per week and five or more meals per week ($p < 0.05$).

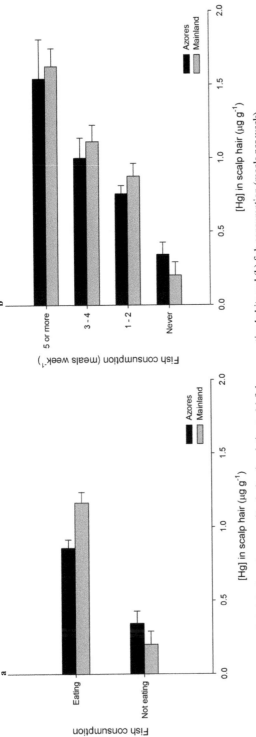

Fig. 6. Hg concentration in hair in the volunteers of both sites in relation to (a) fish consumption habits and (b) fish consumption (meals per week).

Hg Intake Levels, Through the Application of Formulas Established by the World Health Organization

In humans, the ingestion of Hg is related to the concentration of Hg found in the bloodstream (Sherlock et al. 1984). On the other hand, the concentration of Hg in human hair is assumed to be 250 times the concentration in blood (USEPA 1997c).

From the concentration present in the hair, it is possible to estimate the MeHg intake (μg MeHg kg body weight^{-1} day^{-1}) by conversion of the Hg concentration in the hair (μg g^{-1}) to that in the blood (μg L^{-1}), according the following equation a) (USEPA 1997c):

$$\frac{Hghair(\mu gg^{-1})}{250} \times 1000 \tag{a}$$

And the conversion of the Hg blood concentration into MeHg intake can be estimated from hair Hg concentration using the following formula b) (USEPA 1997c):

$$d = \frac{c \times b \times V}{A \times f \times bw} \tag{b}$$

Where

d	=	dose (μg MeHg kg body weight^{-1} day^{-1})	
C	=	mercury concentration in blood (μg l^{-1})	
b	=	elimination rate constant (0.014 per day^{-1})	
V	=	blood volume (5 liters)	
A	=	fraction of the dose absorbed (0.95)	
f	=	the absorbed fraction distributed to the blood (0.05)	
bw	=	body weight (60 kg)	

and to extrapolate the MeHg intake, the following formula was used c) (WHO 2008):

MeHg intake

$$= \frac{Amount\ of\ fishing\ gested\ (gweek^{-1}) \times [MeHg]\ in\ fishing\ gested\ (\mu gg^{-1})}{Kilogram\ body\ weight\ (Kgbw)} \tag{c}$$

Through the MeHg intake obtained with the formula b), the amount of fish consumed per week and body weight, the concentration of MeHg in fish was calculated following the equation d)

[MeHg] in fish (ppm)

$$= \frac{MeHg\ intake\ (\mu gMeHgkg\ body\ weight\ day)}{Amount\ of\ fishing\ gested\ (gweek)} \times 60\ (Kgbw) \tag{d}$$

The potential levels of Hg uptake inferred from PTWIs formulas were evaluated in adolescents of two distinct locations (Angra do Heroísmo and Murtosa), based on the concentrations of mercury present in the human hair.

The levels of MeHg exposure for Azores and Mainland were estimated through formulas a) and b). A body weight of 60 kg was used as denominator of the equation in the USEPA guidelines to calculate the Hg daily dose (USEPA 1997c). In our case study, this value was assumed for both sampling sites, since the average body weight in Azores and Mainland was 60.5 kg and 60.8 kg, respectively.

The exposure level (μg kg^{-1} bw week^{-1}) results showed that adolescents from the Mainland present a higher exposure to the MeHg than the adolescents from Azores, 0.77 μg kg^{-1} bw week^{-1} against 0.57 μg kg^{-1} bw week^{-1}. The difference between sampling sites is statistically significant ($p < 0.001$).

The majority of Azorean adolescents (77.6 percent) had exposure levels below the USEPA RfD, in contrast with the 52.2 percent observed in the Mainland adolescents.

MeHg intake values calculated based on the concentration of Hg present in hair, showed obviously identical trends to Hg concentration in (Fig. 7), increasing with the fish consumption. Similar results were found in other studies such as Bjornberg et al. (2005) and Rubio et al. (2008).

Both sites presented values above the RfD; however, in the Azores these values were only found in volunteers who consumed five or more meals per week, while in the Mainland, the volunteers that had three or more meals of fish per week already exceeded this reference value.

On the other hand, considering fish consumption as the main cause of Hg concentration present in the hair, the application of formula (d) would allow researchers to estimate [MeHg] in the fish.

The calculated MeHg in fish (Fig. 8) show levels of 0.16 μg g^{-1} and 0.13 μg g^{-1} (Azores and Mainland, respectively) for the 1–2 meals of fish per week groups. However, with the increase of fish consumption (3–4 and 5 or more fish meal per week) in both sampling sites, a surprising decrease in the levels of calculated MeHg in fish (< 0.1 μg g^{-1}) was observed.

Higher fish consumption rates led to an increase in the amount (mass) of fish consumed, but the corresponding Hg burden is not linearly assimilated; in this way, MeHg concentrations determined in the ingested fish (source) by volunteers from Azores and Mainland indicate a decrease of Hg concentration in the fish in the highest rates of fish consumption. However, assuming fish as the common source of MeHg for all fish consumption categories, this occurrence could not be acceptable. This is because consuming one fish meal per week or five fish meal per week would have had no effect on the Hg concentration in the fish specimens. Accordingly, considering fish consumption as the main route of Hg in humans, the levels of MeHg in fish should be identical regardless of the category of consumption (1–2, 3–4 and 5 or more than 5 meals per week). Thus, assuming a common level of MeHg in fish for every consumption category and having the MeHg level in fish obtained for the lowest fish meal per week (0.16 μg g^{-1} for Azores and 0.13 μg g^{-1} for Mainland) as reference, the potential levels of Hg in hair would have to be higher than the determined [Hg] (found in hair) that our results had first indicated.

The extrapolated intake (Fig. 9) increased linearly with the fish consumption. The differences between intakes are approximately the same when compared with the real intake for the fish consumption of 1–2 meals per week, approximately 2-fold in

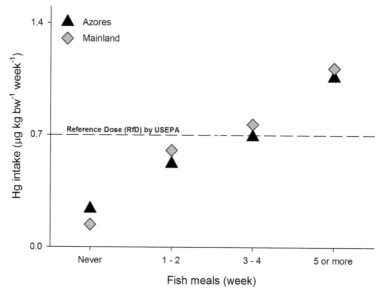

Fig. 7. MeHg intake (μg MeHg kg body weight^{-1} day^{-1}) in volunteers of Azores and Mainland.

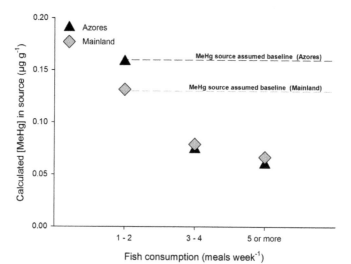

Fig. 8. MeHg concentration in fish ingested in volunteers from Azores and Mainland.

the volunteers that consume 2–3 fish meals per week in both sites and approximately 3-fold and 2-fold in the category of 5 or more fish meals per week in the Azores and the Mainland, respectively.

Metals, such as Hg, are known to induce metallothioneins (MT) (Bucheli and Fent 1995, Yasutake et al. 1998) that have high cysteine content, acting similarly to metal chelators, providing heavy metal tolerance and regulating Hg distribution and retention (Yoshida et al. 1997, Klaassen and Liu 1998, Yoshida et al. 1999). Chelators of Hg (e.g., glutathione and N-acetylcysteine) have been shown to

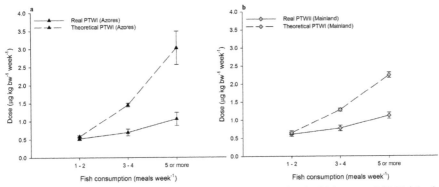

Fig. 9. Real MeHg and potential MeHg intake versus fish consumption for (a) Azores and (b) Mainland.

enhance the clearance of Hg from the blood perhaps by reducing metal uptake by erythrocytes (Gochfeld 1997). According to Gundacker et al. (2007), an increment in fish consumption per week (higher Hg intake), may lead to higher levels of MT in the bloodstream. Since MT are important detoxification proteins, an increase of this type of proteins may result in the reduction of Hg concentration in the blood at the time of hair formation, contributing to the differences observed between Hg concentration measured in the hair of the volunteers and the extrapolated Hg concentration after applying the formula c). Micronutrients such as selenium, methionine, cysteine and vitamin E can also protect against Hg bioactivity, likely via antioxidant responses (e.g., bcl2 gene induction in the kidney, free radical scavenging) trapping Hg and it may also aid to explain Hg detoxification (Drasch et al. 1996, Patrick 2002, Battin and Brumaghim 2009), leading to a lower Hg concentration compared to the expected (extrapolated) values.

Isocurves of the Maximum Number of Fishmeal per week Without Exceeding the MeHg RfD

Epidemiological studies in New Zealand, Seychelles and the Faeroe are the basis of the safety levels of daily Hg exposure established by international agencies such as the USEPA and JECFA (Burger et al. 2001, Rice 2004, Maycock and Benford 2007). USEPA fixed the guideline values for maximum exposure limits (RfD) of 0.1 μg MeHg kg bw^{-1} day^{-1} (USEPA 1997a) and JECFA established PTWI equivalent to 0.19 μg kg^{-1} day^{-1} for MeHg (EFSA 2012).

Uncertainty factors (UF) are used for the calculation of RfD. The USEPA RfD was established by applying a 10-fold Uncertainty Factor (UF) to explain the variation between prolonged healthy human populations exposed to the consumption of contaminated fish (USEPA 1997b).

The reference dose setting established by the USEPA was suggested by Grandjean and Budtz-Jørgensen (2007) based on developmental neurotoxicity studies at exposure levels close to 50 percent below USEPA RfD (Jedrychowski et al. 2007, Lederman et al. 2008, Suzuki et al. 2010). With this adjustment, the reference dose would be reduced from 0.1 to 0.05 μg MeHg kg bw^{-1} day^{-1}.

The level of exposure to MeHg depends on (i) type and amount of ingested fish per unit time (such as day or week); (ii) MeHg concentrations in fish; and (iii) the body weight of the fish consumers. Thus, MeHg intake (μg MeHg kg bw^{-1} week^{-1} for individuals or populations) can be calculated by the following the formula e) (WHO 2008):

MeHg intake

$$= \frac{\textit{Amount of fishing gested (gweek}^{-1}) \times [Hg] \textit{ in fishing gested } (\mu gg^{-1})}{\textit{Kilogram body weight (Kgbw)}} \qquad \text{e)}$$

Therefore, the maximum number of fish meals per week can be calculated using formula f) by considering a reference dose for MeHg as the maximum exposure value, the body weight of 60 kg based on USEPA guidelines (USEPA 1997c), the amount (g) of fish ingested per meal and MeHg concentration present in ingested fish:

$$\textit{Fish meal week} = \frac{(RfD \times bw) \times 7 \textit{ days}}{(\textit{Fish meal size (g)} \times [MeHg] \textit{ in fish}) \div 1000} \qquad \text{f)}$$

The results for the application of the formula f) are shown as trend lines (isocurves) (Fig. 10) indicating the maximum number fish meals per week allowed without exceeding the RfD (0.1 μg MeH g kg bw^{-1} day^{-1} or 0.05 μg MeHg kg bw^{-1} day^{-1}), combining fish meal size of 50 g (Fig. 10a), 100 g (Fig. 10b), 150g (Fig. 10c), 170 g (Fig. 10d), 200 g (Fig. 10e) and 250 g (Fig. 10f) under different MeHg concentrations.

Regarding a RfD of 0.1 μg MeHg kg bw^{-1} day^{-1}, the isocurves demonstrate that the RfD is exceeded if we consume a 50 g fish meal once a week with MeHg concentrations above 0.84 μg g^{-1} (Fig. 9a), whereas when considering a RfD of 0.05 μg MeHg kg bw^{-1} day^{-1} the MeHg concentrations should not exceed 0.42 μg of MeHg g^{-1} of fish (Fig. 9a). As expected, as the portion of fish per meal increases the values of mercury in fish would have to decrease in order to obtain a specific level of exposure. For example, to the RfD of 0.1 μg MeHg kg bw^{-1} day^{-1} and the fish size of 100, 150, 170, 200 and 250 g (Fig. 9b–9f), the MeHg concentration in fish would not have to exceed the 0.42, 0.28, 0.24, 0.21 and 0.17 μg g^{-1}, respectively, following that RfD. Regarding the RfD of 0.05 μg MeHg kg bw^{-1} day^{-1} and for the fish size of 100, 150, 170, 200 and 250 g, the values related to MeHg in fish would obviously have been decreased to half when compared to the values of fish meal size using USEPA RfD.

Conclusion

The concentration of Hg found in the hair of the two populations studied (Azores and Mainland) indicates that individuals with higher fish consumption per week generally have higher concentrations of Hg, which makes fish consumption the main route of exposure for these populations. Overall data indicate relatively low concentration of mercury in hair despite the high fish consumption per capita.

Hg concentration in human hair is assumed to reflect MeHg intake through fish consumption, allowing researchers to evaluate the risk associated with the MeHg

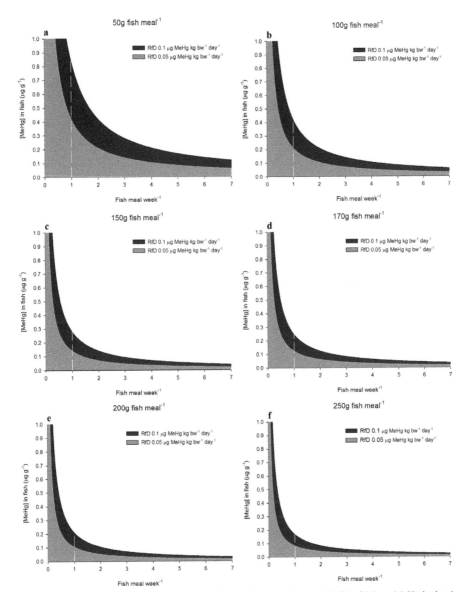

Fig. 10. Fish meal week^{-1} in relation to the [MeHg] in fish (µg g^{-1}) for: (a) RfD of 0.1 µg MeHg kg bw^{-1} day^{-1} and (b) ½ RfD (0.05 µg MeHg kg bw^{-1} day^{-1}).

Color version at the end of the book

intake for the human population. The results of the Hg concentration of adolescents from the Azores and the Mainland indicate that the effective Hg uptake becomes lower than expected as the rate of fish consumption increases. These results accentuate the capacity of the human body to induce a self-protection response, minimising MeHg assimilation probably by detoxification mechanisms, thereby mitigating Hg levels in the blood when experiencing increasing Hg exposure.

Finally, even a single meal (per week) of 50 g fish with 0.84 µg g⁻¹ of MeHg would reach the USEPA RfD maximum levels, despite the 1.0 µg g⁻¹ of MeHg in fish which is being allowed for fish consumption.

Acknowledgements

This work was supported by the Portuguese Science Foundation (FCT) through a PhD grant awarded to Vieira, H.C. (PD/BD/127808/2016). FCT also financed CESAM: UID/AMB/50017/2013.

References

Afonso, C., Lourenço, H.M., Dias, A., Nunes, M.L. and Castro, M. 2007. Contaminant metals in black scabbard fish (*Aphanopus carbo*) caught off Madeira and the Azores. Food Chemistry 101: 120–125.

Agusa, T., Kunito, T., Iwata, H., Monirith, I., Tana, T.S., Subramanian, A. and Tanabe, S. 2005. Mercury contamination in human hair and fish from Cambodia: levels, specific accumulation and risk assessment. Environmental Pollution 134: 79–86.

Agusa, T., Kunito, T., Iwata, H., Monirith, I., Chamnan, C., Tana, T.S., Subramanian, A. and Tanabe, S. 2007. Mercury in hair and blood from residents of Phnom Penh (Cambodia) and possible effect on serum hormone levels. Chemosphere 68: 590–596.

Al-Majed, N.B. and Preston, M.R. 2000. Factors influencing the total mercury and methyl mercury in the hair of the fishermen of Kuwait. Environmental Pollution 109: 239–250.

Andersen, J.L. and Depledge, M.H. 1997. A survey of total mercury and methylmercury in edible fish and invertebrates from Azorean waters. Marine Environmental Research 44: 331–350.

ATSDR. 2013. Sumary data for 2013 priority list of hazardous substances. Agency for Toxic Substances and Disease Registry, Atlanta, GA. Available at http://www.atsdr.cdc.gov/spl/resources/ATSDR_2013_SPL_Detailed_Data_Table.pdf [accessed on February 2015].

Baeyens, W., Leermakers, M., Papina, T., Saprykin, A., Brion, N., Noyen, J., De Gieter, M., Elskens, M. and Goeyens, L. 2003. Bioconcentration and biomagnification of mercury and methylmercury in north sea and scheldt estuary fish. Archives of Environmental Contamination and Toxicology 45: 498–508.

Battin, E.E. and Brumaghim, J.L. 2009. Antioxidant activity of sulfur and selenium: a review of reactive oxygen species scavenging, glutathione peroxidase, and metal-binding antioxidant mechanisms. Cell Biochemistry and Biophysics 55: 1–23.

Björnberg, K.A., Vahter, M., Petersson-Grawé, K., Glynn, A., Cnattingius, S., Darnerud, P.O., Atuma, S., Aune, M., Becker, W. and Berglund, M. 2003. Methyl mercury and inorganic mercury in Swedish pregnant women and in cord blood: influence of fish consumption. Environmental Health Perspectives 111: 637–641.

Björnberg, K.A., Vahter, M., Grawé, K.P. and Berglund, M. 2005. Methyl mercury exposure in Swedish women with high fish consumption. Science of the Total Environment 341: 45–52.

Bucheli, T.D. and Fent, K. 1995. Induction of cytochrome P450 as a biomarker for environmental contamination in aquatic ecosystems. Critical Reviews in Environmental Science and Technology 25: 201–268.

Burger, J., Gaines, K.F. and Gochfeld, M. 2001. Ethnic differences in risk from mercury among savannah river fishermen. Risk Analysis 21: 533–544.

Carrasco, L., Barata, C., García-Berthou, E., Tobias, A., Bayona, J.M. and Díez, S. 2011. Patterns of mercury and methylmercury bioaccumulation in fish species downstream of a long-term mercury-contaminated site in the lower Ebro River (NE Spain). Chemosphere 84: 1642–1649.

Cernichiari, E., Toribara, T.Y., Liang, L., Marsh, D.O., Berlin, M., Myers, G.J., Cox, C., Shamlaye, C.F., Choisy, O. and Davidson, P. 1995. The biological monitoring of mercury in the Seychelles study. Neuro Toxicology 16: 613–628.

Cizdziel, J.V. and Gerstenberger, S. 2004. Determination of total mercury in human hair and animal fur by combustion atomic absorption spectrometry. Talanta 64: 918–921.

Compeau, G.C. and Bartha, R. 1985. Sulfate-reducing bacteria: Principal methylators of mercury in anoxic estuarine sediment. Applied and Environmental Microbiology 50: 498–502.

Coolbaugh, M., Gustin, M. and Rytuba, J. 2002. Annual emissions of mercury to the atmosphere from natural sources in Nevada and California. Environmental Geology 42: 338–349.

Costa, V.L., Lourenço, H.M., Figueiredo, I., Carvalho, L., Lopes, H., Farias, I., Pires, L., Afonso, C., Vieira, A.R., Nunes, M.L. and Gordo, L.S. 2009. Mercury, cadmium and lead in black scabbardfish (Aphanopus carbo Lowe, 1839) from mainland Portugal and the Azores and Madeira archipelagos. Gordo, L.S. (Ed.) (2009). Stock structure and quality of black scabbardfish in the southern NE Atlantic. Scientia Marina (Barcelona) 73(Suppl. 2): 77–88.

Díez, S., Montuori, P., Pagano, A., Sarnacchiaro, P., Bayona, J.M. and Triassi, M. 2008. Hair mercury levels in an urban population from southern Italy: Fish consumption as a determinant of exposure. Environment International 34: 162–167.

Dolbec, J., Mergler, D., Larribe, F., Roulet, M., Lebel, J. and Lucotte, M. 2001. Sequential analysis of hair mercury levels in relation to fish diet of an Amazonian population, Brazil. Science of the Total Environment 271: 87–97.

Dommergue, A., Ferrari, C.P., Planchon, F.A.M. and Boutron, C.F. 2002. Influence of anthropogenic sources on total gaseous mercury variability in grenoble suburban air (France). Science of the Total Environment 297: 203–213.

Dorea, J., Barbosa, A., Ferrari, Í. and De Souza, J. 2003. Mercury in hair and in fish consumed by Riparian women of the Rio Negro, Amazon, Brazil. International Journal of Environmental Health Research 13: 239–248.

Drasch, G., Wanghofer, E., Roider, G. and Strobach, S. 1996. Correlation of mercury and selenium in the human kidney. Journal of Trace Elements in Medicine and Biology 10: 251–254.

EFSA. 2012. Scientific opinion on the risk for public health related to the presence of mercury and methylmercury in food. EFSA J 10.

Egeland, G.M. and Middaugh, J.P. 1997. Balancing fish consumption benefits with mercury exposure. Science 278: 1904–1905.

Failler, P., Van de Walle, G., Lecrivain, N., Himbes, A. and Lewins, R. 2007. Future prospects for fish and fishery products. 4. Fish consumption in the European Union in 2015 and 2030. Part 1. European overview. FAO Fisheries Circular (FAO).

FAO. 2010. Fishery and Aquaculture Statistics 2008. Statistics and Information Service of the Fisheries and Aquaculture Department, FAO Yearbook. FAO Fisheries and Aquaculture Department, Rome.

FAO. 2012. The State of World Fisheries and Aquaculture 2012. Food & Agriculture Organization.

Freire, C., Ramos, R., Lopez-Espinosa, M.-J., Díez, S., Vioque, J., Ballester, F. and Fernández, M.-F. 2010. Hair mercury levels, fish consumption, and cognitive development in preschool children from Granada, Spain. Environmental Research 110: 96–104.

Gochfeld, M. 1997. Factors influencing susceptibility to metals. Environ Health Perspect 105 Suppl. 4: 817–822.

Gochfeld, M. 2003. Cases of mercury exposure, bioavailability, and absorption. Ecotoxicology and Environmental Safety 56: 174–179.

Grandjean, P. and Budtz-Jørgensen, E. 2007. Total imprecision of exposure biomarkers: implications for calculating exposure limits. Am. J. Ind. Med. 50: 712–719. doi:10.1002/ajim.20474.

Gundacker, C., Komarnicki, G., Jagiello, P., Gencikova, A., Dahmen, N., Wittmann, K.J. and Gencik, M. 2007. Glutathione-S-transferase polymorphism, metallothionein expression, and mercury levels among students in Austria. Science of the Total Environment 385: 37–47.

Gustin, M.S., Lindberg, S.E., Austin, K., Coolbaugh, M., Vette, A. and Zhang, H. 2000. Assessing the contribution of natural sources to regional atmospheric mercury budgets. Science of the Total Environment 259: 61–71.

Hajeb, P., Selamat, J., Ismail, A., Bakar, F., Bakar, J. and Lioe, H. 2008. Hair mercury level of coastal communities in Malaysia: a linkage with fish consumption. European Food Research and Technology 227: 1349–1355.

Hajeb, P., Jinap, S., Ismail, A., Fatimah, A.B., Jamilah, B. and Abdul Rahim, M. 2009. Assessment of mercury level in commonly consumed marine fishes in Malaysia. Food Control 20: 79–84.

Harris, H.H., Pickering, I.J. and George, G.N. 2003. The chemical form of mercury in fish. Science 301: 1203.

Holsbeek, L., Das, H.K. and Joiris, C.R. 1996. Mercury in human hair and relation to fish consumption in Bangladesh. Science of the Total Environment 186: 181–188.

Jedrychowski, W., Perera, F., Jankowski, J., Rauh, V., Flak, E., Caldwell, K.L., Jones, R.L., Pac, A. and Lisowska-Miszczyk, I. 2007. Fish consumption in pregnancy, cord blood mercury level and cognitive and psychomotor development of infants followed over the first three years of life: Krakow epidemiologic study. Environment International 33: 1057–1062.

Klaassen, C.D. and Liu, J. 1998. Induction of metallothionein as an adaptive mechanism affecting the magnitude and progression of toxicological injury. Environ. Health Perspect. 106 Suppl. 1: 297–300.

Koenig, S., Solé, M., Fernández-Gómez, C. and Díez, S. 2013. New insights into mercury bioaccumulation in deep-sea organisms from the NW Mediterranean and their human health implications. Science of The Total Environment 442: 329–335.

Kris-Etherton, P.M., Harris, W.S., Appel, L.J. and Committee, f.t.N. 2002. Fish Consumption, Fish Oil, Omega-3 Fatty Acids, and Cardiovascular Disease. Circulation 106: 2747–2757.

Lederman, S.A., Jones, R.L., Caldwell, K.L., Rauh, V., Sheets, S.E., Tang, D., Viswanathan, S., Becker, M., Stein, J.L., Wang, R.Y. and Perera, F.P. 2008. Relation between cord blood mercury levels and early child development in a world trade center cohort. Environmental Health Perspectives 116: 1085–1091.

Liao, P.-Y., Liu, C.-W. and Liu, W.-Y. 2016. Bioaccumulation of mercury and polychlorinated dibenzo-p-dioxins and dibenzofurans in salty water organisms. Environ. Monit. Assess. 188: 1–15.

Lin, C.-J. and Pehkonen, S.O. 1999. The chemistry of atmospheric mercury: a review. Atmospheric Environment 33: 2067–2079.

Lindqvist, O. and Rodhe, H. 1985. Atmospheric mercury—a review. Tellus B 37B: 136–159.

Loppi, S. 2001. Environmental distribution of mercury and other trace elements in the geothermal area of Bagnore (Mt. Amiata, Italy). Chemosphere 45: 991–995.

MacIntosh, D.L., Williams, P.L., Hunter, D.J., Sampson, L.A., Morris, S.C., Willett, W.C. and Rimm, E.B. 1997. Evaluation of a food frequency questionnaire-food composition approach for estimating dietary intake of inorganic arsenic and methylmercury. Cancer Epidemiology Biomarkers & amp; Prevention 6: 1043–1050.

Magalhães, M.C., Costa, V., Menezes, G.M., Pinho, M.R., Santos, R.S. and Monteiro, L.R. 2007. Intra- and inter-specific variability in total and methylmercury bioaccumulation by eight marine fish species from the Azores. Mar. Pollut. Bull. 54(10): 1654–1662. doi:10.1016/j.marpolbul.2007.07.006.

Malm, O., Branches, F.J.P., Akagi, H., Castro, M.B., Pfeiffer, W.C., Harada, M., Bastos, W.R. and Kato, H. 1995. Mercury and methylmercury in fish and human hair from the Tapajós river basin, Brazil. Science of the Total Environment 175: 141–150.

Maycock, B.J. and Benford, D.J. 2007. Risk assessment of dietary exposure to methylmercury in fish in the UK. Human & Experimental Toxicology 26: 185–190.

Megapesca. 2007. Relatório Estatístico—Caracterização do consumo de Pescado nos Açores. Megapesca, Alfeizerão, Portugal.

Mergler, D., Anderson, H.A., Chan, L.H., Mahaffey, K.R., Murray, M., Sakamoto, M. and Stern, A.H. 2007. Methylmercury exposure and health effects in humans: a worldwide concern. Ambio. 36: 3–11.

Mil-Homens, M., Branco, V., Vale, C., Stevens, R., Boer, W., Lebreiro, S., Cato, I. and Abrantes, F. 2008. Historical trends in Hg, Pb and Zn sedimentation in the central shelf area of Portugal. Journal of Iberian Geology 34: 287–298.

Monteiro, L., Isidro, E. and Lopes, H. 1991. Mercury content in relation to sex, size, age and growth in two scorpionfish (*Helicolenus dactylopterus* and *Pontinus kuhlii*) from Azorean waters. Water, Air, & Soil Pollution 56: 359–367.

Monteiro, L.R., Costa, V., Furness, R.W. and Santos, R.S. 1996. Mercury concentrations in prey fish indicate enhanced bioaccumulation in mesopelagic environments. Marine Ecology Progress Series 141: 21–25.

Ocké, M.C., Bueno-de-Mesquita, H.B., Goddijn, H.E., Jansen, A., Pols, M.A., van Staveren, W.A. and Kromhout, D. 1997. The Dutch EPIC food frequency questionnaire. I. Description of the questionnaire, and relative validity and reproducibility for food groups. International Journal of Epidemiology 26: S37–48.

Oken, E., Rifas-Shiman, S.L., Amarasiriwardena, C., Jayawardene, I., Bellinger, D.C., Hibbeln, J.R., Wright, R.O. and Gillman, M.W. 2016. Maternal prenatal fish consumption and cognition in mid childhood: Mercury, fatty acids, and selenium. Neurotoxicology and Teratology 57: 71–78.

Pacyna, E.G., Pacyna, J.M. and Pirrone, N. 2001. European emissions of atmospheric mercury from anthropogenic sources in 1995. Atmospheric Environment 35: 2987–2996.

Passos, C.J., Mergler, D., Gaspar, E., Morais, S., Lucotte, M., Larribe, F., Davidson, R. and Grosbois, S.d. 2003. Eating tropical fruit reduces mercury exposure from fish consumption in the Brazilian Amazon. Environmental Research 93: 123–130.

Patrick, L. 2002. Mercury toxicity and antioxidants: part i: role of glutathione and alpha-lipoic acid in the treatment of mercury toxicity-mercury toxicity. Toxicol. Appl. Pharmacol. 7: 456–471.

Pereira, M., Lillebø, A., Pato, P., Válega, M., Coelho, J., Lopes, C., Rodrigues, S., Cachada, A., Otero, M., Pardal, M. and Duarte, A. 2009. Mercury pollution in Ria de Aveiro (Portugal): a review of the system assessment. Environ. Monit. Assess. 155: 39–49.

Pirrone, N., Cinnirella, S., Feng, X., Finkelman, R., Friedli, H., Leaner, J., Mason, R., Mukherjee, A., Stracher, G. and Streets, D. 2010. Global mercury emissions to the atmosphere from anthropogenic and natural sources. Atmospheric Chemistry and Physics 10: 5951–5964.

Rasmussen, R.S., Nettleton, J. and Morrisey, M.T. 2005. A review of mercury in seafood: special focus on tuna. Journal of Aquatic Food Product Technology 14: 71–100.

Renzoni, A., Zino, F. and Franchi, E. 1998. Mercury levels along the food chain and risk for exposed populations. Environmental Research 77: 68–72.

Rice, D.C. 2004. The US EPA reference dose for methylmercury: sources of uncertainty. Environmental Research 95: 406–413.

Rodrigues, A.F., Freitas, M.D.C., Vieira, B., Soares, P., Sousa, E., Amaral, A.S., Rodrigues, C. and Aptroot, A. 2004. Mercury levels on the eastern flanks of the mid-atlantic ridge (azores region). Materials and Geoenvironment 51: 1336–1339.

Rubio, C., Gutiérrez, Á., Burgos, A. and Hardisson, A. 2008. Total dietary intake of mercury in the Canary Islands, Spain. Food Additives & Contaminants: Part A 25: 946–952.

Shao, D., Kang, Y., Cheng, Z., Wang, H., Huang, M., Wu, S., Chen, K. and Wong, M.H. 2013. Hair mercury levels and food consumption in residents from the Pearl River Delta: South China. Food Chemistry 136: 682–688.

Sherlock, J., Hislop, J., Newton, D., Topping, G. and Whittle, K. 1984. Elevation of mercury in human blood from controlled chronic ingestion of methylmercury in fish. Human Toxicology 3: 117–131.

Siegel, S.M. and Siegel, B.Z. 1984. First estimate of annual mercury flux at the Kilauea main vent. Nature 309: 146–147.

Suzuki, K., Nakai, K., Sugawara, T., Nakamura, T., Ohba, T., Shimada, M., Hosokawa, T., Okamura, K., Sakai, T., Kurokawa, N., Murata, K., Satoh, C. and Satoh, H. 2010. Neurobehavioral effects of prenatal exposure to methylmercury and PCBs, and seafood intake: Neonatal behavioral assessment scale results of Tohoku study of child development. Environmental Research 110: 699–704.

Torres, P., Rodrigues, A., Soares, L. and Garcia, P. 2015. Metal Concentrations in Two Commercial Tuna Species from an Active Volcanic Region in the Mid-Atlantic Ocean. Archives of Environmental Contamination and Toxicology 70: 341–347.

Ullrich, S.M., Tanton, T.W. and Abdrashitova, S.A. 2001. Mercury in the aquatic environment: a review of factors affecting methylation. Critical Reviews in Environmental Science and Technology 31: 241–293.

USEPA. 1997a. Mercury Study Report to Congress Volume IV. An Assessment of Exposure to Mercury in the United States. Office of Air Quality Planning and Standards and Office of Research and Development, Washington, DC.

USEPA. 1997b. Mercury Study Report to Congress Volume V: Health Effects of Mercury and Mercury Compounds.

USEPA. 1997c. Mercury Study Report to Congress Volume VII: Characterization of Human Health and Wildlife Risks from Mercury Exposure in the United States. U.S. Environmental Protection Agency.

Vieira, H.C., Morgado, F., Soares, A.M.V.M. and Abreu, S.N. 2013. Mercury in scalp hair near the mid-atlantic ridge (mar) in relation to high fish consumption. Biological Trace Element Research 156: 29–35.

WHO. 1990. Environmental health criteria 101—Methylmercury. World Health Organization, Geneva.

WHO. 2003. Diet, nutrition and the prevention of chronic diseases: Report of a joint WHO/FAO expert consultation, 28 January–1 February 2002. Geneva: World Heath Organization/Food and Agricultural Organization.

WHO. 2008. Guidance for identifying populations at risk from mercury exposure. Geneva, Switzerland.

Wolkin, A., Hunt, D., Martin, C., Caldwell, K.L. and McGeehin, M.A. 2012. Blood mercury levels among fish consumers residing in areas with high environmental burden. Chemosphere 86: 967–971.

Yasutake, A., Nakano, A. and Hirayama, K. 1998. Induction by mercury compounds of brain metallothionein in rats: Hg0 exposure induces long-lived brain metallothionein. Archives of Toxicology 72: 187–191.

Yoshida, M., Satoh, H. and Sumi, Y. 1997. Effect of ethanol pretreatment on mercury distribution in organs of fetal guinea pigs following in utero exposure to mercury vapor. Toxicology 119: 193–201.

Yoshida, M., Satoh, M., Yasutake, A., Shimada, A., Sumi, Y. and Tohyama, C. 1999. Distribution and retention of mercury in metallothionein-null mice after exposure to mercury vapor. Toxicology 139: 129–136.

You, C.H., Kim, B.G., Kim, Y.M., Lee, S.A., Kim, R.B., Seo, J.W. and Hong, Y.S. 2014. Relationship between dietary mercury intake and blood mercury level in Korea. J. Korean Med. Sci. 29: 176–182.

Contaminants Impact on Marine Turtle Populations Development
An Overview

Patricia Salvarani,[1] *Vania C. Foster,*[2] *Jaime Rendon*[3]
and *Fernando Morgado*[1,*]

INTRODUCTION

There are seven sea turtle species in the world, distributed across almost all the oceans (*Caretta caretta, Chelonia mydas, Dermochelys coriacea, Eretmochelys imbricata, Lepidochelys olivacea, Natator depressus* and *Lepidochelys kempii*). With the exception for *Natator depressus*, all others are catalogued as vulnerable, endangered or critically endangered in the IUCN list (IUCN 2016). Currently, six species of sea turtles are included in the lists of endangered species worldwide (MTSG 1995, Lutcavage et al. 1997) and on the Red List of Threatened Animals (IUCN). According to the Marine Turtle Specialist Group (MTSG), currently, the main threats to sea turtles are coastal development, incidental capture by fisheries, direct use for human consumption, climate change, pollution and biotic threats.

During their life cycle, sea turtles face various challenges in a fight for survival and are subject to many biotic, abiotic and anthropogenic threats. Biotic threats are related to diseases and parasites such as fibro papillomatosis and spiroquidiasis, bacterial or fungi infections in eggs (Aguirre 1995, 2000, Rossi et al. 2009, Rodenbusch et al. 2014), presence of other turtles spawning (Eckrich and Owens 1995), eggs and offspring predation (Fowler 1979) and also vegetation growth around

[1] Department of Biology & CESAM, University of Aveiro, 3810-193 Aveiro, Portugal.
 Email: patysalvarani@hotmail.com
[2] Programa de Pós-graduação em Ecologia e Conservação, Universidade Federal de Mato Grosso do Sul, 79070-900 Campo Grande, Mato Grosso do Sul, Brazil.
[3] Instituto Epomex, Universidad Autónoma de Campeche, Av Augustin de Melgar y Juan de la Barrera s/n, 24039 Campeche, México.
* Corresponding author

the nests, since roots can invade nests and cause offspring mortality (Godfrey et al. 1995). Abiotic threats are related to strong rains, erosions, heat stress and climate change (Hawkes et al. 2009). The eggs absorb moisture from the environment around them and can change the morphological and embryonic and hatchling physiological features (Hewavisenthi and Parmenter 2001). Such adsorbed substances may affect the process of hatching nests and influence the development and survival of the offspring since the early stages of embryonic development are the most vulnerable to toxic exposure (Bishop et al. 1991, Alam and Brim 2000, Hewavisenthi and Parmenter 2001).

In addition to effects on phenology, according to Walther et al. (2002), climate change may have more direct impacts on thermally sensitive organisms, and under future scenarios of climate warming, and given the conservation concern regarding the sea turtle population, understanding of the potential effects of temperature increases on sea turtle populations and their potential to cope with such changes is important. Sea turtles have temperature-dependent sexual determination, where the sex ratio is influenced by the temperature of the sand during incubation. In these organisms, sex is determined by temperature in the middle third period of incubation with males offspring occurring when the eggs are exposed to lower temperatures and females under higher temperatures, within a thermal tolerance range of 25 to 35°C (Ackerman 1997, Hawkes et al. 2007).

Threats associated with human activities are related to population increase (Marcovaldi and Marcovaldi 1999), hunt and egg collection (Márquez and Doi 1973, Wetherall et al. 1993, Gallo 2006, Pupo et al. 2006), erosion and accretion of beaches (Witherington and Martin 2000), artificial lighting (Marcovaldi and Marcovaldi 1999, Witherington and Martin 2000), fishing nets (Spotila et al. 2000, Heppel et al. 2003) and shading (Van De Merwe et al. 2006).

Sea turtles assume a very important ecological role in trophic nets of the coastal ecosystem, either as consumers (seaweed, seagrass, sponges, tunicates, crustaceans, cnidarians) or as prey (eggs, youth and adults). Their movements during spawning and feeding between different habitats (seagrass beds, coral reefs, ocean waters, sandy beaches) are important in energy transfer and nutrient recycling (Bjorndal 1997). As species with slow growth and a long-life cycle, they are often used as models for evolutionary studies of adaptation to different environmental conditions, since they are extremely susceptible to the action of man in all phases of its life cycle (Dodd and Dreslik 2008).

Organochlorine contaminants (OCs) cause a strong impact on the environment due to three basic characteristics: environmental persistence, bioaccumulation and high toxicity. These characteristics may affect the health of marine animals, including sea turtles. The bioaccumulation of these pollutants in tissues and organs, influence the growth and development of natural populations of sea turtles worldwide, leading to mortality in various stages of development (egg, juvenile or adult) (Lake et al. 1994, von Osten 2005, de Andréa 2008, Marcovecchio and Freije 2013, Guerranti et al. 2014).

These contaminants are transferred to the offspring via eggs, with serious implications for embryonic development and health once outside the shell (Russell and Haffner 1999, Alava et al. 2006). Environmental contaminants of chemical origin are chlorinated hydrocarbons synthesised by man, not occurring naturally in the environment and can be divided into two groups: low molecular weight (industrial solvents and chlorofluorocarbon (CFCs)) and high molecular weight (organochlorine pesticides (OCPs) and polychlorinated biphenyls (PCBs)) (Clark 1992). The bioaccumulation of these substances due to transfer in the food chain has become a reason for concern about the health of these communities (Storelli et al. 2003, Alava et al. 2006, da Silva et al. 2014). Because they are threatened with extinction, it is extremely important to understand how sea turtle population responds to impacts and what are the required conservation measures needed in a long-term perspective.

Heavy metals as Hg, Pb, Cd, Cr, As, Se, Fe, Mn, Zn, Cu, Ni and Co in tissues, eggs, and blood can contribute to the contamination of sea turtle populations. Some of them are essential to the life processes of organisms, but others are toxic, even at low concentrations, they accumulated in organisms via food, respiratory pathways or the dermal contact (Sakai et al. 1995, Becker et al. 2002, Innis et al. 2008, Jakimska et al. 2011) that may affect, for example, the reproductive system, gastrointestinal, respiratory, immunologic and renal (Godley et al. 1999, Sakai et al. 2000, Franzellitti et al. 2004, Cortés-Gómez et al. 2014).

Sea Turtles Classification

The sea turtles are classified into two families: Cheloniidae and Dermochelyidae. The first family has six species (*C. caretta, C. mydas, N. depressus, E. imbricata, L. olivacea, L. kempii*), with a carapace covered with plates, variable in numbers for each species. This family is characterized by having a strong skull and secondary palate. Their edges are shaped as nonretractable fins covered with numerous small plates; elongated fingers and tightly stuck for connective tissue. Their claws are reduced to one or two fins (Pritchad et al. 1997). The second family includes only the species *D. coriacea* in which instead of having a shell covered with plates, it has a skin similar to leather. This species is characterized by an extreme reduction of shell and cravats; development of a dorsal surface composed of a mosaic with millions polygonal bones; lack of claws and plates in the shells (there are plates until the juvenile stage); and a skeleton full of fat with extensive areas of vascularised cartilage in the vertebras and fin joints. Its skull does not have nasal bones and its surface mandibular is covered with keratin. This species is considered the largest one among all the species (MOLL et al. 1979).

From seven species listed in the world, most of the species are distributed in the tropical seas (*C. mydas, E. imbricata, C. caretta, L. olivacea, D. coriacea*), and two species have restricted distribution, *N. depressus* in northeast of Australia and *L. kempii* in the Gulf of Mexico and North Atlantic (IUCN 2016) (Table 1). Turtles are identified for many characteristics such as inframarginal scutes (womb), costal scutes (back), size and shape of the head, the shape of heads scutes and the formation and colour of costal scutes.

Table 1. Main characteristics of sea turtles.

Species	Key Features	Habitat	Feeding	Spawning	Length/Weight	Location	Status Iucn
Caretta caretta (Linnaeus 1758)	large size of the head, proportionally larger than the other species	shallow depth, about 20 m	crustaceans, molluscs, jellyfish, hydrozoans, fish eggs	from 110 to 130 eggs	110 cm carapace/150 kg	Florida, South Africa, Greece, Turkey, Indian Ocean, Australia, Japan, United States, Mediterranean, Brazil	Vulnerable
Eremochelys imbricata (Linnaeus 1766)	hull is formed by overlapping scales, corneum beak	shallow depth, about 40 m	fish eggs, crustaceans, molluscs, bryozoans, coelenterates, hedgehogs	from 110 to 180 eggs	110 cm carapace/120 kg	Atlantic, Indo-Pacific, the Caribbean Islands, Australia and Oceanic Islands Brazilian	Critically endangered
Chelonia mydas (Linnaeus 1758)	color is gray-green; the belly is white	shallow depth, about 20 m	puppies are omnivorous, becoming herbivorous in adulthood	from 110 to 130 eggs	120 cm carapace/230 kg	Globally distributed and generally found in tropical and subtropical waters	Endangered
Lepidochelys olivácea (Escholtz 1829)	one of the smallest species, greenish gray	shallow depth, about 20 m	fishes, molluscs, crustaceans, bryozoans, jellyfish, fish eggs	from 100 to 110 eggs	70 cm carapace/70 kg	Central America, Mexico, India, Suriname, French Guiana, Brazil	Vulnerable
Dermochelys coriácea (Linnaeus 1758)	color is black with white spots, carapace has seven longitudinal keels	depth between 50 and 80 m	ctenophores and cnidarians	from 80 to 90 eggs	180 cm carapace/700 kg	French Guiana Pacific and Costa Rica, Atlantic, South Africa	Vulnerable
Lepidochelys kempii (Garman 1880)	endemic to the Gulf of Mexico, least of all species, arribadas	depth about 50 m	crabs, fishes, jellyfish, molluscs	from 100 to 120 eggs	60–70 cm carapace/45 kg	Gulf of Mexico	Critically endangered
Natator depressus (Garman 1880)	carapace flattened, oval and smooth	depth of about 10 m	squid; sea cucumbers; soft corals; molluscs	from 50 to 70 eggs	100 cm carapace/90 kg	Australia	Data deficient

Ecotoxicological Studies on Sea Turtles

Bioaccumulation of heavy metals in sea turtles

Sea turtle characteristics offer many specific advantages as pollution in dicators by heavy metals and other kinds of marine environment contaminants, as their long life spans and migratory cycles provide more time to integrate environmental exposure. The concentration of a metal in the tissues of a predator may be higher than in the tissues of its victims, because of bioaccumulation, resulting from trophic transfer factor (DeForest et al. 2007, Jakimska et al. 2011). Metal bioaccumulation in an animal depends on many biotic factors such as size, weight, age, gender, diet, metabolism and the position in the food chain and abiotic factors such as metals distribution in their environmental, salinity, temperature, pH, habitat type, water and the interactions with other metals (de Souza 2008).

The organisms that live near the coast are the most susceptible to exposure to all types of pollutants, due to proximity to areas of human occupation. The bioaccumulation of contaminants in the tissues of these animals is considered a major problem for growth and development factors (Jakimska et al. 2011). The metals present in an animal may have no effect on its health, but they can also be harmful. Some metals in small concentrations allow the metabolic system to function normally, but if these levels are exceeded, they become toxic and can cause death (Bianchini 2008).

Many works have been conducted on determining heavy metals in tissues and sea turtle organs (Caurant et al. 1999, Godley 1999, Franzellitti 2004, Maffucci et al. 2005, Lam et al. 2006, Storelli et al. 2008, Andreani et al. 2008, Silva 2011). Agusa et al. (2008) carried out analyses of concentrations of total arsenic (As) in liver, kidney, muscle and stomach content of green turtles (*C. mydas*) and hawksbill turtle (*E. imbricata*), where the concentrations in hawksbill turtles were higher than those in green turtles, indicating that hawksbill turtles may have a specific accumulation mechanism for As.

Anan et al. (2001), carried out analyses in the liver, kidney, and muscle to determine the concentration of 18 trace elements in turtles in Japan. The accumulation features of trace elements in the three tissues were similar between green and hawksbill turtles, and no gender differences in trace element accumulation in liver and kidney were found for most of the elements. Sakai et al. (1995), carried out analyses on sea turtle eggs to determine the concentration of heavy metals concentration and thus develop a non-lethal technique method. Also, Hg concentrations also were measured in keratinized fragments (shells) and in internal tissues of green turtles, is the use of shell fragments considered a non-invasive and non-lethal technique (Bezerra et al. 2012, 2013). Lam et al. (2004) analyzed the concentrations of 19 individuals trace elements and heavy metals in various tissues and organs to establish a baseline dataset for chemical contaminants in green turtles nesting in South China, thus provide data for comparison with previous studies. The results showed that As levels in different turtle tissues of South China were similar to those reported from Japan and Hawaii. Other studies focused in tissues and organs of sea turtle population focusing heavy metals and organochlorine compounds concentrations in the Pacific Ocean (Sakai et al. 2000, Lam et al. 2004, Malarvannan et al. 2011), Mediterranean Sea (McKenzie et al. 1999, Storelli et al. 2005, Storelli et al. 2014) and in the Atlantic Ocean (Casal et al. 2007, Monagas et al. 2008, Oros et al. 2009, Rossi 2014).

Bioaccumulation of persistent organic pollutants (POPs) in sea turtles

POPs are chemicals that do not break down easily in the environment, being resistant to chemical, photolytic and biological degradation (Clark 1992). As a result of their generalized use, ability to be transported long-distance and due to their bioaccumulative and biomagnifiable characteristics in the food chain, these compounds are present in all the sea systems of the world (Hamann et al. 2010). Bioaccumulation of these toxic substances has become a cause of concern for the likelihood of its possible transfer through the food chain, causing an impact on many species of animals in the marine environmental (Marcotrigiano and Storelli 2003, Keller et al. 2004, Ogata et al. 2009). The pollution of the marine system is one of the investigation priorities by researchers who work in areas related to turtle biology and its conservation (Hamann et al. 2006).

Pursuant to a decision in the Stockholm Convention on Persistent Organic Pollutants, control measures focused on reducing and eliminating releases of 12 POPs which are grouped into three categories. The list includes eight pesticides (aldrin, chlordane, DDT, dieldrin, endrin, heptachlor, mirex, and toxaphene), two industrial chemicals (PCBs and hexachlorobenzene, also used as a pesticide) and two involuntary by-products from the industrial combustion process (dioxins and furans).

Organochlorine contaminants, such as OCPs and PCBs, due to their toxicity and higher persistence in the environment may be absorbed and accumulated by living organisms and through the bloodstream, can accumulate in the fat tissue (Corsolini et al. 2000). Many studies have reported contamination levels in sea turtles in the whole world in the past decade (Cobb and Wood 1997, Keller et al. 2006, Alava et al. 2006, Orós et al. 2009, D'Ilio et al. 2011, Camacho et al. 2014, De Andrés et al. 2016).

Gardner et al. (2003), measured organochlorine residues in three species of sea turtles from the Baja California peninsula, Mexico. Seventeen of 21 organochlorine pesticides analysed were detected in the 35 tissue samples, and PCBs were detected in all but one of the nine turtles studied. The researchers concluded that the levels of organochlorines detected in the study were low, possibly to the feeding habits of the predominant species collected that are herbivorous or by their samples have been collected in an unindustrialized region.

Studies that analysed pollutants concentration in sea turtles using tissues collected from dead animals (Malarvannan et al. 2011, Guerranti et al. 2014, Storelli et al. 2014), as adipose tissue (Lazar et al. 2011) or in more than one tissue (Corsolini et al. 2000a, Miao et al. 2001, Gardner et al. 2003, Keller et al. 2004b, Monagas et al. 2008, da Silva 2009, Oros et al. 2009) are represented in the Table 2.

Certain levels of POPs may have negative effects on sea turtle health parameters, as shown by some studies (Cobb et al. 1997, De Andres et al. 2016, Keller et al. 2004, Storelli et al. 2014). These contaminants are maternally transferred and may adversely affect normal embryonic development and hatchling sizes. Guirlet et al.

Table 2. Published studies on persistent organic pollutant concentrations in sea turtle tissues.

Location	Species	Tissue (n)	ΣPCBs	ΣDDTs	Reference
Hawaii	C. mydas	liver/adipose tissue (3)	52.1 ± 664.7*		Miao et al. (2001)
Mexico	L. olivacea	liver (1) kidney (1) adipose (1) muscle (1)	58.1 63.4 18.4 16.9	10.4 18.3 5.1 8.6	Gardner et al. (2003)
Florida	C. caretta	adipose tissue (44)	2.01 ± 2.96	452 ± 643	Keller et al. (2004b)
Japan	C. mydas E. imbricata C. caretta	liver (5) liver (3) liver (2)	0.04 ± 0.92 2.5 ± 13 1.3 ± 3.8	0.008 ± 0.27 0.56 ± 9.0 2.0 ± 3.8	Malarvannan et al. (2011)
Eastern Adriatic Sea	C. caretta	yellow fat (27)	474 ± 547	112.5 ± 72.3	Lazar et al. (2011)
Southeast Coast of Brazil	C. mydas	adipose tissue (51)/hepatic tissue (64)		Adipose (8.11 ± 13.72); Hepatic (4.25 ± 8.53)	Sarmiento (2013)
Southeast Italy	C. caretta	liver (12)	Males (1255.39 ± 702.10); Female (1497.16 ± 1671.50)		Storelli et al. (2014)

PCBs (ng/g^{-1} lipid weight); *female.

(2010) examined the OCs, pesticides (DDTs and HCHs), PCBs and the temporal variation of blood and eggs concentrations in leatherback turtles, where all OCs detected in leatherback blood were detected in eggs, suggesting a maternal transfer of OCs. This transfer was shown to depend on female blood concentration for ΣDDTs and for the most prevalent PCB congeners since significant relationships were found between paired blood–egg concentrations.

van de Merwe et al. (2010a) investigated the maternal transfer and effects of POPs on embryonic development in a green sea turtle, *C. mydas* population in Peninsular Malaysia. They found a significant correlation between increasing egg POP concentration and decreasing hatchling mass: length ratio. POPs may, therefore, have subtle effects on the development of *C. mydas* eggs, which may compromise offshore dispersal and predator avoidance. Similar results have been found in the study by Stewart et al. (2011), where concentrations were lower than those shown to have acute toxic effects in other aquatic reptiles but may have sub-lethal effects on the hatchling body condition and health.

It is for this reason that fat tissues and blood are being used to verify a possible maternal transfer in female turtles as shown by some studies (Cobb et al. 1997, McKenzie et al. 1999, Alava et al. 2006, D'Ilio et al. 2011, García-Besné et al. 2014). Blood samples were successfully used to measure organochlorine pollutant concentrations, being considered a non-lethal collect technique in different species of sea turtle (Table 3). Other analyses were also carried out through blood tests with the Comet assay in erythrocytes of *C. caretta* resulting in a useful method to detect

Table 3. Published studies on persistent organic pollutant concentrations in blood and eggs of sea turtles.

Location	Species	Matrix (n)	ΣPCBs	ΣDDTs	ΣChlordane	ΣHCHs	ΣOCPs	ΣPOPs (OC+ PCB+PBDE)	ΣPAHs	Reference
Merritt Island, Florida	Cc, Cm	eggs (Cm-2/Cc-9)	78	66*						Clark and Krynitsky (1980)
Merritt Island, Florida	Cc	eggs (56) dw		99						Clark and Krynitsky (1985)
South Carolina	Cc	eggs (16) lw	1188 ± 311							Cobb (1997)
Hernon Island, Queensland	Cm	eggs (15)		1.7 ± 0.3*						Podreka et al. (1998)
Mar Medit., Escocia	Cc, Cm, Dc	eggs (1)	Cc-89/ Cm-6.1	Cc-155/ Cm-4.3						McKenzie et al. (1999)
Northwest Florida	Cc	eggs (20) dw		753–800						Alam and Brim (2000)
North Carolina	Cc	blood (12) ww	5.14 ± 3.95a	0.58 ± 0.30	0.26 ± 0.18					Keller et al. (2004a)
North Carolina	Cc	blood (48) ww	5.56 ± 5.28	0.65 ± 0.68	0.22 ± 0.20		6.55 ± 6.14			Keller et al. (2004)
North Carolina	Cc, Lk	blood (Cc-44/ Lk-10) lw	Cc-0.002 ± 0.003/ Lk-0.98 ± 0.001	Cc-0.30 ± 0.57/Lk- 0.17 ± 0.15	Cc-0.10 ± 0.15/Lk-0.07 ± 0.08					Keller et al. (2004b)
Southern Flórida	Cc	eggs (22) ww	144 ± 280	50.2 ± 92.4	25.5 ± 46.7	0.258 ± 0.508				Alava et al. (2006)
North Carolina	Cc	blood (27) ww	6.25 ± 1.14	0.721 ± 0.152*	0.253 ± 0.048					Keller et al. (2006)

Location	Species	Sample (n)							Reference
Cape Cod, Massachusetts	Lk	blood (31)		ND (n =13)/ <10 (n =5)*					Innis et al. (2008)
Australia	Cm	eggs (10)	PCB 18-0.016 ± 0.0004		trans-0.02 ± 0.0004		endolsufan I-0.20 ± 0.005		van de Merwe et al. (2009a)
Malaysia	Cm	eggs (55) ww	0.47 ± 0.083	0.083 ± 0.018	0.057 ± 0.009	0.069 ± 0.009	0.39 ± 0.04	1.09 ± 0.43	van de Merwe et al. (2009)
French Guiana	Dc	eggs/blood (38) ww	B-1.26 ± 0.71/E- 6.98 ± 5.02	B-0.31 ± 0.22/E-1.44 ± 1.26		B-0.15 ± 0.16/E- 0.41 ± 0.26			Guirlet et al. (2010)
Gulf of Mexico	Lk, Cm	blood (Cm-9/Lk-46) ww	Cm-0.53 ± 0.70/ Lk-0.42 ± 0.36	Cm-0.12 ± 0.11/Lk- 0.68 ± 0.65	Cm-0.01 ± 0.020/Lk- 0.11 ± 0.10	Cm-0.014 ± 0.011/ Lk-0.015 ± 0.014			Swarthout et al. (2010)
Australia	Cm	blood (16) ww	0.68 ± 0.15					0.93 ± 0.17	van de Merwe et al. (2010)
Malaysia	Cm	eggs (33)/blood (11) ww	E-0.55 ± 0.05/B- 0.58 ± 0.09	N.A.		E-0.17 ± 0.007 **/ B-0.50 ± 0.06		E-1.29 ± 0.07/ B-1.38 ± 0.19	van de Merwe et al. (2010a)
Southeastern United States	Cc	eggs—WF-11/ EF-24/NC-9 Lw	WF-32.4 ± 14.1/ EF-372 ± 148/ NC-1460 ± 493	WF-23.8 ± 7.1/EF-136 ± 56/NC- 694 ± 251	WF-20.8 ± 9.6/EF-113 ± 31/NC- 375 ± 146	WF-0.449 ± 0.017/EF- 1.21 ± 0.49/ NC-3.15 ± 1.39			Alava et al. (2011)

Table 3 contd....

...Table 3 contd.

Location	Species	Matrix (n)	ΣPCBs	ΣDDTs	ΣChlordane	ΣHCHs	ΣOCPs	ΣPOPs (OC+ PCB+PBDE)	ΣPAHs	Reference
San Diego Bay	Cm	blood (20)		0.736 ± 0.097*	0.554 ± 0.073 (n=12)	0.915 ± 0.092 (γHCH)				Komoroske (2011)
Eastern, Florida	Cc	blood (19) ww	0.005 ± 0.002					0.005 ± 0.002		Ragland et al. (2011)
Eastern, Florida	Dc	eggs (34)	8.45 ± 3.1	1.87 ± 0.4		<LOD				Stewart et al. (2011)
Canary Islands/Cape Verde	Cc	blood (IC-162/ CV-205)							IC-6.38 ± 4.95/ CV-5.95 ± 4.88	Camacho et al. (2012)
Canary Islands/Cape Verde	Cc	blood (IC-162/CV-197)	IC-3.77 ± 5.90/ CV-0.15 ± 0.40	IC-0.78 ± 1.22/CV-0.21 ± 0.30			IC-1.98 ± 2.58/ CV-0.77 ± 1.21			Camacho et al. (2013)
Cape Verde	Cc	blood (50)	0.17 ± 0.19	0.075 ± 0.07*			0.095 ± 0.08		1.64 ± 1.14	Camacho et al. (2013a)
South Carolina	Cc	eggs (10) lw b	659 ± 216	325 ± 185	94.9 ± 41.2					Keller (2013)
Cape Verde	Cm, Ei	Blood (Cm-21/Ei-13)	Cm-0.53 ± 1.05/ Ei-0.19 ± 0.56				Cm-0.33 ± 0.75/Ei-0.2 ± 0.37		Cm-12.06 ± 0.61/ Ei-2.95 ± 2.90	Camacho et al. (2014)
Canary Islands	Cc	blood (56)	27.06 ± 42.09				1.16 ± 1.79		6.00 ± 8.26	Camacho et al. (2014a)

Location	Species	Sample						Reference
Mexico	Cm, Ei	eggs/blood (30) lw		Cm-B-3.38 ± 3.08/E-38.72 ± 0 // Ei – B-3.55 ± 3.52/E-410.5 ± 379.7*				García-Besné et al. (2014)
Hawaii	Cm	blood (13) wwc	0.144 ± 0.165	0.013 ± 0.008*	0.013 ± 0.004			Keller et al. (2014)
Adriatic Sea/ Canary Islands	Cc	blood (MA-35/IC-30)		MA-1.15 ± 0.96/C-0.17 ± 0.33			MA-20.1 ± 30.89/IC-6.29 ± 4.54	Bucchia et al. (2015)
Caribbean	Cc, Cm	eggs (Cm-11/Ei-4) Ww	Cm-4.57 ± 0.47/Ei-1.72	Cm-0.17 ± 0.04/Ei-0.19	Cm-0.17 ± 0.007/Ei-0.47			Dyc et al. (2015)
Costa Rica	Dc	eggs (18) lw	102.3 ± 93.3			113.9 ± 96.5		De Andrés et al. (2016)

Loggerhead (Cc), green (Cm), leatherback (Dc), Kemp's ridley (Lk), olive ridley (Lo), hawksbill (Ei), flatback (Nd) sea turtles; Mean ± SD in ng/g; * for only p,p´-DDE; ** only γHCH; dw - dry weight; lw – lipid weight; ww - wet weight; ND = not detected; N.A. = not available; < LOD = below detection; B – blood; E – egg; a - whole blood - technique A; b - Botany Bay Island, South Carolina; c - Kailua Bay; MA - Adriatic Sea; IC- Canary Islands; CV = Cape Verde; WF = Western Florida; EF = Eastern Florida; NC = North Carolina.

possible genotoxic effects, increasing knowledge about the state of ecotoxicological healthy of this species (Caliani et al. 2014).

Biochemical Biomarkers in Sea Turtles

Biochemical biomarkers are essential for assessing health risks from exposure to potentially chemical toxic products (Timbrell 1998). Biomarkers are used in a broad meaning to include any measure which shows an integration between a biological system and a potential danger which may be chemical, physical and biological (Organization 1993). It may be used for various purposes depending on study aim and on chemical exposure and can be classified into three types: exposure biomarkers, affect biomarkers and susceptibility (Amorin 2003, von Osten 2005). The use of biochemical biomarkers is being used by some studies, to analyse the level of several indicators of oxidative and pollutant stress in the tissue and blood of the sea turtle population (Yoshida 2012, Labrada-Martagón et al. 2011, Richardson et al. 2009).

According to Agarwal et al. (2005), oxidative stress is a situation created by an imbalance between oxidants and antioxidants, in favour of oxidants. In certain physiological, environmental, psychological and social situations, our organism goes to alert state, when oxygen consumption and enzymatic activities of some oxygenases and oxidases are high. Enzymes from oxidative stress represented by catalase (CAT), superoxide dismutase (SOD), glutathione reductase (GR) and peroxidase (GPx) are necessary for the maintenance of life as they are associated with the detoxification process of compounds formed in living beings and are considered important for allowing the survival of environmental impacted organisms (Cogo et al. 2009). An analysis of the activities of these enzymes allows environmental control and works as a contamination alert signal. Researchers applied these analyses in blood (Table 4) and in several tissues of sea turtle population (Table 5).

A need for sublethal detection of toxic effects on the organisms led to biomarkers indicators. They can be used as early indicators of environmental changes before happening permanent damages happens in an ecosystem (Huggett et al. 1992). Use of biochemical biomarkers in monitoring programs of sea turtle offers some advantages since they are normally the first to be changed and show good sensibility to contaminants, relative specificity and low-cost analysis when compared to chemical analysis (Stegeman and Hahn 1994).

Glutathione-S-transferase (GST), for instance, develops a critical role in the electrophiles exogenous and endogenous detoxification and acts just as well as products of oxidative stress. Considering the knowledge available on the natural and anthropogenic threats to sea turtles, the chemical and environmental effects and biochemical mechanisms related such as GST detoxification catalyzed is probably least understand. In the study carried out by Richardson et al. (2009), GST activity was characterized in four species of sea turtles with varied life histories and feeding strategies. Results of this study provided information about differences regarding the biotransformation potential and the possible impacts of the studied contaminants on the health and biotransformation ability of sea turtles. The results of the study indicate that hawksbill, loggerhead, olive ridley, and green sea turtles possess functional GST enzymes with similar kinetic parameters; however, rates of catalytic

Table 4. Published studies on antioxidant enzyme activity (GST; CAT; Mn-SOD; t-SOD; GR; GPx; U mg Hb–1) in blood of sea turtles.

Location	Species	Matrix (n)	GST	CAT	Mn-SOD	t-SOD	GR	GPx	Reference
Baja California, Mexico	*Cm*	blood (PA-39/BM-11)	PA-0.99 ± 0.17*/ BM-0.62 ± 0.11*	PA-223.29 ± 19.77*/ BM-92.22 ± 14.07*	PA-0.27 ± 0.04*/ BM-1.99 ± 0.37*	PA-0.43 ± 0.06*/ BM-2.95 ± 0.32*		PA-12.39 ± 0.97*/ BM-12.50 ± 2.68*	Labrada-Martagón et al. (2011)
Brazil	*Cm*	blood(27)**	78.39 ± 5.73	3.66 ± 0.35		74.66 ± 7.56	0.65 ± 0.12	3.52 ± 0.16	Yoshida (2012)
Southwest Florida	*Lk*	blood (13)				48.0 ± 10.3***			Perrault et al. (2014)

Mean ± SD; *SE = standard error; **group 2 (n=10); ***SOD: U/ml; PA-Punta Abreojos; BM-Bahia Magdalena; kemp's ridley *(Lk)*, green turtle *(Cm)*; GST (glutathione S-transferase); CAT (catalase); t-SOD, Mn-SOD (total and mitochondrial superoxide dismutase); GR (glutathione reductase); GPx (glutathione peroxidase).

Table 5. Published studies on antioxidant enzyme activity (GST; CAT; t-SOD; U mg Hb−1) in tissue of sea turtles.

Location	Species	Tissue (n)	Enzymes	GST	CAT	t-SOD	Reference
Baja California, Mexico	*C. mydas agassizii*	heart, kidney, liver, lung, pectoral muscle (19)	GST*/SOD*/CAT*	h (102.6 ± 18.1); k (457.7 ± 49.4); li (1032.2 ± 147.5); lu (68.6 ± 9.2); pm (210.3 ± 17.5)	h (8.9 ± 1.2); k (63.6 ± 8.5); li (78.1 ± 12.2); lu (3.7 ± 0.4); pm (5.6 ± 0.8)	h (22.05 ± 3.7); k (3.1 ± 3.4); li (35.1 ± 4.7); lu (12.1 ± 1.5); pm (24.1 ± 2.5)	Valdivia et al. (2007)
Baja California, Mexico	*Cc, Lo, Ei, Cm*	liver (*Cc*-4/*Lo*-4/*Ei*-3/*Cm*-4)	GST**	*Cc*-7.52 (3.89)/*Lo*-4.68 (1.26)/*Ei*-6.16 (1.16)/*Cm*-7.70 (1.89)			Richardson et al. (2009)
Baja California, Mexico	*Cc, Lo, Cm*	liver (*Cc*-4/*Lo*-4/*Cm*-4)	PCB/GST	***			Richardson et al. (2010)

*Mean ± S.E.M.; **Mean ± SD, catalytic efficiency (Vmax/Km) as nmol/min/mg protein/μM; ***no significant differences were observed in cytosolic GST activity; Loggerhead (*Cc*), green turtle (*Cm*), olive ridley (*Lo*), hawksbill (*Ei*); GST (glutathione S-transferase); CAT (catalase); SOD (superoxide dismutase).

activities using class-specific substrates show inter and intraspecies variation. GST from the herbivorous green sea turtle shows 3–4.5 fold higher activity with the substrate ethacrynic acid than the carnivorous olive ridley sea turtle.

Hematologic and biochemical standards analyses, including oxidative stress indicators were carried out in a subspecies green turtle group of Oriental Pacific (*C. mydas agassizii*) from a relatively undisturbed habitat in Mexico. Antioxidant enzymes, SOD, CAT, and GST were determined as a defense mechanisms indicator against reactive oxygen species (ROS) in the study carried out by Valdivia et al. (2007), where overall levels of all found enzymes were within ranges reported for other reptile species. Nevertheless, these authors suggest differences in oxidative metabolism among tissues. For example, liver and muscle showed the highest SOD activity, while kidney had the highest CAT and GST activities. This study has achieved information on an oxidative stress indicators base, providing important insights on the evaluation of health wildlife, especially for threatened species.

Perrault et al. (2014), analysed plasma brevetoxin concentrations and superoxide dismutase (SOD) in sea turtles Ridley Kemp at Florida. No significant correlations were observed between plasma brevetoxin concentrations and plasma proteins or SOD activity, however, alpha-globulins tended to increase with increasing brevetoxin concentrations in the bloom group, where Smaller (carapace length and mass) bloom turtles had higher plasma brevetoxin concentrations than larger bloom turtles.

In order to determine potential contaminant effects in young turtles, *C. mydas* from East Pacific, Labrada-Martagón et al. (2010), has analysed environmental concentrations organochlorine of pesticides, trace metals and correlation to body condition, measuring the activity of antioxidant enzymes and lipid peroxidation levels, using spectrophotometric analyse, where the results indicate that correlations between antioxidant enzyme activities and concentration of xenobiotics suggest physiological sensitivity of East Pacific green turtles to chemicals. Analysis of hematological and biochemical parameters, including oxidative stress indicators, is a valuable tool in assessing wildlife health.

Hematological biochemical parameters in sea turtles

In the clinical investigation of reptiles, blood samples can be easily obtained and are of great diagnostic value, where hematological evaluation has great importance. Serum biochemistry represents an important tool for monitoring the health and physiological status of sea turtles, due to the increasing need to evaluate their health status, for their possible maintenance as healthy animals in captivity as well as rehabilitation of free-living individuals (Pires et al. 2006).

The comparison of the data obtained is, however, limited due to possible differences among populations, as well as variations in the techniques used and factors such as age, size, seasonality, health, habitat, and diet, which may also influence in hematological parameters. On the other hand, the descriptions of the morphological characteristics of blood cells of marine chelonians are limited (Pires et al. 2009).

Due to chemical pollution exposure, the sea turtle population face of numerous environmental challenges, which may contribute to an immune system failure resulting in increasing diseases. Therefore, it is necessary to implement a complete evaluation of the immune status of these threatened animals that are in danger of extinction, and these hematological analysis are being successfully used in several studies and are used to evaluate immune functions in a range of aquatic species (Rousselet et al. 2013).

In the last years, blood sampling has been used as a non-destructive technique to monitor and evaluate clinically the condition of the organism (Table 6). As in the results obtained by Keller et al. (2004), the study provides the first evidence, although strictly correlative, that OCs contaminant may be affecting sea turtle health. Although the concentrations of OCs are relatively low compared with other species, significant correlations between OCs levels and health indicators as homeostasis of proteins, carbohydrates, and ions were observed, according to the above mentioned study.

Hamann et al. (2006), presented comparative data on population demographics and biochemical blood parameters for green sea turtles *C. mydas* in their foraging grounds in the Gulf of Carpentaria, Australia. Based on the biochemical analysis of blood, green turtles appeared to be healthy. However, their mean levels of glucose and magnesium were generally lower than the ranges observed in other studies of clinically healthy green turtles. This study was the first to describe the population structure and provide a biochemical assessment of *C. mydas* for this population and it will provide an important dataset for assessing future impacts such as cyclone damage to seagrass communities.

Plasma samples collected from fed and fasting *C. mydas* turtles to analyses Triglycerides, β-hydroxybutyrate, and glycerol concentration using spectrophotometric assays, for determining recent feeding history using a single blood sample (Hamann et al. 2006). Serum triglyceride and glycerol concentrations decreased during fasting periods, while serum ß-hydroxybutyrate concentration increased during fasts. For triglyceride and glycerol, this decrease apparently occurred in the first five days of fasting and was unaltered by further fasting (Price et al. 2013). Blood analyses were also measured in studies with *L. kempii* to measure plasma corticosterone, glucose, and testosterone concentrations, to determine the effects of acute handling stress in Cedar Keys, Florida, showing that mean plasma corticosterone and glucose concentrations increased significantly with time (Gregory et al. 2011). No significant difference was observed over time for mean testosterone concentrations. Approximately half of the turtles demonstrated an increase in plasma testosterone after 60 min of captivity while the others demonstrated a decrease (Gregory et al. 2011).

As shown by Innis et al. (2010), health evaluations of 19 leatherback turtles in the northwestern Atlantic were undertaken, where relatively high blood concentrations of selenium and cadmium in all turtles were found. Blood samples too were successfully used to measure heavy metals and metalloids concentration due to vital functions carried out by blood cells and their susceptibility to intoxication (Innis et al. 2008, Jerez et al. 2010, van de Merwe et al. 2010).

Hematological values and plasma biochemistry in the *D. coriacea* species were determined in some studies (Deem et al. 2006, Perrault et al. 2012) and the OCs concentrations found in leatherback in these studies were lower than concentrations

Table 6. Published studies on hematological parameters in blood matrices of sea turtles.

Species	Location	Blood analysis (n)	Variables	Reference
	Cayman Islands	Hematological parameter (51)	PVC; Hb; RBC; WBC	Wood and Ebanks (1984)
	Bahamas	Blood biochemical (100)/Packed cell volume (106)	PVC; GLC; Na; K; Cl; CO2; IB; BUN; CR; BUN/CR; UA; Ca; P; TP; Alb; Glob; A/G; IC; TB; ALP; LDH; AST; ALT; CHOL; TC; Fe	Bolten and Bjorndal (1992)
	Hawaii	Morphology and blood cytochemistry (26)	Lymp; MO; Het; Eos; Bas; WBC	Work et al. (1998)
	Australia	Blood biochemical	CR; U; GLC; ALT; AST; CPK; TP; Alb; Glob; A/G; Ca; P; Mg; TSI; UA	Hamann et al. (2006)
	Brazil	Blood biochemical (33)	ALT; AST; ALP; CPK; CR; U; UA; TP; Alb; Glob; A/G; GLC; TC; P; Cl; Na; K;	Leite (2007)
Chelonia mydas	Brazil	Hematological parameter (60)	Hct; RBC; Hb; RBC; MCH; MCHC; WBC; Mo; Lymp; Het; Eos; Bas; THB	de Deus Santos et al. (2009)
	Australia	Hematological parameter/Blood biochemical (290)	PCV; THB; Lymp; Het; Eos; Mo; Bas; WBC/Alb; ALP; AST; Ca; Cl; CR; CPK; Glob; GLC; LDH; Mg; P; K; TP; Na; TB; U; UA	Flint et al. (2010)
	USA	Hematological parameter/Blood biochemical (41)	WBC; Het; Lymp; Eos; Bas; Mo; Hct/ ALP; ALT; CR; LDH; TP; Alb; Glob; CHOL; GLC; Ca; P; Ca; K; Na; UA; U; Cl	Komoroske et al. (2011)
	Cayman Islands	Blood biochemical (5)	GLYC; TRIG; BUTY	Price et al. (2013)
	Atol Palmyra	Hematological parameter/Blood biochemical (157)	Bas; Eos; Het; Lymp; Mo; WBC; PCV/CPK; ALP; AST; LDH; TP; Alb; Glob; CHOL; TRIG; Ca; GLC; K; P; Na; UA	McFadden et al. (2014)

Table 6 contd. ...

...Table 6 contd.

Species	Location	Blood analysis (n)	Variables	Reference
	USA	Hematological parameter/Blood biochemical (48)	RBC; Hb; Hct; Het; Lymp; Mo; Eos; Azu; Bas; Het/GLS; TP; Alb; Glob; A/G; BUN; UA; CR; SBR; AST; ALP; LDH; CPK;	Keller et al. (2004)
	USA	Plasma Protein Electrophoresis (41)	plasma protein fractions	Gicking et al. (2004)
	Brazil	Hematological parameter/Plasma Protein (8)	PVC; Hb; RBC; WBC; Het; Eos; Bas; Lymp; Mo; THB; TP	Pires et al. (2006)
	Spain	Hematological parameter (35)	Ery; Het; Eos; Lymp; Mo; THB	Casal and Orós (2007)
	Japan	Hematological parameter/Blood biochemical (5)	RBC; WBC; PVC; Het; Bas; Lymp; Mo/SBR; GOT; GPT; GGT; ALP; Amy; LDH; CPK; TP; Alb; CR; BUN; UA; GLC; TC; CHOL; Ca; P; K; Na; Cl	Kakizoe et al. (2007)
	Spain	Ultrastructural characterization of blood cell (15)	Ery; Het; Eos; Lymp; Mo; THB	Casal et al. (2007)
Caretta caretta	Brazil	Hematological parameter/Blood biochemical (27)	Hct; RBC; Hb; THB; MCH; MCHC; WBC; Het; Lymp; Eos/TP; Alb; Glob; A/G; GLC; CHOL; TC; CR; UA; AST; ALP	Pires et al. (2009)
	USA	Hematological parameter/Blood biochemical (83)	PCV; RBC; WBC; Het; Lymp; Mo; Eos; Bas; Azu/GLC; Na; K; Cl; CO_2; U; CR; TP; Alb; Glob; CHOL; TC; Ca; P; UA; ALT; AST; LDH; CPK; Amy; Lip; GGT	Deem et al. (2009)
	Spain	Hematological parameter/Blood biochemical (103)	PCV; RBC; THB; WBC; Het; Lymp; Eos; Bas; Mo/TP; Alb; Glob; CR; UA; U; SBR; CHOL; TC; GLC; Ca; AST; ALT; ALP; LDH	Casal et al. (2009)
	USA	Hematological parameter/Blood biochemical (85)	PVC; WBC; Het; Lymp; Eos; Bas; Mo/TP; Alb; Glob; Alb; CR; UA; BUN; GLC; SBR; CHOL; ALP; CPK; AST; ALT; GGT; Amy; Ca; Na; K; Cl	Rousselet et al. (2013)
	Africa	Hematological parameter/Blood biochemical (50)	PVC; RBC; WBC; THB; Het; Lymp; Eos; Bas; Mo/TP; Alb; Glob; GLC; CR; UA; U; CHOL; SBR; TC; ALT; AST; ALP; LDH; GGT; CPK; Amy; Lip; Na; K; Cl; Ca; P; Mg	Camacho et al. (2013)

	USA	Hematological parameter (100)	Ery; Het; Eos; Lymp; Mo	Cannon (1992)
Lepidochelys kempii	USA	Hematological parameter/ Blood biochemical (176)	WBC; Hct; Het; Lymp; Mo; Eos; Bas/ALT; AST; CPK; LDH; GGT; Alb; TP; Glob; SBR; BUN; CR; CHOL; GLC; Ca; P; Cl; K; Na; UA; A/G;	Innis et al. (2009)
	Mexico	Hematological parameter (44)	corticosterone; GLC, and testosterone	Gregory et al. (2011)
	Africa	Hematological parameter/ Blood biochemical (35)	PCV; RBC; WBC; Het; Lymp; Mo; Eos/GLC; Na; K; CO2; BUN; CR; TP; CHOL; TC; Ca; P; UA; ALT; AST; LDH; CPK; Amy; Lip; GGT	Deem et al. (2006)
Dermochelys coriacea	Canada	Hematological parameter/ Blood biochemical (12)	Lymp; Het; Eos; Mo; Bas; WBC/ALT; AST; ALP; LDH; CPK; Alb; CR; CHOL; GLC; Ca; P; K; UA; TP; Glob; Cu; Se; Hg; A; E	Innis et al. (2010)
	USA	Hematological parameter/ Blood biochemical (60–70)	PVC; RBC; WBC; THB; Het; Lymp; Eos; Bas; Mo/TP; GLC; CR; UA; BUN; CHOL; ALT; AST; ALP; LDH; CPK; Amy; Lip; Na; K; Cl; Ca; P;	Perrault et al. (2012)
Lepidochelys kempii/ Chelonia mydas	Mexico	Hematological parameter/ Blood biochemical (*Lk* 46); *Cm* (9)	WBC; Lymp; Neo/ TP; GLC; Alb; Glob; BUN; UA; AST; CPK; Na; K; Ca; P; Cl	Swarthout et al. (2010)

PCV = packed cell volume; Hb = hemoglobin level; RBC = red blood cell count; WBC = white blood cell count; MCH = Mean corpuscular hemoglobin concentration; Amy = Amylase; Alb = Albumin; ALP = Alkaline phosphatase; AST = Aspartate Aminotransferase; ALT = Alanine Aminotransferase; BUN = Urea nitrogen; BUN/CR = creatinine ratio; Ca = Calcium; Cl = Chloride; CR = Creatinine; CPK = Creatine phosphokinase; CO2 = Carbon dioxide; CHOL = Cholesterol; Fe = Iron; Glob = Globulin; GLC = Glucose; IC = Ionized calcium; IB = Ion balance; K = Potassium; Lip = Lipase; LDH = Lactic dehydrogenase; Mg = Magnesium; Na = Sodium; P = Phosphorus; TP = Total protein; TB = Total bilirubin; A/G = Albumin/globulin ratio; TC = Triglycerides; TSI = Total serum iron; U = Urea; UA = Uric acid; SBR = Bilirubin; GOT = glutamate oxaloacetic transaminase; GPT = glutamate pyruvate transaminase; GGT = gama -glutamyl transferase; Cu = copper; Se = selenium; Cad = Cadmium; Hg = mercurium; A = vitamin A; E = vitamin E; Neo = Neutrophis; Ery = Erythrocyte; Lymp = Lymphocyte; Mo = Monocyte; Eos = Eosinophiles; Bas = Basophiles; WCC = Total WBC; Hct = Hematocrit; Het = Heterophiles; THB= Thrombocytes; TRIG = triglycerides; BUTY = ß-hydroxybutyrate; GLYC = glycero; Azu = Azuroph; *Lepidochelys kempii* (*Lk*); *Chelonia mydas* (*Cm*).

measured in other sea turtles, which might be due to the lower trophic position (diet based on gelatinous zooplankton) and to the location of their foraging and nesting grounds. All OCs detected in leatherback blood were detected in eggs, suggesting a maternal transfer of OCs in the studies carried out by Guirlet et al. (2008). This method was also used by Páez-Osuna et al. (2010) to validate the maternal transfer of lead (Pb) via egg-laying.

Ley-Quiñónez et al. (2011) determined baseline concentrations of zinc, cadmium, copper, nickel, selenium, manganese, mercury and lead in the blood of 22 clinically healthy, loggerhead turtles (*C. caretta*), captured for several reasons in Mexico. That concluded that blood is an excellent tissue to measure in relatively non-invasive way baseline values of heavy metals. In the study located in the Republic of Cape Verde, almost all of the samples showed detectable levels of the 11 elements (Cu, Mn, Pb, Zn, Cd, Ni, Cr, As, Al, Hg and Se), strengthening the usefulness of blood for the monitoring of the levels of contaminating elements and their adverse effects on blood parameters in sea turtles (Camacho et al. 2013a).

Blood tests were carried out in sea turtles from the *C. caretta* species with a goal to obtain values from the hemogram standards and plasma protein of this species (Camacho et al. 2013, Rousselet et al. 2013). Nonlethal fat biopsies and blood samples were collected from live turtles for OC contaminant analysis, and concentrations were compared with clinical health assessment data, including hematology, plasma chemistry, and body condition. Blood concentrations of ∑Chlordanes were negatively correlated with red blood cell counts, hemoglobin, and hematocrit, indicative of anemia. Positive correlations were observed between most classes of OC contaminants and white blood cell counts and between mirex and ∑TCDD-like PCB concentrations and the heterophil: lymphocyte ratio, suggesting modulation of the immune system. These correlations suggest that OC contaminants may be affecting the health of loggerhead sea turtles (Keller et al. 2004).

A total of 5 millilitres of blood was collected from captive *C. caretta* sea turtles in order to obtain hemoglobin and plasma protein parameters, whereof the analysed variables such as erythrocyte count, mean globular volume, the mean globular hemoglobin and total and differential counts of leukocytes presented different values for the species in question when compared with the literature consulted (Pires et al. 2006, 2009). Hematologic characteristics and plasma chemistry values of juvenile loggerhead turtles (*C. caretta*), was analysed, and the results of these studies were used to establish a hematology and blood chemistry baseline for captive juvenile loggerhead turtles and will aid in their medical management (Casal et al. 2007, Kakizoe et al. 2007, Deem et al. 2009). de Deus Santos et al. (2009), has set young *C. mydas* turtles hematological values in captivity, concluding his results vary in relation to other authors, due to different methodologies with the same species. Hamann et al. (2006) found mean levels of glucose and magnesium were generally lower than the ranges observed in other studies of clinically healthy green turtles.

Wood and Ebanks (1984), analysed distinguishable blood cells, and Differential counts of white blood cells showed age dependent differences and the hemoglobin level and packed cell volume of the green sea turtle apparently increase with age. Similar to the study reported by Bolten and Bjorndal (1992) where there was a significant correlation of the body size to the blood parameters measured. Thus, research dealing with this issue will have increasing importance due to the use of this non-lethal technique, which helps in with wildlife contaminant monitoring. The information obtained will be used to monitor any changes caused by potentially inorganic pollutants and may help in the clinical diagnosis of diseases that affect these sea turtles (D'Ilio et al. 2011, Ikonomopoulou et al. 2011, Cortés-Gómez et al. 2014, McFadden et al. 2014, Villa et al. 2015).

Conclusions

Sea turtle population conservation strategies must include research, strengthening environmental education, local management strategies, interaction with local fishermen, encouraging turtle safe release when accidentally caught in their fishing net, discussion programs for wildlife conservation threatened by extinction, and voluntaries training courses. Worldwide there are many studies that address the issue of contaminants and their effects on sea turtles and the conservation plans that include management programs for the protection of nests and nesting females, the nest translocation to protected hatcheries and protection against poaching. It is also important to emphasize the information exchange between science, policy, and public participation in the design and implementation of conservation actions. Because contaminants are maternally transferred to eggs or by bioaccumulation, information about contamination by heavy metals, POPs or PAHs in nesting females are crucial for assessing the state of health of marine turtles. Also, help to evaluate and prevent malformations in embryos and hatchlings. Despite there being conservation programs for sea turtles, still, there is still a need to strengthen these programs and to raise awareness in the communities surrounding major beaches in order to involve them in the work of conservation protection in the long term.

Acknowledgements

This work was supported by Coordination for the Improvement of Higher Education Personnel (CAPES Brazil) (1201/2013-01).

References

Ackerman, R.A., Lutz, P. and Musick, J. 1997. The nest environment and the embryonic development of sea turtles. The Biology of Sea Turtles 1: 83–106.
Agarwal, A., Gupta, S. and Sharma, R. 2005. Oxidative stress and its implications in female infertility–a clinician's perspective. Reproductive Biomedicine Online 11: 641–650.
Agusa, T., Takagi, K., Kubota, R., Anan, Y., Iwata, H. and Tanabe, S. 2008. Specific accumulation of arsenic compounds in green turtles (*C. mydas*) and hawksbill turtles (*Eretmochelys imbricata*) from Ishigaki Island, Japan. Environmental Pollution 153: 127–136.
Aguirre, A.A., Balazs, G.H., Spraker, T.R. and Gross, T.S. 1995. Adrenal and hematological responses to stress in juvenile green turtles (*Chelonya mydas*) with and without fibropapillomas. Physiological Zoology 831–854.
Aguirre, A.A. and Balazs, G. 2000. Blood biochemistry values of green turtles, *Chelonia mydas*, with and without fibropapillomatosis. Comparative Haematology International 10: 132–137.

Alam, S. and Brim, M. 2000. Organochlorine, PCB, PAH, and metal concentrations in eggs of loggerhead sea turtles (*Caretta caretta*) from northwest Florida, USA. Journal of Environmental Science & Health Part B 35: 705–724.

Alava, J.J., Keller, J.M., Kucklick, J.R., Wyneken, J., Crowder, L. and Scott, G.I. 2006. Loggerhead sea turtle (*Caretta caretta*) egg yolk concentrations of persistent organic pollutants and lipid increase during the last stage of embryonic development. Science of The Total Environment 367: 170–181.

Alava, J.J., Keller, J.M., Wyneken, J., Crowder, L., Scott, G. and Kucklick, J.R. 2011. Geographical variation of persistent organic pollutants in eggs of threatened loggerhead sea turtles (*Caretta caretta*) from southeastern United States. Environmental Toxicology and Chemistry 30: 1677–1688.

Amorim, L.C.A. 2003. Os biomarcadores e sua aplicação na avalição da exposição aos agentes químicos ambientais. Rev Bras Epidemiol, 158–170.

Anan, Y., Kunito, T., Watanabe, I., Sakai, H. and Tanabe, S. 2001. Trace element accumulation in hawksbill turtles (*Eretmochelys imbricata*) and green turtles (*Chelonia mydas*) from Yaeyama Islands, Japan. Environmental Toxicology and Chemistry 20: 2802–2814.

Andreani, G., Santoro, M., Cottignoli, S., Fabbri, M., Carpenè, E. and Isani, G. 2008. Metal distribution and metallothionein in loggerhead (*Caretta caretta*) and green (*Chelonia mydas*) sea turtles. Science of The Total Environment 390: 287–294.

Becker, P.H., González-Solís, J., Behrends, B. and Croxall, J. 2002. Feather mercury levels in seabirds at South Georgia: influence of trophic position, sex and age. Marine Ecology Progress Series 243: 261–269.

Bezerra, M., Lacerda, L., Lima, E. and Melo, M. 2013. Monitoring mercury in green sea turtles using keratinized carapace fragments (scutes). Marine Pollution Bulletin 77: 424–427.

Bezerra, M.F., Lacerda, L.D., Costa, B.G. and Lima, E.H. 2012. Mercury in the sea turtle *Chelonia mydas* (Linnaeus, 1958) from Ceará coast, NE Brazil. Anais da Academia Brasileira de Ciências 84: 123–128.

Bianchini, A., Lauer, M.M., Nery, L.E.M., Colares, E.P., Monserrat, J.M. and dos Santos Filho, E.A. 2008. Biochemical and physiological adaptations in the estuarine crab Neohelice granulata during salinity acclimation. Comparative Biochemistry and Physiology Part A: Molecular & Integrative Physiology 151: 423–436.

Bishop, C.A., Brooks, R.J., Carey, J.H., Ng, P., Norstrom, R.J. and Lean, D.R. 1991. The case for a cause-effect linkage between environmental contamination and development in eggs of the common snapping turtle (*Chelydra s. serpentina*) from Ontario, Canada. Journal of Toxicology and Environmental Health, Part A Current Issues 33: 521–547.

Bjorndal, K.A., Lutz, P. and Musick, J. 1997. Foraging ecology and nutrition of sea turtles. The Biology of Sea Turtles 1: 199–231.

Bolten, A.B. and Bjorndal, K.A. 1992. Blood profiles for a wild population of green turtles (*Chelonia mydas*) in the southern Bahamas: size-specific and sex-specific relationships. Journal of Wildlife Diseases 28: 407–413.

Bucchia, M., Camacho, M., Santos, M.R.D., Boada, L.D., Roncada, P., Mateo, R., Ortiz-Santaliestra, M.E., Rodríguez-Estival, J., Zumbado, M., Orós, J., Henríquez-Hernández, L.A., García-Álvarez, N. and Luzardo, O.P. 2015. Plasma levels of pollutants are much higher in loggerhead turtle populations from the Adriatic Sea than in those from open waters (Eastern Atlantic Ocean). Science of The Total Environment 523: 161–169.

Caliani, I., Campani, T., Giannetti, M., Marsili, L., Casini, S. and Fossi, M.C. 2014. First application of comet assay in blood cells of Mediterranean loggerhead sea turtle (*Caretta caretta*). Marine Environmental Research 96: 68–72.

Camacho, M., Boada, L.D., Orós, J., Calabuig, P., Zumbado, M. and Luzardo, O.P. 2012. Comparative study of polycyclic aromatic hydrocarbons (PAHs) in plasma of Eastern Atlantic juvenile and adult nesting loggerhead sea turtles (*Caretta caretta*). Marine Pollution Bulletin 64: 1974–1980.

Camacho, M., Luzardo, O.P., Boada, L.D., Jurado, L.F.L., Medina, M., Zumbado, M. and Orós, J. 2013. Potential adverse health effects of persistent organic pollutants on sea turtles: evidences from a cross-sectional study on Cape Verde loggerhead sea turtles. Science of The Total Environment 458:283–289.

Camacho, M., Oros, J., Boada, L., Zaccaroni, A., Silvi, M., Formigaro, C., López, P., Zumbado, M. and Luzardo, O. 2013a. Potential adverse effects of inorganic pollutants on clinical parameters of loggerhead sea turtles (*Caretta caretta*): results from a nesting colony from Cape Verde, West Africa. Marine Environmental Research 92: 15–22.

Camacho, M., Boada, L.D., Orós, J., López, P., Zumbado, M., Almeida-González, M. and Luzardo, O.P. 2014. Monitoring organic and inorganic pollutants in juvenile live sea turtles: results from a study of *Chelonia mydas* and *Eretmochelys imbricata* in Cape Verde. Science of The Total Environment 481: 303–310.

Camacho, M., Orós, J., Henríquez-Hernández, L.A., Valerón, P.F., Boada, L.D., Zaccaroni, A., Zumbado, M. and Luzardo, O.P. 2014a. Influence of the rehabilitation of injured loggerhead turtles (*Caretta caretta*) on their blood levels of environmental organic pollutants and elements. Science of The Total Environment 487: 436–442.

Cannon, M.S. 1992. The morphology and cytochemistry of the blood leukocytes of Kemp's ridley sea turtle (*Lepidochelys kempi*). Canadian Journal of Zoology 70: 1336–1340.

Casal, A., Freire, F., Bautista-Harris, G., Arencibia, A. and Orós, J. 2007. Ultrastructural characteristics of blood cells of juvenile loggerhead sea turtles (*Caretta caretta*). Anatomia, Histologia, Embryologia 36: 332–335.

Casal, A. and Orós, J. 2007. Morphologic and cytochemical characteristics of blood cells of juvenile loggerhead sea turtles (*Caretta caretta*). Research in Veterinary Science 82: 158–165.

Casal, A.B., Camacho, M., López-Jurado, L.F., Juste, C. and Orós, J. 2009. Comparative study of hematologic and plasma biochemical variables in Eastern Atlantic juvenile and adult nesting loggerhead sea turtles (*Caretta caretta*). Veterinary Clinical Pathology 38: 213–218.

Caurant, F., Bustamante, P., Bordes, M. and Miramand, P. 1999. Bioaccumulation of cadmium, copper and zinc in some tissues of three species of marine turtles stranded along the French Atlantic coasts. Marine Pollution Bulletin 38: 1085–1091.

Clark, R. Marine Pollution, 1992. Clarendon Press, Oxford.

Clark, Jr., D. and Krynitsky, A. 1980. Organochlorine residues in eggs of loggerhead and green sea turtles nesting at Merritt Island, Florida—July and August 1976. Pesticides Monitoring Journal 14: 7–10.

Clark, D.R. and Krynitsky, A.J. 1985. DDE residues and artificial incubation of loggerhead sea turtle eggs. Bulletin of Environmental Contamination and Toxicology 34: 121–125.

Cobb, G.P. and Wood, P.D. 1997. PCB concentrations in eggs and chorioallantoic membranes of loggerhead sea turtles (*Caretta caretta*) from the Cape Romain National Wildlife Refuge. Chemosphere 34: 539–549.

Cogo, A.J., Siqueira, A.F., Ramos, A.C., Cruz, Z.M. and Silva, A.G. 2009. Utilização de enzimas do estresse oxidativo como biomarcadoras de impactos ambientais. Natureza on line 7: 37–42.

Corsolini, S., Aurigi, S. and Focardi, S. 2000. Presence of polychlorobiphenyls (PCBs) and coplanar congeners in the tissues of the Mediterranean loggerhead turtle *Caretta caretta*. Marine Pollution Bulletin 40: 952–960.

Corsolini, S. and Focardi, S. 2000a. Bioconcentration of polychlorinated biphenyls in the pelagic food chain of the Ross Sea. Pages 575–584 Ross sea ecology. Springer.

Cortés-Gómez, A.A., Fuentes-Mascorro, G. and Romero, D. 2014. Metals and metalloids in whole blood and tissues of Olive Ridley turtles (*Lepidochelys olivacea*) from La Escobilla Beach (Oaxaca, Mexico). Marine Pollution Bulletin 89: 367–375.

D'ilio, S., Mattei, D., Blasi, M., Alimonti, A. and Bogialli, S. 2011. The occurrence of chemical elements and POPs in loggerhead turtles (*Caretta caretta*): an overview. Marine Pollution Bulletin 62: 1606–1615.

da Silva, C.C., Varela, A.S., Barcarolli, I.F. and Bianchini, A. 2014. Concentrations and distributions of metals in tissues of stranded green sea turtles (*Chelonia mydas*) from the southern Atlantic coast of Brazil. Science of The Total Environment 466: 109–118.

da Silva, J. 2009. Ocorrência de pesticidas organoclorados e bifenilos policlorados em tartarugas marinhas *Chelonia mydas*. Universidade de São Paulo.

de Andréa, M.M. 2008. Bioindicadores ecotoxicológicos de agrotóxicos.

De Andrés, E., Gómara, B., González-Paredes, D., Ruiz-Martín, J. and Marco, A. 2016. Persistent organic pollutant levels in eggs of leatherback turtles (*Dermochelys coriacea*) point to a decrease in hatching success. Chemosphere 146: 354–361.

de Deus Santos, M.R., Ferreira, L.S., Batistote, C., Grossman, A. and Bellini, C. 2009. Valores hematológicos de tartarugas marinhas *Chelonia mydas* (Linaeus, 1758) juvenis selvagens do Arquipélago de Fernando de Noronha, Pernambuco, Brasil. Brazilian Journal of Veterinary Research and Animal Science 46: 491–499.

de Souza Spinosa, H., Górniak, S.L. and Neto, J.P. 2008. Toxicologia aplicada à medicina veterinária. Manole.

Deem, S.L., Norton, T.M., Mitchell, M., Segars, A., Alleman, A.R., Cray, C., Poppenga, R.H., Dodd, M. and Karesh, W.B. 2009. Comparison of blood values in foraging, nesting, and stranded loggerhead turtles (*Caretta caretta*) along the coast of Georgia, USA. Journal of Wildlife Diseases 45: 41–56.

DeForest, D.K., Brix, K.V. and Adams, W.J. 2007. Assessing metal bioaccumulation in aquatic environments: the inverse relationship between bioaccumulation factors, trophic transfer factors and exposure concentration. Aquatic Toxicology 84: 236–246.

Dodd, C.K. and Dreslik, M.J. 2008. Habitat disturbances differentially affect individual growth rates in a long-lived turtle. Journal of Zoology 275: 18–25.

Dyc, C., Covaci, A., Debier, C., Leroy, C., Delcroix, E., Thomé, J.-P. and Das, K. 2015. Pollutant exposure in green and hawksbill marine turtles from the Caribbean region. Regional Studies in Marine Science 2: 158–170.

Eckrich, C.E. and Owens, D.W. 1995. Solitary versus arribada nesting in the olive ridley sea turtles (*Lepidochelys olivacea*): a test of the predator-satiation hypothesis. Herpetologica, 349–354.

Fowler, L.E. 1979. Hatching success and nest predation in the green sea turtle, *Chelonia mydas*, at Tortuguero, Costa Rica. Ecology 60: 946–955.

Flint, M., Morton, J.M., Limpus, C.J., Patterson-Kane, J.C., Murray, P.J. and Mills, P.C. 2010. Development and application of biochemical and haematological reference intervals to identify unhealthy green sea turtles (*Chelonia mydas*). The Veterinary Journal 185: 299–304.

Franzellitti, S., Locatelli, C., Gerosa, G., Vallini, C. and Fabbri, E. 2004. Heavy metals in tissues of loggerhead turtles (*Caretta caretta*) from the northwestern Adriatic Sea. Comparative Biochemistry and Physiology Part C: Toxicology & Pharmacology 138: 187–194.

Gallo, B.M., Macedo, S., Giffoni, B.d.B., Becker, J.H. and Barata, P.C. 2006. Sea turtle conservation in Ubatuba, southeastern Brazil, a feeding area with incidental capture in coastal fisheries. Chelonian Conservation and Biology 5: 93–101.

García-Besné, G., Valdespino, C. and Rendón-von Osten, J. 2015. Comparison of organochlorine pesticides and PCB residues among hawksbill (*Eretmochelys imbricata*) and green (*Chelonia mydas*) turtles in the Yucatan Peninsula and their maternal transfer. Marine Pollution Bulletin 91: 139–148.

Gardner, S.C., Pier, M.D., Wesselman, R. and Juárez, J.A. 2003. Organochlorine contaminants in sea turtles from the Eastern Pacific. Marine Pollution Bulletin 46: 1082–1089.

Gicking, J.C., Foley, A.M., Harr, K.E., Raskin, R. and Jacobson, E. 2004. Plasma protein electrophoresis of the Atlantic loggerhead sea turtle, *Caretta caretta*. Journal of Herpetological Medicine and Surgery Volume 14.

Godley, B.J., Thompson, D.R. and Furness, R.W. 1999. Do heavy metal concentrations pose a threat to marine turtles from the Mediterranean Sea? Marine Pollution Bulletin 38: 497–502.

Godfrey, M.H. and Barreto, R. 1995. Beach vegetation and seafinding orientation of turtle hatchlings. Biological Conservation 74: 29–32.

Gregory, L.F. and Schmid, J.R. 2001. Stress responses and sexing of wild Kemp's ridley sea turtles (*Lepidochelys kempii*) in the northeastern Gulf of Mexico. General and Comparative Endocrinology 124: 66–74.

Guerranti, C., Baini, M., Casini, S., Focardi, S.E., Giannetti, M., Mancusi, C., Marsili, L., Perra, G. and Fossi, M.C. 2014. Pilot study on levels of chemical contaminants and porphyrins in *Caretta caretta* from the Mediterranean Sea. Marine Environmental Research 100:33–37.

Guirlet, E., K. Das, and M. Girondot. 2008. Maternal transfer of trace elements in leatherback turtles (*Dermochelys coriacea*) of French Guiana. Aquatic Toxicology 88:267–276.

Guirlet, E., K. Das, J.-P. Thomé, and M. Girondot. 2010. Maternal transfer of chlorinated contaminants in the leatherback turtles, *Dermochelys coriacea*, nesting in French Guiana. Chemosphere 79:720–726.

Hamann, M., C. S. Schäuble, T. Simon, and S. Evans. 2006. Demographic and health parameters of green sea turtles *Chelonia mydas* foraging in the Gulf of Carpentaria, Australia. Endangered Species Research 2:81–88.

Hamann, M., M. Godfrey, J. Seminoff, K. Arthur, P. Barata, K. Bjorndal, A. Bolten, A. Broderick, L. Campbell, and C. Carreras. 2010. Global research priorities for sea turtles: informing management and conservation in the 21st century. Endangered Species Research 11:245–269.

Hawkes, L., A. Broderick, M. Godfrey, and B. Godley. 2007. Investigating the potential impacts of climate change on a marine turtle population. Global Change Biology 13:923–932.

Hawkes, L. A., A. C. Broderick, M. H. Godfrey, and B. J. Godley. 2009. Climate change and marine turtles. Endangered Species Research 7:137–154.

Heppell, S., L. Crowder, D. Crouse, S. Epperly, and N. B. Frazer. 2003. Population models for Atlantic loggerheads: past, present, and future.

Hewavisenthi, S. and C. J. Parmenter. 2001. Influence of incubation environment on the development of the flatback turtle *(Natator depressus)*. Copeia 2001:668–682.

Huggett, R. J. 1992. Biomarkers: biochemical, physiological, and histological markers of anthropogenic stress. CRC.

Ikonomopoulou, M. P., H. Olszowy, C. Limpus, R. Francis, and J. Whittier. 2011. Trace element concentrations in nesting flatback turtles (*Natator depressus*) from Curtis Island, Queensland, Australia. Marine environmental research 71:10–16.

Innis, C., M. Tlusty, C. Perkins, S. Holladay, C. Merigo, and E. S. Weber III. 2008. Trace metal and organochlorine pesticide concentrations in cold-stunned juvenile Kemp's ridley turtles (*Lepidochelys kempii*) from Cape Cod, Massachusetts. Chelonian Conservation and Biology 7:230–239.

Innis, C. J., J. B. Ravich, M. F. Tlusty, M. S. Hoge, D. S. Wunn, L. B. Boerner-Neville, C. Merigo, and E. S. Weber III. 2009. Hematologic and plasma biochemical findings in cold-stunned Kemp's ridley turtles: 176 cases (2001–2005). Journal of the American Veterinary Medical Association 235:426–432.

Innis, C., C. Merigo, K. Dodge, M. Tlusty, M. Dodge, B. Sharp, A. Myers, A. McIntosh, D. Wunn, and C. Perkins. 2010. Health evaluation of leatherback turtles (*Dermochelys coriacea*) in the Northwestern Atlantic during direct capture and fisheries gear disentanglement. Chelonian Conservation and Biology 9:205–222.

IUCN, 2016, The IUCN Red List of Threatened Species. Version 2016.3. <www.iucnredlist.org>. Downloaded on 10 June 2016.

Jakimska, A., P. Konieczka, K. Skóra, and J. Namieśnik. 2011. Bioaccumulation of metals in tissues of marine animals, part I: the role and impact of heavy metals on organisms. Pol. J. Environ. Stud 20:1117–1125.

Jerez, S., M. Motas, R. Á. Cánovas, J. Talavera, R. M. Almela, and A. B. del Río. 2010. Accumulation and tissue distribution of heavy metals and essential elements in loggerhead turtles (*Caretta caretta*) from Spanish Mediterranean coastline of Murcia. Chemosphere 78:256–264.

Kakizoe, Y., K. Sakaoka, F. Kakizoe, M. Yoshii, H. Nakamura, Y. Kanou, and I. Uchida. 2007. Successive changes of hematologic characteristics and plasma chemistry values of juvenile loggerhead turtles (*Caretta caretta*). Journal of Zoo and Wildlife Medicine 38:77–84.

Keller, J. M., J. R. Kucklick, M. A. Stamper, C. A. Harms, and P. D. McClellan-Green. 2004. Associations between organochlorine contaminant concentrations and clinical health parameters in loggerhead sea turtles from North Carolina, USA. Environmental health perspectives:1074–1079.

Keller, J., J. Kucklick, and P. McClellan-Green. 2004a. Organochlorine contaminants in loggerhead sea turtle blood: extraction techniques and distribution among plasma and red blood cells. Archives of Environmental Contamination and Toxicology 46: 254–264.

Keller, J.M., Kucklick, J.R., Harms, C.A. and McClellan-Green, P.D. 2004b. Organochlorine contaminants in sea turtles: correlations between whole blood and fat. Environmental Toxicology and Chemistry 23: 726–738.

Keller, J.M., McClellan-Green, P.D., Kucklick, J.R., Keil, D.E. and Peden-Adams, M.M. 2006. Effects of organochlorine contaminants on loggerhead sea turtle immunity: comparison of a correlative field study and in vitro exposure experiments. Environmental Health Perspectives, 70–76.

Keller, J.M. 2013. Forty-seven days of decay does not change persistent organic pollutant levels in loggerhead sea turtle eggs. Environmental Toxicology and Chemistry 32: 747–756.

Keller, J.M., Balazs, G.H., Nilsen, F., Rice, M., Work, T.M. and Jensen, B.A. 2014. Investigating the potential role of persistent organic pollutants in Hawaiian green sea turtle fibropapillomatosis. Environmental Science & Technology 48: 7807–7816.

Komoroske, L.M., Lewison, R.L., Seminoff, J.A., Deheyn, D.D. and Dutton, P.H. 2011. Pollutants and the health of green sea turtles resident to an urbanized estuary in San Diego, CA. Chemosphere 84: 544–552.

Labrada-Martagón, V., Méndez-Rodríguez, L.C., Gardner, S.C., Cruz-Escalona, V.H. and Zenteno-Savín, T. 2010. Health indices of the green turtle (*Chelonia mydas*) along the Pacific coast of Baja California Sur, Mexico. II. Body condition index. Chelonian Conservation and Biology 9: 173–183.

Labrada-Martagón, V., Rodríguez, P.A.T., Méndez-Rodríguez, L.C. and Zenteno-Savín, T. 2011. Oxidative stress indicators and chemical contaminants in East Pacific green turtles (*Chelonia mydas*) inhabiting two foraging coastal lagoons in the Baja California peninsula. Comparative Biochemistry and Physiology Part C: Toxicology & Pharmacology 154: 65–75.

Lake, J.L., Haebler, R., McKinney, R., Lake, C.A. and Sadove, S.S. 1994. PCBs and other chlorinated organic contaminants in tissues of juvenile Kemp's Ridley Turtles (*Lepidochelys kempi*). Marine Environmental Research 38: 313–327.

Lam, J.C., Tanabe, S., Chan, S.K., Yuen, E.K., Lam, M.H. and Lam, P.K. 2004. Trace element residues in tissues of green turtles (*Chelonia mydas*) from South China Waters. Marine Pollution Bulletin 48: 174–182.

Lam, J.C., Tanabe, S., Chan, S.K., Lam, M.H., Martin, M. and Lam, P.K. 2006. Levels of trace elements in green turtle eggs collected from Hong Kong: evidence of risks due to selenium and nickel. Environmental Pollution 144: 790–801.

Lazar, B., Maslov, L., Romanić, S.H., Gračan, R., Krauthacker, B., Holcer, D. and Tvrtković, N. 2011. Accumulation of organochlorine contaminants in loggerhead sea turtles, *Caretta caretta*, from the eastern Adriatic Sea. Chemosphere 82: 121–129.

Leite, A.T.M. 2007. Determinação do perfil bioquímico de tartarugas-verdes (*Chelonia mydas*) juvenis selvagens no litoral sul do Brasil.

Ley-Quiñónez, C., Zavala-Norzagaray, A., Espinosa-Carreon, T.L., Peckham, H., Marquez-Herrera, C., Campos-Villegas, L. and Aguirre, A. 2011. Baseline heavy metals and metalloid values in blood of loggerhead turtles (*Caretta caretta*) from Baja California Sur, Mexico. Marine Pollution Bulletin 62: 1979–1983.

Lutcavage, M.E., Plotkin, P., Witherington, B. and Lutz, P.L. 1997. Human impacts on sea turtle survival. The Biology of Sea Turtles 1: 387–409.

Maffucci, F., Caurant, F., Bustamante, P. and Bentivegna, F. 2005. Trace element (Cd, Cu, Hg, Se, Zn) accumulation and tissue distribution in loggerhead turtles (*Caretta caretta*) from the Western Mediterranean Sea (southern Italy). Chemosphere 58: 535–542.

Malarvannan, G., Takahashi, S., Isobe, T., Kunisue, T., Sudaryanto, A., Miyagi, T., Nakamura, M., Yasumura, S. and Tanabe, S. 2011. Levels and distribution of polybrominated diphenyl ethers and organochlorine compounds in sea turtles from Japan. Marine Pollution Bulletin 63: 172–178.

Marcotrigiano, G. and Storelli, M. 2003. Heavy metal, polychlorinated biphenyl and organochlorine pesticide residues in marine organisms: risk evaluation for consumers. Veterinary Research Communications 27: 183–195.

Marcovaldi, M.Â. and Dei Marcovaldi, G.G. 1999. Marine turtles of Brazil: the history and structure of Projeto TAMAR-IBAMA. Biological Conservation 91: 35–41.

Marcovecchio, J. and Freije, R. 2013. Procesos químicos en Estuarios. Universidad Tecnológica Nacional.

Márquez, R. and Doi, T. 1973. Ensayo teórico sobre el análisis de la población de tortuga prieta, *Chelonia mydas* (Caldwell), en aguas del Golfo de California, México. Bull. Tokai Reg. Fish. Res. Lab 73: 1–22.

McFadden, K.W., Gómez, A., Sterling, E.J. and Naro-Maciel, E. 2014. Potential impacts of historical disturbance on green turtle health in the unique & protected marine ecosystem of Palmyra Atoll (Central Pacific). Marine Pollution Bulletin 89: 160–167.

McKenzie, C., Godley, B., Furness, R. and Wells, D. 1999. Concentrations and patterns of organochlorine contaminants in marine turtles from Mediterranean and Atlantic waters. Marine Environmental Research 47: 117–135.

Miao, X.-S., Balazs, G.H., Murakawa, S.K. and Li, Q.X. 2001. Congener-specific profile and toxicity assessment of PCBs in green turtles (*Chelonia mydas*) from the Hawaiian Islands. Science of the Total Environment 281: 247–253.

Monagas, P., Oros, J., Araña, J. and Gonzalez-Diaz, O. 2008. Organochlorine pesticide levels in loggerhead turtles (*Caretta caretta*) stranded in the Canary Islands, Spain. Marine Pollution Bulletin 56: 1949–1952.

MOLL, E., Harless, M. and Morlock, H. 1979. Turtles, perspectives and research. Turtles, Perspectives and Research.

MTSG (Marine Turtle Specialist Group). 1995. A Global Strategy for the Conservation of Marine Turtles. IUCN, Gland, Switzerland.

Ogata, Y., Takada, H., Mizukawa, K., Hirai, H., Iwasa, S., Endo, S., Mato, Y., Saha, M., Okuda, K. and Nakashima, A. 2009. International Pellet Watch: Global monitoring of persistent organic pollutants (POPs) in coastal waters. 1. Initial phase data on PCBs, DDTs, and HCHs. Marine Pollution Bulletin 58: 1437–1446.

Organization, W.H. 1993. Biomarkers and Risk Assessment: Concepts and Principles.

Oros, J., Gonzalez-Diaz, O. and Monagas, P. 2009. High levels of polychlorinated biphenyls in tissues of Atlantic turtles stranded in the Canary Islands, Spain. Chemosphere 74: 473–478.

Páez-Osuna, F., Calderón-Campuzano, M., Soto-Jiménez, M. and Ruelas-Inzunza, J. 2010. Lead in blood and eggs of the sea turtle, *Lepidochelys olivacea*, from the Eastern Pacific: concentration, isotopic composition and maternal transfer. Marine Pollution Bulletin 60: 433–439.

Perrault, J.R., Miller, D.L., Eads, E., Johnson, C., Merrill, A., Thompson, L.J. and Wyneken, J. 2012. Maternal health status correlates with nest success of leatherback sea turtles (*Dermochelys coriacea*) from Florida. PLoS One 7: e31841.

Perrault, J.R., Schmid, J.R., Walsh, C.J., Yordy, J.E. and Tucker, A.D. 2014. Brevetoxin exposure, superoxide dismutase activity and plasma protein electrophoretic profiles in wild-caught Kemp's ridley sea turtles (*Lepidochelys kempii*) in southwest Florida. Harmful Algae 37: 194–202.

Pires, T.T., Rostan, G. and Guimarães, J.E. 2006. Hemograma e determinação da proteína plasmática total de tartarugas marinhas da espécie *Caretta caretta* (Linnaeus, 1758), criadas em cativeiro, Praia do Forte, Município de Mata de São João-Bahia. Brazilian Journal of Veterinary Research and Animal Science 43: 348–353.

Pires, T.T., Rostan, G., de Bittencourt, T.C.C. and Guimarães, J.E. 2009. Hemograma e bioquímica sérica de tartarugas cabeçudas (*Caretta caretta*) de vida livre e mantidas em cativeiro, no litoral norte da Bahia. Brazilian Journal of Veterinary Research and Animal Science 46: 11–18.

Podreka, S., Georges, A., Maher, B. and Limpus, C.J. 1998. The environmental contaminant DDE fails to influence the outcome of sexual differentiation in the marine turtle *Chelonia mydas*. Environmental Health Perspectives 106: 185.

Price, E.R., Jones, T.T., Wallace, B.P. and Guglielmo, C.G. 2013. Serum triglycerides and ß-hydroxybutyrate predict feeding status in green turtles (*Chelonia mydas*): evaluating a single blood sample method for assessing feeding/fasting in reptiles. Journal of Experimental Marine Biology and Ecology 439: 176–180.

Pritchard, P.C., Lutz, P. and Musick, J. 1997. Evolution, phylogeny, and current status. The biology of sea turtles 1: 1–28.

Pupo, M.M., Soto, J.M. and Hanazaki, N. 2006. Captura incidental de tartarugas marinhas na pesca artesanal da Ilha de Santa Catarina, SC. Biotemas 19: 63–72.

Ragland, J.M., Arendt, M.D., Kucklick, J.R. and Keller, J.M. 2011. Persistent organic pollutants in blood plasma of satellite-tracked adult male loggerhead sea turtles (*Caretta caretta*). Environmental Toxicology and Chemistry 30: 1549–1556.

Richardson, K.L., Gold-Bouchot, G. and Schlenk, D. 2009. The characterization of cytosolic glutathione transferase from four species of sea turtles: Loggerhead (*Caretta caretta*), green (*Chelonia mydas*), olive ridley (*Lepidochelys olivacea*), and hawksbill (*Eretmochelys imbricata*). Comparative Biochemistry and Physiology Part C: Toxicology & Pharmacology 150: 279–284.

Richardson, K.L., Lopez Castro, M., Gardner, S.C. and Schlenk, D. 2010. Polychlorinated biphenyls and biotransformation enzymes in three species of sea turtles from the Baja California Peninsula of Mexico. Archives of Environmental Contamination and Toxicology 58: 183–193.

Rodenbusch, C., Baptistotte, C., Werneck, M., Pires, T., Melo, M., de Ataíde, M., dos Reis, K., Testa, P., Alieve, M. and Canal, C. 2014. Fibropapillomatosis in green turtles *Chelonia mydas* in Brazil: characteristics of tumors and virus. Diseases of Aquatic Organisms 111: 207–217.

Rossi, S., Zwarg, T., Sanches, T.C., Cesar, M.d.O., Werneck, M.R. and Matushima, E.R. 2009. Hematological profile of *Chelonia mydas* (Testudines, Cheloniidae) according to the severity of fibropapillomatosis or its absence. Pesquisa Veterinária Brasileira 29: 974–978.

Rossi, S. 2014. Analise da atividade de leucócitos e de bifenilas policloradas aplicada ao estudo da fibropapilomatose em *Chelonia mydas* (Testudines, Cheloniidae) (Linnaeus 1758). PhD thesis. Escola Superior de Agricultura Luiz de Queiroz–Centro de Energia Nuclear na Agricultura, Universidade de Sao Paulo, Piracicaba, Brazil.

Rousselet, E., Levin, M., Gebhard, E., Higgins, B.M., DeGuise, S. and Godard-Codding, C.A. 2013. Evaluation of immune functions in captive immature loggerhead sea turtles (*Caretta caretta*). Veterinary Immunology and Immunopathology 156: 43–53.

Russell, R.W., Gobas, F.A. and Haffner, G.D. 1999. Maternal transfer and in ovo exposure of organochlorines in oviparous organisms: a model and field verification. Environmental Science & Technology 33: 416–420.

Sakai, H., Ichihashi, H., Suganuma, H. and Tatsukawa, R. 1995. Heavy metal monitoring in sea turtles using eggs. Marine Pollution Bulletin 30: 347–353.

Sakai, H., Saeki, K., Ichihashi, H., Suganuma, H., Tanabe, S. and Tatsukawa, R. 2000. Species-specific distribution of heavy metals in tissues and organs of loggerhead turtle (*Caretta caretta*) and green turtle (*Chelonia mydas*) from Japanese coastal waters. Marine Pollution Bulletin 40: 701–709.

Sarmiento, A.M.S. 2013. Determinação de pesticidas organoclorados em tecidos de tartarugas-verdes (*Chelonia mydas*) provenientes da costa sudeste do Brasil: estudo da ocorrência em animais com e sem fibropapilomatose. Universidade de São Paulo. Dissertação (Mestrado em Patologia Experimental e Comparada) - Faculdade de Medicina Veterinária e Zootecnia, University of São Paulo, São Paulo, 2013. Disponível em: <http://www.teses.usp.br/teses/disponiveis/10/10133/tde-16072014-151420/>. Acesso em: 2016-04-2

Silva, L.M. 2011. Metais pesados em tecidos de *Chelonia mydas* encalhadas no Litoral do Rio Grande do Sul, Brasil.

Spotila, J.R., Reina, R.D., Steyermark, A.C., Plotkin, P.T. and Paladino, F.V. 2000. Pacific leatherback turtles face extinction. Nature 405: 529–530.

Stegeman, J.J. and Hahn, M.E. 1994. Biochemistry and molecular biology of monooxygenases: current perspectives on forms, functions, and regulation of cytochrome P450 in aquatic species. Aquatic Toxicology: Molecular, Biochemical, and Cellular Perspectives 87: 206.

Stewart, K.R., Keller, J.M., Templeton, R., Kucklick, J.R. and Johnson, C. 2011. Monitoring persistent organic pollutants in leatherback turtles (*Dermochelys coriacea*) confirms maternal transfer. Marine Pollution Bulletin 62: 1396–1409.

Storelli, M. and Marcotrigiano, G. 2003. Heavy metal residues in tissues of marine turtles. Marine Pollution Bulletin 46: 397–400.

Storelli, M., Storelli, A., D'Addabbo, R., Marano, C., Bruno, R. and Marcotrigiano, G. 2005. Trace elements in loggerhead turtles (*Caretta caretta*) from the eastern Mediterranean Sea: overview and evaluation. Environmental Pollution 135: 163–170.

Storelli, M., Barone, G., Storelli, A. and Marcotrigiano, G. 2008. Total and subcellular distribution of trace elements (Cd, Cu and Zn) in the liver and kidney of green turtles (*Chelonia mydas*) from the Mediterranean Sea. Chemosphere 70: 908–913.

Storelli, M.M. and Zizzo, N. 2014. Occurrence of organochlorine contaminants (PCBs, PCDDs and PCDFs) and pathologic findings in loggerhead sea turtles, *Caretta caretta*, from the Adriatic Sea (Mediterranean Sea). Science of The Total Environment 472: 855–861.

Swarthout, R.F., Keller, J.M., Peden-Adams, M., Landry, A.M., Fair, P.A. and Kucklick, J.R. 2010. Organohalogen contaminants in blood of Kemp's ridley (*Lepidochelys kempii*) and green sea turtles (*Chelonia mydas*) from the Gulf of Mexico. Chemosphere 78: 731–741.

Timbrell, J.A. 1998. Biomarkers in toxicology. Toxicology 129: 1–12.

Valdivia, P.A., Zenteno-Savín, T., Gardner, S.C. and Aguirre, A.A. 2007. Basic oxidative stress metabolites in eastern Pacific green turtles (*Chelonia mydas* agassizii). Comparative Biochemistry and Physiology Part C: Toxicology & Pharmacology 146: 111–117.

Van De Merwe, J., Ibrahim, K. and Whittier, J. 2006. Effects of nest depth, shading, and metabolic heating on nest temperatures in sea turtle hatcheries. Chelonian Conservation and Biology 5: 210–215.

van de Merwe, J.P., Hodge, M., Olszowy, H.A., Whittier, J.M., Ibrahim, K. and Lee, S.Y. 2009. Chemical contamination of green turtle (*Chelonia mydas*) eggs in peninsular Malaysia: implications for conservation and public health. Environmental Health Perspectives 117: 1397.

van de Merwe, J.P., Hodge, M., Whittier, J.M. and Lee, S.Y. 2009a. Analysing persistent organic pollutants in eggs, blood and tissue of the green sea turtle (*Chelonia mydas*) using gas chromatography with tandem mass spectrometry (GC-MS/MS). Analytical and Bioanalytical Chemistry 393: 1719–1731.

van de Merwe, J.P., Hodge, M., Olszowy, H.A., Whittier, J.M. and Lee, S.Y. 2010. Using blood samples to estimate persistent organic pollutants and metals in green sea turtles (*Chelonia mydas*). Marine Pollution Bulletin 60: 579–588.

van de Merwe, J.P., Hodge, M., Whittier, J.M., Ibrahim, K. and Lee, S.Y. 2010a. Persistent organic pollutants in the green sea turtle *Chelonia mydas*: Nesting population variation, maternal transfer, and effects on development. Marine Ecology Progress Series 403: 269–278.

Villa, C., Finlayson, S., Limpus, C. and Gaus, C. 2015. A multi-element screening method to identify metal targets for blood biomonitoring in green sea turtles (*Chelonia mydas*). Science of the Total Environment 512: 613–621.

von Osten, J.R. 2005. Uso de biomarcadores en ecosistemas acuáticos. Golfo de México: contaminación e impacto ambiental: diagnóstico y tendencias, 121.

Walther, G.-R., Post, E., Convey, P., Menzel, A., Parmesan, C., Beebee, T.J., Fromentin, J.-M., Hoegh-Guldberg, O. and Bairlein, F. 2002. Ecological responses to recent climate change. Nature 416: 389.

Wetherall, J., Balazs, G., Tokunaga, R. and Yong, M.Y. 1993. Bycatch of marine turtles in North Pacific high-seas driftnet fisheries and impacts on the stocks. North Pacific Commission Bulletin 53: 519–538.

Witherington, B.E. and Martin, R.E. 2000. Understanding, Assessing, and Resolving Light-Pollution Problems on Sea Turtle Nesting Beaches.

Wood, F.E. and Ebanks, G.K. 1984. Blood cytology and hematology of the green sea turtle, *Chelonia mydas*. Herpetologica, 331–336.

Work, T.M., Raskin, R.E., Balazs, G.H. and Whittaker, S. 1998. Morphologic and cytochemical characteristics of blood cells from Hawaiian green turtles. American Journal of Veterinary Research 59: 1252–1257.

Yoshida, E.T.E. 2012. Avaliação da influência da ingestão de lixo plástico nos indicadores de estresse oxidativo no sangue de tratarugas verdes (*Chelonia mydas*).

Index

About the Editors

Bernardo Duarte graduated in 2008 in Cellular Biology and Biotechnology (Faculty of Sciences of the University of Lisbon), followed by a Master degree in 2011 in Management and Conservation of Natural Resources (University of Évora) and by the PhD in Biology and Ecology (Faculty of Sciences of the University of Lisbon) in 2016. He is an Investigator of the Faculty of Sciences of the University of Lisbon and of MARE – Marine and Environmental Sciences Centre. In the last 15 years developed his research mostly focusing on contaminant biogeochemistry, ecotoxicology and ecophysiology of marine organisms. This work resulted in more than 70 publications in international ISI Journals (h-index 18), 13 book chapters and over 1,300 citations. He is routinely reviewing papers for several international journals in these fields of expertise and acts as a reviewer for several funding bodies in USA, South Africa and Europe. He is presently Reviewing Editor (Marine Pollution section) at Frontiers in Marine Science and Frontiers in Environmental Science. He currently leads several projects focusing on bio-optical techniques and its application in the development of new ecotoxicological tests.

Isabel Caçador graduated in Biology in 1978, finished her PhD in 1995 and her Habilitation in 2008. She is presently a Professor in Plant biology Department at the Faculty of Sciences of the University of Lisbon, Portugal and a senior investigator at MARE – Marine and Environmental Sciences Centre. Her major expertise is in marine and estuarine ecology and ecotoxicology. Her 40 years of expertise in the field have resulted in 200 publications, of which 121 publications in international peer reviewed journals, resulting in over 3,100 citations and h-index of 29. In these areas she has obtained international recognition in the arbitration of scientific articles and books, collaborations in European projects, establishment of cooperation actions, invitation to participate in international juries of merit to award prizes and also, have seen award winning work of which was co-author. She participated in more than 50 national and international projects, being coordinator of 16. Isabel Caçador is also the author of about 170 international communications (40 by invitation). Additionally, she supervised more than 15 PhD thesis. She is regularly reviewing papers for over 50 international journals and has been evaluator for several funding agencies.

Color Plate Section

Chapter 7

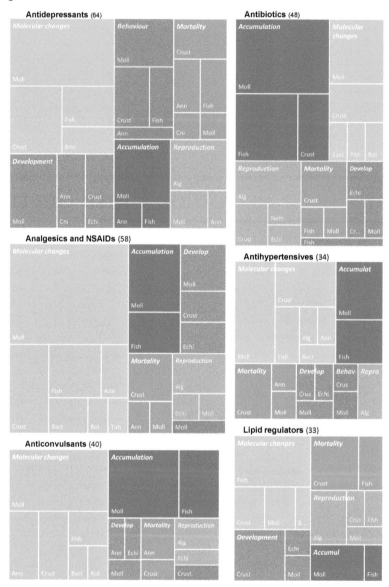

Fig. 2. Tree map representation of studies on the effects of pharmaceutical exposure in coastal and marine organisms per therapeutic class, biological endpoints and major taxonomic groups. Therapeutic classes are antidepressants, analgesics and non-steroid anti-inflammatories (NSAIDs), anticonvulsants, antibiotics, antihypertensives and lipid regulators. Biological endpoints and respective abbreviations are molecular changes, accumulation (accumul), development (develop), mortality, reproduction (repro) and behavior (behav). Major taxonomic groups and respective abbreviations are fish, tunicates (tun), echinoderms (echi), mollusks (moll), crustaceans (crust), rotifers (rot), annelids (ann), nematods (nem), cnidarians (cni), algae (alg) and bacteria (bact). Individual box sizes are proportional to number of entries, and total number of entries per therapeutic class is shown (*n*). Note that a single study may have multiple entrances per therapeutic class (total number of studies 124).

Chapter 8

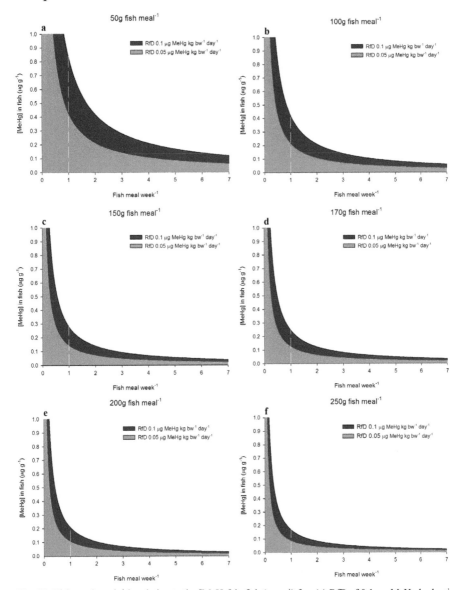

Fig. 10. Fish meal week^{-1} in relation to the [MeHg] in fish (µg g^{-1}) for: (a) RfD of 0.1 µg MeHg kg bw^{-1} day^{-1} and (b) ½ RfD (0.05 µg MeHg kg bw^{-1} day^{-1}).

9 780367 779238